Air Quality and Human Health

Pratap Kumar Padhy • Soumya Niyogi
Pulak Kumar Patra • Markus Hecker
Editors

Air Quality and Human Health

Springer

Editors
Pratap Kumar Padhy
Department of Environmental Studies
Visva-Bharati University
Santiniketan, West Bengal, India

Pulak Kumar Patra
Department of Environmental Studies
Visva-Bharati University
Santiniketan, West Bengal, India

Soumya Niyogi
Department of Biology and Toxicology Centre
University of Saskatchewan
Saskatoon, SK, Canada

Markus Hecker
School of Environment and Sustainability and Toxicology Centre
University of Saskatchewan
Saskatoon, SK, Canada

ISBN 978-981-97-1362-2 ISBN 978-981-97-1363-9 (eBook)
https://doi.org/10.1007/978-981-97-1363-9

This Springer imprint is published by the registered company Springer Nature Singapore Pte Ltd.
The registered company address is: 152 Beach Road, #21-01/04 Gateway East, Singapore 189721, Singapore

Paper in this product is recyclable.

Contents

About the Editors

Pratap Kumar Padhy is a Professor and Head at the Department of Environmental Studies, Institute of Science, Visva-Bharati, Santiniketan. Prof. Padhy is working on various aspects of Air Pollution and Health. He has published many research papers in internationally refereed journals and several chapters in books. He has supervised 12 doctoral students and 3 postdoctoral scholars. He has also undertaken four research projects, including one international collaborative project. He was awarded the President of India Gold Medal for General Proficiency from JNU from 1995 to 1996.

Soumya Niyogi is a Professor at the Biology & Toxicology Centre, University of Saskatchewan Saskatoon, Canada. He is an Environmental Toxicologist with over 25 years of research experience specializing in the toxicology of metals/metalloids and polyaromatic hydrocarbons (PAHs). He has published more than 80 peer-reviewed papers and presented at more than 200 conferences. He has undertaken many sponsored research projects and supervised Ph.D. and M.Sc. students. He is an editorial board member in some reputed journals and a frequent reviewer in many journals.

Pulak Kumar Patra is an Associate Professor at the Department of Environmental Studies, Institute of Science, Visva-Bharati, Santiniketan. He has more than 20 years of teaching and research experience in Environment Science. His research interests include environmental geoscience, hydrogeochemistry, and medical geology. He has published more than 50 research papers in journals and edited books. He has also authored three books and edited two books. He has supervised eight Ph.D. students.

Markus Hecker is a Professor & Canada Research Chair at the School of Environment and Sustainability, University of Saskatchewan. His research interest includes ecotoxicology, endocrine disruption, environmental DNA, exposure and hazard characterization of environmental contaminants, mechanistic toxicology, and toxicogenomics. He serves as the editorial board member of many reputed journals. Markus has authored over 170 peer-reviewed papers, review articles, editorials, and book chapters. He currently serves on the board of directors of the Society of Toxicology and Chemistry (North America).

1 Introduction

Pratap Kumar Padhy, Soumya Niyogi,
Pulak Kumar Patra, and Markus Hecker

1.1 Introduction

Air quality and human health are closely linked. Two thousand five hundred years ago, Hippocrates, the father of medicine, mentioned the critical role of the environment (especially the air we breathe and the water we drink) in our health (Balasubramanian et al. 2021). Following the eighteenth-century industrial revolution in Europe, air quality deteriorated noticeably due to the rise of industrial and fireplace coal burning. The cause-effect relationship between air quality and human health was scientifically established. It led some countries to make laws for controlling emissions, with some success. The tragedy of the great London smog in December 1952 woke the world to the repercussions of air pollution and paved the way for landmark legislations such as the Clean Air Act (Goodsite et al. 2021). Since then, scientific knowledge of the harmful consequences of air pollution on human health has grown many times. Pollution abatement technology has also progressed. Powerful techniques are now available for data collection and processing. Research in this field has grown enormously with the availability of much more data and the development of data analysis methods. Air quality and human health science encompass a variety of disciplines, and a multidisciplinary approach is needed to address this pressing global problem.

Air pollution and its health consequences are a problem all around the globe. According to WHO data, 99% of the world's population breathes air that contains high levels of contaminants and exceeds WHO guideline limits. While air pollution has been improving in some regions, it is getting worse mostly in low- and middle-income nations including India. Rapid economic expansion, fossil fuel-based energy generation, as well as a lack of expertise and resources to combat pollution are the main causes of higher levels of air pollution in these countries.

The two most widely cited and regularly updated estimates for the death toll from air pollution come from the World Health Organization (WHO) and the Institute of Health Metrics and Evaluation's Global Burden of Disease study (Roser 2021). Their latest estimation of air pollution-induced annual deaths are 7 million and 6.7 million respectively. However, there is no consensus on these estimations. They also do not

P. K. Padhy (✉) · P. K. Patra
Department of Environmental Studies, Visva-Bharati University, Santiniketan, West Bengal, India
e-mail: pkpadhy@visva-bharati.ac.in

S. Niyogi
Department of Biology and Toxicology Centre, University of Saskatchewan, Saskatoon, SK, Canada

M. Hecker
School of Environment and Sustainability and Toxicology Centre, University of Saskatchewan, Saskatoon, SK, Canada

P. K. Padhy et al. (eds.), *Air Quality and Human Health*,
https://doi.org/10.1007/978-981-97-1363-9_1

discriminate among the deaths caused by different air pollutants. In general, more recent studies find a higher death toll than earlier studies. It is not because of worsening air pollution, but due to the fact that, recent scientific evidence suggests that the health impacts of exposure to pollution are more significant than previously thought. Research on the effects of air pollution is still in its early stage. More scientific research is needed to arrive at greater certainty. Traditionally, air pollution was linked mainly to the respiratory system and haematological health. Recent studies provide evidence of a wide range of health issues due to air pollution. The often-complex nature of air pollution due to its diverse chemical constituents makes it challenging to pinpoint the specific causes of health risks in exposed populations.

The nature of the air pollution problem varies widely from country to country, also from one region to another within a country. It is not only dictated by the pollutants released or emitted into the air from different sources, but also by the prevailing weather conditions. The range of pollutants produced is also wide. Introducing new methods of detecting and quantifying these pollutants and establishing rural and urban air pollution networks to monitor regional and local patterns has provided invaluable information that has been used to build up a profile of air pollutant problems.

A better understanding of atmospheric chemistry, new methods for pollutant monitoring, detailed epidemiology, and the use of controlled exposure studies to examine single and mixtures of pollutants provide a firm framework against which to devise new legislation to control this intractable problem.

1.2 Source and Nature of Air Pollutants

Air pollution comes from a wide variety of sources. A pollutant is usually understood to be a substance which, between the point of its emission into the atmosphere and its ultimate removal, causes harm to a target, i.e. an ecosystem or human health. The classical pollutants are sulphur dioxide (SO_2), nitrogen oxides (NO_x), carbon monoxide (CO), volatile organic compounds (VOCs), and particulate matter (PM). Air pollution may be caused by natural processes or by human activities. The largest source of natural air pollution is airborne dust in the world's deserts. A second major natural source is wildfire smoke in forests and grasslands. Additional natural sources are sea spray, pollen, and volcanoes. It is possible to reduce natural air pollution to some degree. It is certainly possible to reduce exposure to it.

Anthropogenic sources of air pollution originate in a variety of human activities: the production of electricity in coal power plants; the burning of solid fuels (wood, charcoal, coal, crop waste and dung) for cooking and heating in billions of poor households; agriculture; industry; and road transport. These pollutions can be reduced with policy measures and technological innovations.

Quantitative information on the emission rate of air pollutants from different sources is vital for developing appropriate abatement strategies. The best means of improving air quality is to identify the major sources of both primary and the precursors of secondary pollutants. Emissions of air pollutants are estimated from knowledge of the processes that form them. This is relatively straightforward for some pollutants, as emissions are mainly dependent on the composition of the fuel. Information on fuel consumption and composition and knowledge of the presence and type of pollution abatement technology is sufficient to estimate emissions from many sources. For other pollutants, emissions depend on combustion conditions, such as temperature, pressure, and pollution abatement technology, and are thus more difficult to estimate.

Annual emission inventories, a database of the emissions of primary pollutants from all known sources for a defined geographical area and year, are being routinely compiled at the national level and increasingly at the urban level. National emission inventories are used both within a country for prioritising pollution control measures and internationally in negotiating pollution control treaties.

1.3 Air Pollution Monitoring

Any estimation of the public health impact of air pollution exposure must depend heavily upon estimates of air pollutant concentrations. Such estimates may be derived either by measurement or by numerical modelling and are often a combination of the two. Measurements of air pollutant concentrations are customarily carried out at fixed-site monitoring stations. The number of measuring stations and their networking has increased many fold. Still, they are concentrated in urban areas. Model predictions are used to interpolate or provide data for areas where monitoring data are lacking. Meteorological information is vital for such modelling. Weather, air quality, and health watch warning systems may all be supported by meteorological modelling, which also serves as the foundation for real-time air quality forecasts.

Provided the boundary conditions of the model are known, i.e. the concentrations of air pollution entering the model domain from outside, then a well-conceived model should be capable of predicting both annual average and short-term concentrations within the urban area. Monitoring still has an important role to play since monitoring data will be used to validate the model's predictions for current air quality, evaluate trends in air pollutant concentrations, and determine the effectiveness of the control strategy once it is implemented. Another use of monitoring data is informing the public of air pollution concentrations and, where necessary, providing the basis for short-term remedial action.

Satellite-based sensors are increasingly being used to monitor the air quality. Geographic Information System (GIS)-based land use regression models are becoming popular to estimate long-term exposure where monitor density is insufficient (Adams and Kanaroglou 2016). These models incorporate land use information to provide additional knowledge about emission source potential to improve spatial variation modelling.

Where there are several measures of varying cost and ease of implementation, pollution control authorities need to know which measure, or combination of measures, will be most cost-effective in improving air quality to the desired level. Thus, emission control approaches for managing air quality better require a proper understanding of the fate and transport of pollutants. Many simulation models are now available for visualising/predicting the effectiveness of the measures in different situations.

Measurements of air pollutant concentrations are best considered in the context of the entire air quality management process. The major forces driving air pollution control and the policies and measures through which that air pollution control is achieved can be represented in the flow chart (Fig. 1.1) as proposed by Harrison (1999).

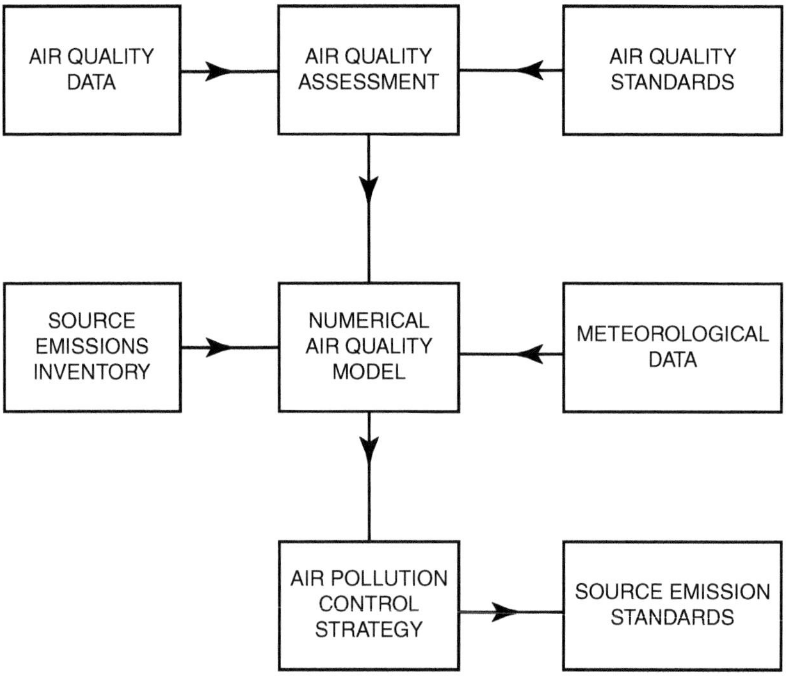

Fig. 1.1 Key components of an air quality management strategy. (Source: Harrison, R.M., 1999, Measurements of Concentrations of Air Pollutants *In* Stephen T. Holgate, Jonathan M. Samet, Hillel S. Koren, Robert L. Maynard editors Air Pollution and Health, Academic Press, 1999) (Reprinted with permission from Elsevier)

1.4 Air Pollution Meteorology and Atmospheric Chemistry

Air pollution arises from the competition between emission processes, which increase pollutant concentrations, and dispersion and advection processes in the atmosphere, which reduce them. Weather and climate are critical in the relationship between air pollution and health. Atmospheric stability controls the extent to which pollutants may be dispersed and diffused, while wind transports and disperses pollutants. Meteorological modelling may provide the basis for real-time air quality forecasting and health watch warning systems. Because people can be sensitive to both weather and air pollution, particular weather and air quality configurations may engender distinct health outcomes. Unravelling these complex interactive effects is a challenge. The exact nature of the synergistic impacts of weather extremes and harsh air quality conditions on health is firmly established.

An essential category of air pollutants consists of substances not emitted into the atmosphere in any significant quantity but formed by chemical reactions. These latter pollutants are termed secondary pollutants to distinguish them from the primary or emitted pollutants. Nitrogen dioxide (NO_2), ozone (O_3), and suspended particulate matter are three major secondary pollution problems with potential public health impacts in urban and industrialised population centres worldwide. The distinction between primary and secondary pollutants is vital for understanding air pollution and devising pollution abatement strategies. There is a proportional relationship between emissions and ambient concentrations for primary pollutants. However, for a secondary pollutant, reducing emissions of the precursor does not lead to a proportionate reduction in its ambient concentration. In some circumstances, reducing precursor emissions may lead to increased concentrations of some pollutants. This is most notable in the case of urban ozone, where reducing local emissions of nitrogen oxides can lead to an increase in local ozone concentrations.

Patterns of atmospheric aerosols have changed enormously since coal burning was widespread, and the particles were mainly coal-derived carbonaceous particles associated with sulphuric acid and other irritant compounds. In terms of the presentation in the air of potentially polluting entities to the human body, the numbers of particles are, therefore, several orders of magnitude smaller than the number of molecules of other pollutants in the air simultaneously. Therefore, the individual particles' structure and composition are crucial in understanding the mechanism(s) by which particles produce adverse health effects. The significance of evolving patterns of particulate and gaseous pollution, particularly for human health, is a major challenge for atmospheric and medical science.

1.5 Exposure Assessment

Accurate estimates of human exposure to inhaled air pollutants are necessary for a realistic appraisal of the risks these pollutants pose and to design and implement strategies to control and limit those risks (Monn 2001). Except in occupational settings, such estimates are usually based on measurements of pollutant concentrations outside (ambient) air, recorded with outdoor fixed-site monitoring. For non-carcinogenic air pollutants, much of our knowledge of public health impacts is derived from community-based epidemiological studies which relate health outcomes to concentrations of air pollution measured at fixed-point outdoor monitoring stations. Exposure-response relationships derived from such studies can readily be combined with other outdoor fixed point concentration measurements and used in a predictive manner to estimate public health impacts of air pollution exposure. Therefore, routine outdoor measurements, instead of personal exposure assessments, have a crucial role in evaluating the health effects of air pollution (Sexton and Ryan 1988). This has been a very active field in recent years, with burgeoning research activity and a drive by both national governments and international organisations.

It demonstrates clearly that inhaled particles can profoundly impact physiology and biochemistry, and it is therefore unsurprising that epidemiological studies show significant impacts of PM exposure. The methodology and approach used in standard setting depend very much on the nature of the pollutant, and this is brought out clearly by case studies of individual pollutants. The focus is on how epidemiological evidence can be used to evaluate the risk from particulate matter exposure.

Personal exposure depends on many factors such as activity patterns, microenvironmental living conditions, deposition, clearance, retention and disposition of the particular matter, and the mechanisms of damage and repair.

The current understanding of the health effects of air pollutants, especially of particulate matters (PM), is limited. Most of the epidemiological evidence linking adverse health effects to PM exposure is not conclusive. The cellular and molecular mechanisms by which PM and other air pollutants prompt toxic effects in humans are yet to be fully understood. Biochemical and molecular biomarker assessments can provide important insights into the cellular perturbations that lead to the onset of adverse health effects during chronic exposure to air pollutants. The advent of next-generation, systems biology molecular techniques (e.g. omics technologies) provides exciting new avenues to study the effects of air pollution on human health.

1.6 Designing of the Book

With contributions from well-known experts from diverse research fields (environmental science, toxicology, geology, public health science, biology, physics, chemistry, and geospatial technology), the book deliberates on the health risks of air pollution and explores its control and mitigation strategies. The recent developments and challenges in diverse aspects of air quality and human health, starting from air pollution monitoring, identification of the source, especially in aerosols, specific health issues to biomarkers, and management issues, are discussed in the following 14 chapters.

The second chapter summarises the existing Global Burden of Disease Study (GBD) studies and presents a roadmap for the way forward. The authors suggest India should invest resources to systematically collate health data, design cohorts, and improve exposure assessment to generate exposure-response functions. Lab-based studies and biomarker analysis can help detect early indications of air pollution health impacts.

Over the years, remote sensing combined with geographic information systems known as geospatial technology-based assessment of air quality has emerged as a popular tool wherein air quality standards can be studied across ample space (regional units and global scales) and time scales. The third paper presents a cross-sectional descriptive bibliometric analysis of publications on geospatial technology-based air quality assessment. Air particulate medical geology is another emerging field of research focusing on atmospheric particle cycles and their complex pathways to assess their health risks. The fourth chapter dealing with this field priorities two research areas: the use of interfacial sciences tools for quantitative assessment of in situ surface, chemical, and mineralogical characteristics of atmospheric particulates; and to understand what role trace elements play at the level of cells in living things. Chapter 5 stresses the importance of land-use regression models that air quality data from limited monitoring stations using predictor variables usually obtained through geographic information systems (GIS).

Chapter 6 deals with the toxicity of particulate pollutants, especially the emerging pollutants microplastics and their impact on mitochondrial health. Chapters 7 and 10 are based on case studies from India on trends of air pollutants and their health effects. Though the overall trend of pollutants has seen a gradual decline, the health effects of air pollution in India are in a grave situation. The pattern and trend of the burden of disease due to particulate matter in the world and India are presented in Chap. 8. Chapter 9 provides a detailed overview of various health effects caused by outdoor air pollution.

Chapter 11 critically examines the recent progress in developing and using both targeted and non-target biomarker approaches to assess the impacts of air pollution on human health. Recent developments in 'omics' biotechnologies and AI tools have revolutionised the way we can probe entire biological systems, allowing for the identification and development of novel and relevant biomarkers which will support human health risk assessment of air pollution.

Chapter 12 deals with air pollutants from power plants and briefly describes their sources and their impact on human health. Best available technologies (BAT) and best environmental practices (BEP) to manage different pollutants are also discussed. Oxidative stress, inflammation and subsequent genotoxic processes are the predominant drivers of adverse health outcomes such as pulmonary and cardiovascular diseases and cancer associated with air pollution. The oxidative damage caused by PM-induced toxicity and its impact on various organ systems is discussed in Chap. 13.

Chapter 14 provides a broad overview of the major components of air pollution, their sources and modes of transportation, concentration in different regions of India, reaction mechanisms, and health effects, along with the major gap areas.

The last chapter (Chap. 15) explains the available management strategies for air pollution. Emission control through public policy formulation has been instrumental in reducing particulate air pollutant load. Innovative technological interventions can help to moderate its adverse health effect.

Thus, we hope the book provides an up-to-date overview of the modern methods and tools used in air quality monitoring and human health risk assessment. It addresses all aspects of air quality, covering indoor-outdoor air pollution; gaseous and particulate pollutants; characterisation of source and pathways of air pollutants; and the modelling and assessment of health risks (respiratory, epidemiological and toxicological) with regional and global perspectives. It also addresses air quality management issues. The lucid explanation of the role of oxidative stress mechanisms and molecular biomarkers (genomics, proteomics) may be considered as inputs into the development of cancer therapeutics.

Along with providing a scientific basis for air pollution, this book will help readers appreciate the environmental determinants of public health and apply research evidence to improve the quality of life. It also delineates future research initiatives and policy actions needed to protect human health from air pollution, locally and globally. The book will be of great educational value and is recommended for consultation and teaching. The multidisciplinary nature of the book and the wide range of issues covered may provide the reader with access to up-to-date information on all matters relating to ambient air pollution. It will therefore, be helpful to those interested in monitoring air pollution and those wishing to know more about its impact on indices of human health.

References

Adams MD, Kanaroglou PS. Mapping real-time air pollution health risk for environmental management: combining mobile and stationary air pollution monitoring with neural network models. J Environ Manag. 2016;168:133–41. https://doi.org/10.1016/j.jenvman.2015.12.012.

Balasubramanian S, Domingo NGG, Hunt ND, Gittlin M, Colgan KK, Marshall JD, Robinson AL, Azevedo IML, Thakrar SK, Clark MA, Tessum CW, Adams PJ, Pandisand SN, Hill JD. The food we eat, the air we breathe: a review of the fine particulate matter-induced air quality health impacts of the global food system. Environ Res Lett. 2021;16:103004. https://doi.org/10.1088/1748-9326/ac065f.

Goodsite ME, Johnson MS, Hertel O, editors. Air pollution sources, statistics and health effects. New York, NY: Springer; 2021. 531 p.

Harrison RM. Measurements of concentrations of air pollutants. In: Holgate ST, Samet JM, Koren HS, Maynard RL, editors. Air pollution and health, vol. 1999. New York: Academic Press; 1999. p. 63–81. https://doi.org/10.1016/B978-012352335-8/50080-6. ISBN 9780123523358.

Monn C. Exposure assessment of air pollutants: a review on spatial heterogeneity and indoor/outdoor/personal exposure to suspended particulate matter, nitrogen dioxide and ozone. Atmos Environ. 2001;35:1–32.

Roser M. Data review: how many people die from air pollution? Our world in data; 2021.

Sexton K, Ryan PB. Assessment of human exposure to air pollution: methods, measurements, and models. In: Watson AY, Bates RR, Kennedy D, editors. Air pollution, the automobile, and public health. Washington, DC: National Academies Press (US); 1988.

Impacts of Air Pollution on Human Health: Present Status in India and Future Directions

2

Pallavi Joshi, Ekta Chaudhary, and Sagnik Dey

2.1 Introduction

Over the years, an extensive body of scientific evidence established the negative health effects associated with exposure to polluted air, using various epidemiological study designs. The Global Burden of Disease (GBD) study evolved a uniform framework to assess and track the long-term effects of exposure to air pollution across the globe. Based on the GBD assessment (GBD 2019 Risk Factors Collaborators 2020), air pollution exposure is ranked fourth on the list of risks globally causing disability-adjusted life years (DALY) and mortality (GBD 2019 Risk Factors Collaborators 2020).

The GBD study revealed that low- and middle-income countries (LMICs) experience a disproportionate burden of air pollution-related health burden compared to high-income countries. LMICs face multiple challenges in managing air quality and associated health burdens. The ground-based monitoring in LMICs is highly inadequate for mapping regional-scale air quality and exposure (Martin et al. 2019). A higher proportion of solid fuel use for domestic purposes results in high household exposure in LMICs. This, in turn, increases the risk of exposure misclassification in health burden estimates. Patchy health data in LMICs make the health burden estimates more uncertain.

The GBD study addressed these critical gaps using geospatial modelling and other relevant techniques and developed exposure-response functions for ischemic heart disease (IHD), chronic obstructive pulmonary disease (COPD), acute lower and upper respiratory tract infections (ARI), stroke, diabetes, and neonatal deficiencies such as preterm birth and low birth weight by compiling cohort studies from around the world (Burnett et al. 2014, 2018). The evolution of exposure-response functions and their applications in estimating excess mortality risks has been summarized in Burnett and Cohen (2020). In terms of air pollutants, exposure metrics consider the ambient fine particulate matter ($PM_{2.5}$), household $PM_{2.5}$, and ambient ozone.

India, home to 1.4 billion population, has one of the highest $PM_{2.5}$ exposure globally (Cohen et al. 2017). Air pollution is ranked second after maternal and child malnutrition and was estimated to cause 1.67 million (95% uncertainty interval, UI: 1.42–1.92) deaths in 2019 in India, causing a loss of up to 1.36% of the gross domestic product (Pandey et al. 2021). In addition to the GBD study, where mortality and morbidity

P. Joshi · E. Chaudhary
Centre for Atmospheric Sciences, IIT Delhi, New Delhi, India

S. Dey (✉)
Centre for Atmospheric Sciences, IIT Delhi, New Delhi, India

Arun Duggal Centre of Excellence for Research in Climate Change and Air Pollution (CERCA), IIT Delhi, New Delhi, India
e-mail: sagnik@cas.iitd.ac.in

P. K. Padhy et al. (eds.), *Air Quality and Human Health*,
https://doi.org/10.1007/978-981-97-1363-9_2

burden was estimated at state levels, numerous studies were carried out in the last few decades in India to understand the health impacts emanating from air pollution exposures, using indigenous health data and standard epidemiological study designs. This chapter summarizes the current state-of-the-art knowledge of air pollution health impacts in India, outlines the critical challenges, and shows the ways forward to accelerate India's progress to meet the United Nations' Sustainable Development Goal (SDG) 3 of promoting healthy lives and well-being.

2.2 Air Pollution Health Effects and Epidemiological Study Designs

The assessment of any health endpoints in relation to air pollution exposure relies on data from health institutions or civic bodies responsible for the collation of vital statistics. The most common methods to estimate the occurrence of health outcomes in environmental research are based on the events in which (1) the subjects have sought medical attention (e.g., emergency hospital visits or outpatient visits, or hospital admissions), (2) death events recorded by civic bodies and certified by medical professionals, with or without cause of death, (3) self-reported or bio-monitored health irregularities and illnesses recorded in small/large scale health surveys. The use of medical record data, however, has inherent limitations, such as recording errors, variability in medical coding, lack of information on the onset/duration of events, and biases due to varied health-seeking behaviour. These limitations impact the accuracy of the health outcome assessments and introduce bias in the estimation air pollution-related health consequences or any other risk factor. The alternatives to surpass the above limitations are by way of bio-monitoring/human tissue sampling, which is expensive and only viable for a very small sample population and is not realistic at a population scale.

The existing body of epidemiological research records health effects from both long- and short-term exposure to ambient air pollution (Liu et al. 2013; Sehgal et al. 2015; Kesavachandran et al. 2015; Cohen et al. 2017; Haque and Singh 2017). Short-term exposure and related health effects are measured at weekly/hourly/daily temporal scale, whereas long-term exposure and health effects are captured over monthly or annual scales. (Beverland et al. 2012). The short-term exposures manifest in acute illnesses or episodes such as exacerbation of asthma, breathing troubles, and cardiovascular disease (CVD) conditions such as ischemia, arrhythmias, and cardiac failure. The short-term effects are recorded in terms of emergency hospital visits, hospitalizations, and even fatal outcomes generally in finer temporal windows (Kumar et al. 2004; Gupta 2008; Patankar and Trivedi 2011; Maji et al. 2015, 2018). The longer duration and more intense exposures are exhibited in terms of chronic and disabling conditions such as compromised lung/heart function, tuberculosis, cough/wheezing, allergies, cardiovascular complaints, and even premature deaths (Kumar and Foster 2007; Guttikunda and Jawahar 2012; Guttikunda and Goel 2013; Ghosh and Mukherji 2014; Tobollik et al. 2015; Gawande et al. 2016; Rajak and Chattopadhyay 2020).

The type of epidemiological study designs is extremely crucial for accurate assessment of air pollution-related health risks. Their selection primarily depends on exposure and health data and determining factors for the results, their synthesis, interpretation, and policy use. There are multiple analysis methods used in air pollution health studies, ranging from case-crossover, case-control, time-series, Bayesian epidemiological studies, and cohort designs to examine associations between pollutant exposures and health outcomes of interest. Generally, the case-crossover and time-series studies utilize local and fine-scale temporal exposure data. They are generally good at assessing the effects of regional pollutants that exhibit greater spatial homogeneity.

It is important to highlight that time-series studies using exposure and health data resolved at fine spatial scales like zip code level show modestly stronger associations (Joshi et al. 2021). Such studies mainly focus on the short-term and acute effects of exposure to ambient air pollution.

Cross-sectional studies utilizing improved spatially resolved health and exposure data rely on ecological assessments and can provide critical evidence of association in the absence of cohort studies. On the other hand, there are cohort-based studies, where a group of subjects is followed, and subsequent health evaluations are conducted to determine participants' exposure characteristics driven by temporal and spatial variations in exposure, retrospectively or prospectively. The advantage of cohort studies is that they are capable of establishing true causal effects but expensive in terms of logistics and financial resources.

2.3 GBD India Study

Health burden attributable to air pollution depends on three factors—relative risk (RR) of a disease corresponding to an exposure level compared to the theoretically minimum exposure level (TMEL), population, and its age distribution (RRs vary with age for some diseases), and background disease rates from all possible risk factors. The GBD India study compiled the data for the period of 1990 onwards and tracked the progress in air pollution health burden (Balakrishnan et al. 2019). Figure 2.1 shows the state-level DALY rate attributable to air pollution (normalized to 100,000 population) from the GBD India study 2019. The data revealed a large disparity in DALY rates attributable to air pollution in 2019, ranging from 1664 (95% UI: 1347–2042) in Arunachal Pradesh to 5002 (95% UI: 4227–5848) in Uttar Pradesh. For ambient $PM_{2.5}$ exposure, the lowest DALY rate was found for Arunachal Pradesh (593, 95% UI: 389–823), while the highest rate was found for Uttar Pradesh (3106, 95% UI: 2387–3824). For household $PM_{2.5}$, the DALY rate varied from 19 (95% UI: 9–37) in Delhi to 2513 (95% UI: 1769–3228) in Chhattisgarh.

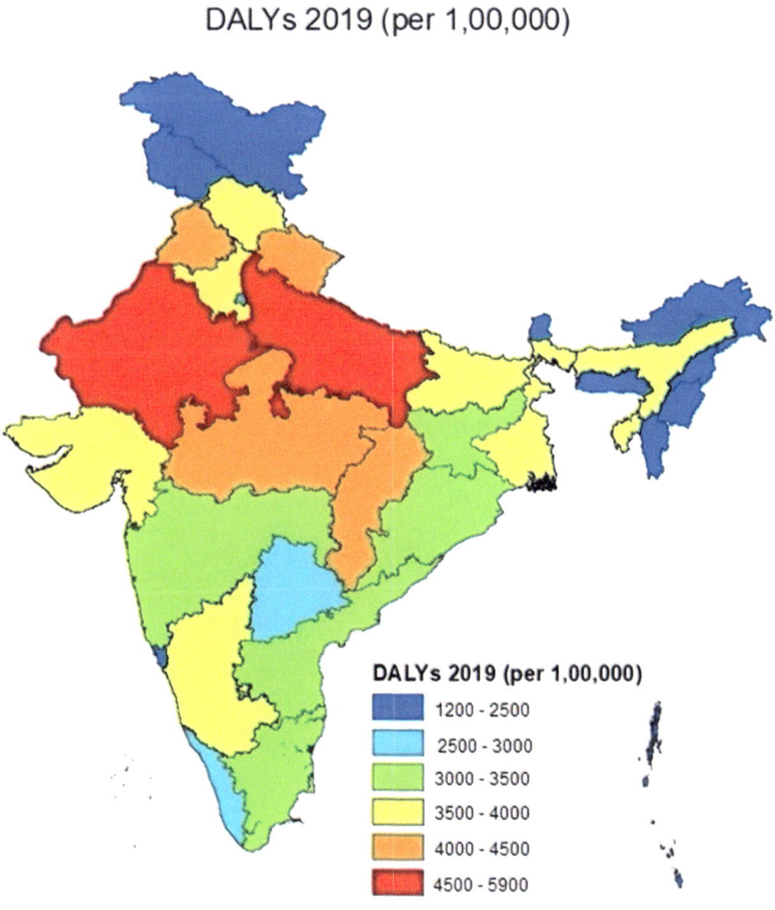

Fig. 2.1 DALYs attributable to air pollution, covering both sexes and all age groups in India, 2019. (Data source: https://vizhub.healthdata.org/gbd-compare/)

The temporal changes in the mortality rates attributable to ambient and household air pollution (Fig. 2.2) further revealed that the age-standardized death rates resulting from household level $PM_{2.5}$ decreased by 72.3% from 215.5 in 1990 to 59.6 in 2019. During the same time, the death rates attributable to ambient $PM_{2.5}$ rose by 57.5% from 60.7 in 1990 to 95.6 in 2019. A large share of DALYs, around 39.5% of the total DALYs, were due to lung diseases, including COPD (22.7%) and ARI (15.5%). Out of the rest, about 38.6% of the total was CVDs, including IHD (24.9%) and stroke (13.7%), neonatal deficiencies (14.5%), diabetes mellitus (5.5%), and cataracts (1.5%). Across the total DALY burden attributable to all risk factors, air pollution was responsible for 11.5%, out of which 6.7%, 4.5%, and 0.7% could be attributed to ambient-household $PM_{2.5}$ and ozone exposure, respectively. In 2017, air pollution-related losses in life expectancy were estimated at 0.9 (0.8–1.1) and 0.7 (0.6–0.8) for ambient and household $PM_{2.5}$, respectively (Balakrishnan et al. 2019).

The GBD India study demonstrated that the health burden and exposure are not linearly proportional. Therefore, an improvement in air quality by the same margin in socio-demographically diverse regions will not result in the same health benefit. For example, the change in ambient $PM_{2.5}$-related mortality burden in India, for the period 2000–2015 is depicted in Fig. 2.3. However, if the other factors remained unaltered and the exposure only changed, the air pollution burden would have increased only by 17.4%. The population rise contributed to 33.6% of the increase in burden alone, and the shift of the age structure towards older age contributed to another 14.5% increase in the burden. This was partially compensated by the reduction of the burden by 14.2% due to an overall reduction in baseline mortality. The net result is an increase in air pollution-linked mortality burden by 47.9% between 2000 and 2015 (Chowdhury et al. 2020). Such analysis implies that the air pollution health burden can be reduced further by accelerating the reduction of baseline mortality by improving the healthcare infrastructure.

In addition to the GBD study, there are a host of epidemiological studies in India which utilized pre-existing exposure-risk functions for various pollutants and health endpoints. The majority of these studies utilize curated software like the Air Q/Air Q+ developed by the WHO European Centre for Environment Health, Bilthoven

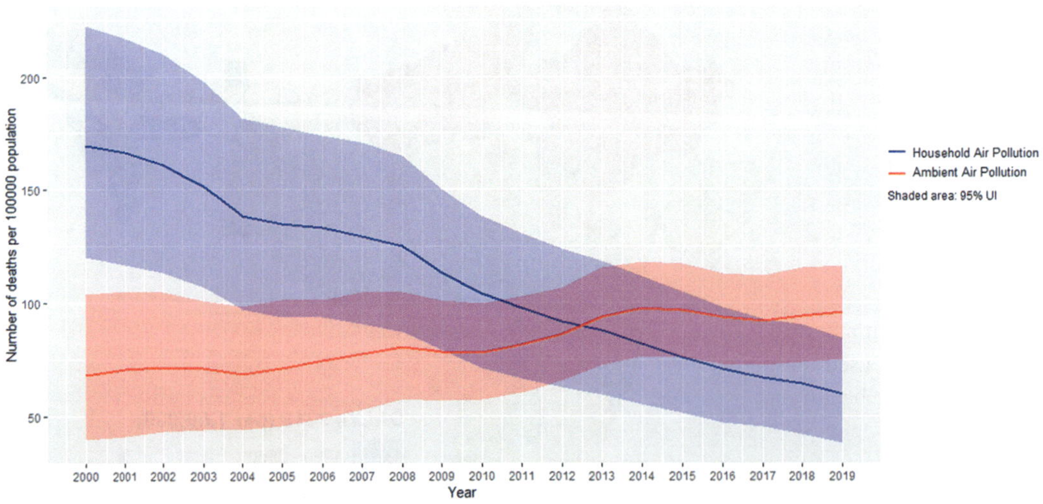

Fig. 2.2 Age-standardized death rates attributable to air pollution, India 2000–2019. (Data Source: GBD (https://vizhub.healthdata.org/gbd-compare/))

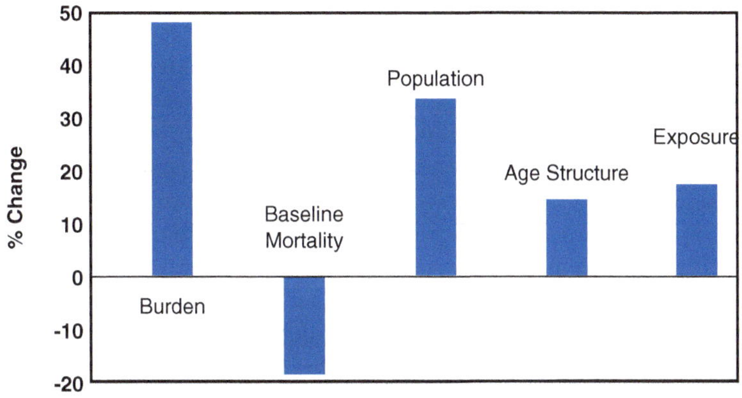

Fig. 2.3 The relative changes in mortality burden attributable to ambient $PM_{2.5}$ exposure in India between 2000 and 2015 and the changes in burden due to the changes in specific factors. (The statistics are taken from Chowdhury et al. 2020, Creative Commons Attribution BY 4.0)

Division. Such software packages are based on the Risk of Mortality/Morbidity due to the Air Pollution (RiMAP) model, modelled to estimate air pollutant impact on a defined population group in a given region and time point. The underlying approach to any risk assessment for human impact is population attributable factor (PAF), the fraction of excess morbidity or mortality in a given population due to atmospheric pollutants, calculated using a pre-existing risk function for exposure and excess health risk, factoring in for the possible confounding factors in that association (Maji et al. 2016). The advantage of using these packages is that they serve as tools for quick analysis to support policymaking for air quality management. The other advantages are that they are versatile and capable of estimating air pollution exposures at short (i.e., daily) and long (i.e., yearly) temporal scales for various health endpoints and list of pollutants and for even sources like household air pollution. The risk functions based on non-indigenous studies and data, however, pose a question of true applicability for the Indian demography (Sacks et al. 2020).

Given these limitations, such simple tools have repeatedly been used on Indian city-level data to demonstrate the health and economic impact of poor air quality. Studies based in Mumbai and Delhi found excess all-cause mortality, respiratory and cardiovascular mortality, also hospital admissions due to COPD and CVD to be more strongly associated with PM_{10}, and $PM_{2.5}$ than gaseous pollutants (Maji 2017a; Afgan and Patidar 2020). Another study in Nagpur used the same model to establish a long-term association (two decades) between COPD risk and PM exposure. Similar studies incorporating multicity analyses have found this excess mortality and morbidity to be on an increasing trend over the years (Maji et al. 2016). The model has also been put to use to establish a strong association between the excess cases of COPD hospital admissions correlated with $PM_{2.5}$ exposures (Manojkumar and Srimuruganandam 2021). Trend analysis in most studies showed peak particulate matter impacts during post-monsoon and winter seasons (Sasmitaa et al. 2022; Kumar and Middey 2022).

2.4 Cohort Studies

The number of epidemiological studies causally linking air quality and health is very limited in India. The Tamil Nadu Air Pollution and Health Effects (TAPHE) cohort study conducted by Balakrishnan et al. (2015) investigated the impact of fine particulate matter ($PM_{2.5}$) exposure in a rural-urban regional setting. The study finding revealed that an increment of 10 μg/m^3 in $PM_{2.5}$ exposure during the gestation period was associated with a 4 g (CI: 1.08–6.76) decrease in birth-weight and subsequently led to a 2% increase in the low birth-weight prevalence (OR: 1.02; CI: 1.00–1.04) (Balakrishnan et al. 2018). Another cohort—Ambient and Indoor Air Pollution in Pregnancy on the Risk of Low Birth Weight and Ensuing Effects in Infant (APPLE)—showed the impact of prenatal exposure to indoor and outdoor air pollutants during pregnancy on low birth weight (Shriyan et al. 2020). In 2015, Choi et al. showed that households cooking with clean fuel like LPG than households using kerosene experienced a lower incidence of respiratory complaints and illnesses among women and children (Choi et al. 2015).

It is important to note that the pieces of evidence demonstrating adverse health effects due to household air pollution primarily arise from the use of solid fuels such as crop residue, wood, and animal dung, particularly in rural areas. Women are very frequently exposed to high levels of particulate matter (PM) in indoor environments, especially when the kitchens are poorly ventilated and not separated from the rest of the living space. The majority of (non-working) women in India, as well as in most LMICs, spend most of their time indoors. As a result, a difference in exposure levels is expected, and the duration of time spent in the kitchen is likely to vary across India. In order to understand the health benefits of using clean fuel for household activities in four LMICs, Johnson et al. (2020) established a 'Household Air Pollution Intervention Network (HAPIN)' trial that includes South India. The data collected during HAPIN showed the impact of personal exposure to indoor air pollution on gestational blood pressure among women dependent on solid fuels for cooking purposes in rural Tamil Nadu (Ye et al. 2022). The Delhi Air Pollution and Health Effects (DAPHNE) was the only cohort focusing on the impact of in-utero exposure to ambient $PM_{2.5}$ on birthweight in India (https://www.urbanair-india.org/daphne) so far. DAPHNE also examined the ambient $PM_{2.5}$ exposure on respiratory outcomes of adolescents in Delhi (Mueller et al. 2022).

While there are substantial findings from developed countries indicating the health impacts of fine particulate matter, there remains a dearth of evidence from low- and middle-income countries' populations. Tonne et al. (2017) designed an observational cohort study titled "Cardiovascular Health Effects of Air Pollution in Telangana, India" (CHAI). The objective of the CHAI study was to assess the relationship between particulate matter and markers of atherosclerosis/cardiovascular diseases. A retrospective cohort study by Liao et al. (2022) provided evidence of the association between exposure—both during pregnancy and after delivery—and infant survival at the national level in India. A prospective cohort study conducted in Delhi, India, from 2010 to 2016 found a significant correlation between high ambient $PM_{2.5}$ exposure levels and an increased risk of systolic blood pressure and the occurrence of hypertension (Prabhakaran et al. 2020).

2.5 Cross-Sectional and Case-Control Studies

More recently, several cross-sectional studies strengthened the indigenous evidence of health impacts in India using the health data from the National Family Health Survey (NFHS). Goyal and Canning (2021) showed that the adjusted OR of low birthweight increases non-linearly from 1.09 (CI: 0.95, 1.26) to 1.40 (CI: 1.12, 1.75) for ambient $PM_{2.5}$ exposure 39.3–77.3 μg/m^3 respectively. A population-based cross-sectional study by Siddique et al. (2010) showed the influence of vehicular exposure on the behavioural activities of school children in Delhi. The study involved 969 school children across various regions within Delhi-NCR for attention deficit hyperactivity disorder (ADHD) screening and found that the children residing in polluted settings had a higher ADHD prevalence. Another study from the Delhi NCR region showed the association between ambient particulate matter exposure with lung function among the population of the age group 18–50 years (Kesavachandran et al. 2013). Agricultural biomass burning has majorly contributed to air pollution in India. Singh et al. (2021a, b) focused on four North Indian states and showed that the adult population living in close proximity to high-intensity biomass burning (primarily crop burning) is more likely to be hypertensive, while Singh and Dey (2021) demonstrated that exposure to such high-intensity burning events during pre- and post-natal period impact the height later in life. Child stunting (Spears et al. 2019), child anaemia (Mehta et al. 2021), child ARI (Odo et al. 2022), and child mortality (deSouza et al. 2022) are also found to be significantly associated with ambient $PM_{2.5}$ exposure.

Similarly, certain cross-sectional studies showed the association of household air pollution (HAP) exposure with reduced birth weight (Wylie et al. 2014; Islam and Mohanty 2021), low IQ (Brabhukumr et al. 2020), and childhood stunting (Islam et al. 2021). Agrawal (2012) found that there is a significantly higher risk of asthma in adult women living in households using unclean fuels than those living in households using cleaner fuels. James et al. (2020) reported that more than two-thirds of women using biomass fuel for cooking were positively associated with self-reported health symptoms such as respiratory, cardiovascular, and adverse obstetric outcomes. All these studies focusing on the health impacts of HAP concluded that reducing household unclean fuel use and enhancing the use of improved stove technology may decrease the chances of health effects due to HAP.

A cross-sectional study by Chaudhary et al. (2022) estimated the impact of ambient air pollutants on anaemia burden among Indian women of reproductive age (WRA; 15–49 years). They found out that with every 10 μg/m^3 increase in $PM_{2.5}$, there was a 7.23% (UI: 6.8%, 7.6%) increase in anaemia burden and a decrease in haemoglobin level by 0.4 g/L (UI: 0.5, 0.4) among WRA. Figure 2.4 represents the exposure-response relationships of ambient $PM_{2.5}$ with anaemia prevalence and haemoglobin level (from NFHS-4) among WRA in India developed in this study, which can be utilized to estimate the health benefits of exposure reduction on a population scale.

A case-control study conducted in Delhi, West Bengal, and Uttarakhand showed an urban disadvantage over rural in terms of risk of reduced lung function both in boys and girls (Siddique et al. 2010). The case (exposed) group included 5671 children from Delhi, whereas the control group included 2245 children from rural settings of West Bengal and Uttarakhand. Another case-control study conducted in India showed the differential risks of HAP from wood and coal burning associated with hypopharyngeal and lung cancer (Sapkota et al. 2008). This study showed that the participants who always relied on coal for cooking purposes had a higher risk of lung cancer than the individuals who never used coal.

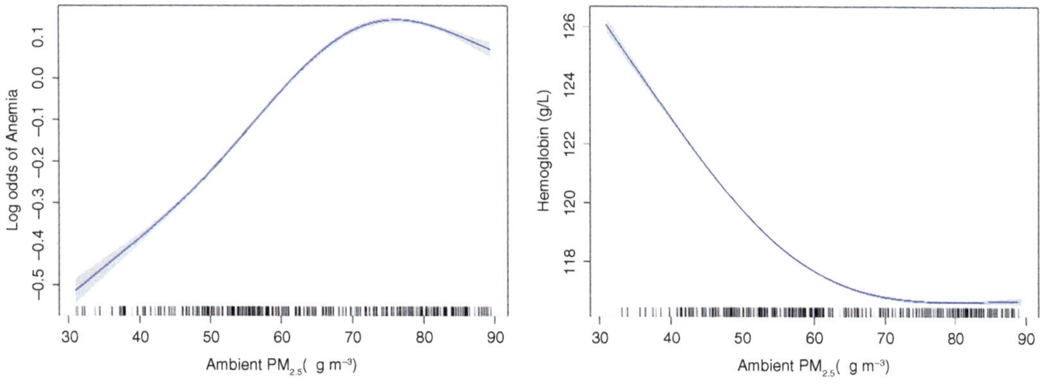

Fig. 2.4 The exposure-response curve between chronic exposure to $PM_{2.5}$ (2007–2016) and log odds of anaemia (left panel). The right panel represents a significant decrease in haemoglobin level with increased $PM_{2.5}$ exposure. (The statistics are taken from Chaudhary et al. 2022, Creative Commons Attribution BY 4.0)

2.6 Time-Series Studies

Many city-level time-series studies have featured an association between various air pollutants with cardiorespiratory or all-cause mortality and hospital admissions in India. The city-level studies hold importance as the evidence can be crucial for air quality management and policy development at a fine spatial scale. An impetus to time series studies for evidencing the impact of air pollution on mortality was by way of the Health Effects Institute call for multicity—Public Health and Air Pollution in Asia (PAPA) study in India. This research call initiated pioneering studies estimating the excess mortality due to acute, i.e., daily, air pollution in two cities: Delhi and Chennai. These two studies became a reference point for all subsequent short-term impact studies until now. Till recently, all time-series studies in India were restricted to PM_{10} as the primary pollutant. Lately, a few studies estimated trends in short-term exposure effects of $PM_{2.5}$ on all-cause non-accidental mortality (Singh et al. 2021; Joshi et al. 2021). Due to gaps in ground-based pollutant data, there were studies that also utilized proxies of respirable suspended particulate matter in terms of haze and visibility to assess the impact on short-term natural mortality (Kumar et al. 2010). The time series design is increasingly being put to use for a variety of health endpoints, the latest being daily COVID-19 infections in association with ambient air pollutants $PM_{2.5}$, PM_{10}, O_3, NO_2, SO_2, and CO levels in NCT Delhi (Singh 2021). As compared to $PM_{2.5}$, gaseous pollutant NO_2 showed a stronger effect on daily all-cause mortality in highly polluted urban centres such as Delhi and Mumbai (Maji et al. 2017a, b).

In such time series studies assessing short-term health impacts, the most popularly used method is the generalized additive model (GAM). The GAM-based regressions have the advantage that they allow the accounting of non-parametric smooth functions, which is important to model non-linear dependence of deaths/hospital admissions etc., on metrological factors such as temperature and relative humidity. There is also a scope to include dependent variables from a wide variety of distributions to model the effects of pollutant predictor variables on them. The other advantages of GAM design include: (a) each additive term is estimated using a univariate smoother, avoiding multicollinearity, (b) the results are easily interpretable to understand the unit change in the dependent variable with corresponding independent variables (Singh 2021). Figure 2.5 shows the dose-response curves of Delhi for ambient $PM_{2.5}$ typically calculated using the GAM model (Joshi et al. 2021).

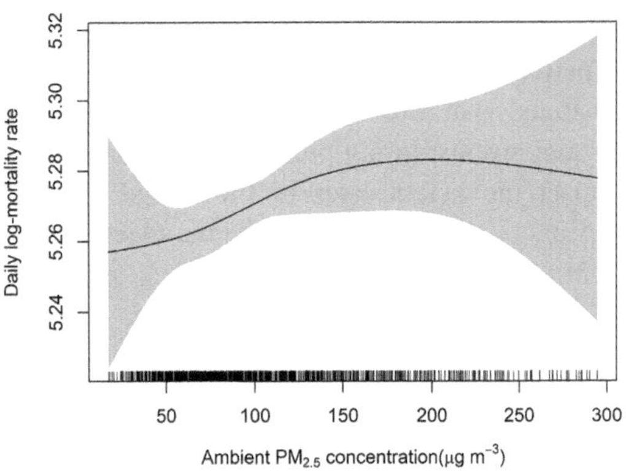

Fig. 2.5 Dose-response relationship 3-day cumulative exposure of $PM_{2.5}$ for Delhi. (The statistics are taken from Joshi et al. 2021, Creative Commons Attribution BY 4.0)

2.7 Challenges and Way Forward

Although a plethora of epidemiologic studies have been conducted in recent times in India, several challenges still remain.

2.7.1 Accurate Exposure Attribution

The robustness of any health burden estimation depends to a major extent on the representativeness of the attributed exposure on a personal level. In prospective cohorts, personal exposure is measured using portable sensors. However, scaling up personal exposure measurement at a population level is not realistic. For ambient exposure, measurements from continuous ground monitors are usually used. In India, the ground-based monitoring network is disproportionately distributed, and the mean distance to the nearest monitor is more than 80 km (Martin et al. 2019). Therefore, ambient air pollution exposure relying on such measurements has large uncertainty.

Advancements in modelling techniques allow estimating $PM_{2.5}$ (and other pollutants) exposure at a large spatial scale using satellite remote sensing data products. For particulate matters, aerosol optical depth (AOD) is converted to either $PM_{2.5}$ or PM_{10} using statistical (Ma et al. 2022) or a scaling-factor-based approach (Donkelaar et al. 2010). Over the years, satellite-derived particulate matter retrieval became standardized (Donkelaar et al. 2015; Shaddick et al. 2018), and local calibration further improved the accuracy of such products (Dey et al. 2012). Recently, a national database of particulate matter under 2.5 μg, was created for India at 1 km spatial and 24-h temporal resolution (Dey et al. 2020), and the data has been used ever since in many epidemiological studies. The other alternative is machine learning (Liao et al. 2022) or land use regression modelling (Morley and Gulliver 2018) to estimate exposure.

It must be noted that the GBD estimates are increasingly based on exposures generated through a hybrid approach. The uncertainty in exposure could vary by a large margin. Figure 2.6 depicts that accuracy is much less in LMIC

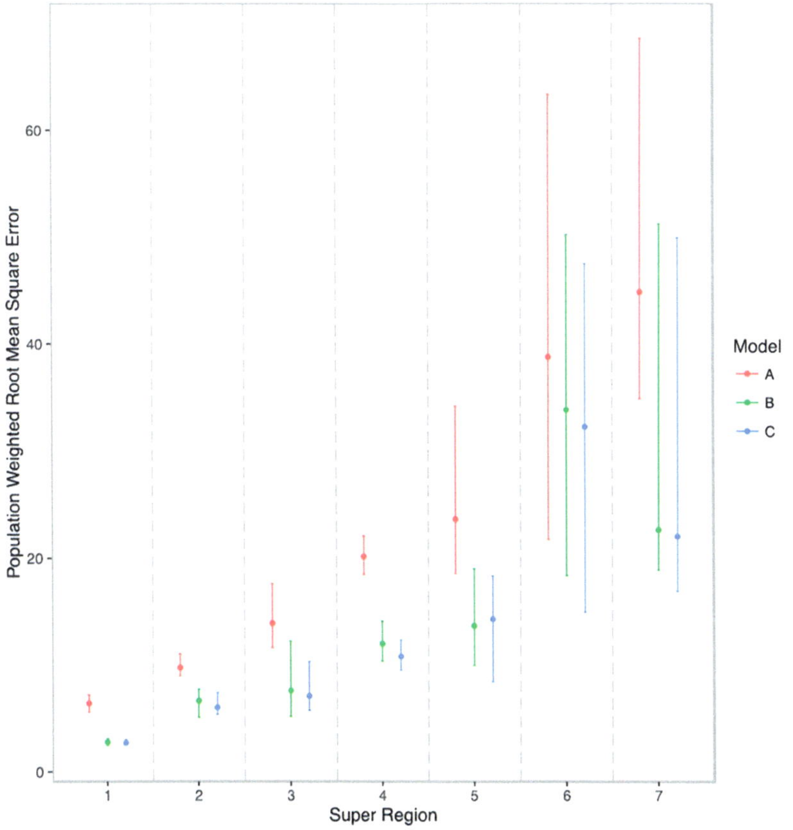

Fig. 2.6 The root means square error of predicted $PM_{2.5}$, results from three models across super-regions. The dots depict the median of the distribution and the vertical lines denote the range of values; from 25 simulations. Model A shows concentrations derived from satellite and chemical transport data; Model B is a variation of Model A, with local population numbers, and Model C is upgraded Model B, along with particle species and monitors elevation. The super regions are: (1) High-Income countries, (2) Central Europe, Eastern Europe, and Central Asia, (3) Latin America and Caribbean, (4) South East Asia, East Asia, and Oceania, (5) North Africa and the Middle East, (6) Sub-Saharan Africa, and (7) South Asia. (The figure is adapted from Ostro et al. 2018)

regions where ground-level monitoring is sketchy or even absent. Significant efforts are required to further improve ambient exposure in India.

Nonetheless, all these approaches opened up the possibility of carrying out retrospective analysis with past cohort data.

However, these methods cannot be applied to household exposure assessment. In India, household exposure is modelled based on the concentration derived from the solid fuel use prevalence and limited indoor measurements to adjust for the exposure window (Balakrishnan et al. 2019). More actual measurements in varying indoor microenvironments are required to further improve household exposure assessment in the future. The ongoing cohorts can be utilized for this purpose.

2.7.2 Reconstructing Personal Exposure and Exposure Apportionment

Health impacts are manifested due to air pollution exposure in both ambient and household modes. Burden estimates independently for either ambient or household $PM_{2.5}$ are affected by exposure misclassification or exposure overlap. The ideal way forward would be reconstructing personal exposure for a target population. Data need to be collected systematically for indoor concentrations representative of various types of microenvironments (e.g., urban neighbourhoods, offices, educational institutions, commercial locations, etc.), simultaneously with ambient concentrations across the major air sheds of India. The outdoor-indoor ratio will allow the reconstruction of personal exposure based on the time-activity profiles of the population that can be collected through surveys. The limited personal exposure data collected through various cohorts can be used as benchmarks to evaluate personal exposure reconstruction.

For air quality management, source apportionment is the foremost requirement. However, exposure may drastically vary due to socio-demographic determinants. India should focus on exposure apportionment to prioritize sectoral interventions for maximum health benefits. For ambient exposure, chemical transport models are a great tool to provide sectoral contributions at a regional scale. The data can be integrated with population distribution for better exposure apportionment. For household exposure, this can be achieved in a hybrid mode—first traditional sampling of PM, chemical speciation in the lab, and application of tools like positive matrix factorization (Jain et al. 2020) or chemical mass balance (Chow et al. 2015) to determine source characterization, and then integrate with models predicting the influx of ambient air indoor. Such an approach will allow health burden estimates tagged to sources and an understanding of the health burden disparity in view of contributions to emissions and exposure.

2.7.3 Moving Beyond $PM_{2.5}$ Mass and Elucidating the Biological Pathway

Recent global evidence suggests health effects beyond cardiovascular and lung diseases, including metabolic and neurodegenerative diseases (Thurston et al. 2017). These health endpoints have not been accounted in the GBD estimates so far, and in doing so, further evidence from local Indian cohort studies needs accounting. Multicentric cohorts are the need of the hour to guide the integrated exposure-risk function used in the GBD at the high exposure level.

The GBD framework also has certain underlying uncertainties for the LMICs such as India, which interestingly contribute to one-half of the total global mortality. Firstly, owing to a large and growing ratio of youthful population in the total demographic scene, there may be masking of the actual cumulative cardiopulmonary effects from air pollution exposures in the current GBD estimates. The actual effects are likely to be observable only years later (Apte et al. 2015). Secondly, the GBD exposure-risk functions assume uniform toxicity across the individual species. Studies (Ostro et al. 2007; Janssen et al. 2011; Lippmann et al. 2013; Chowdhury et al. 2022) suggest that the health impacts could be different for different $PM_{2.5}$ species. In India, the source characterization drastically varies with region and season. India should adopt a dedicated program to carry out epidemiological studies to unravel the differential impacts of individual $PM_{2.5}$ species tagged to key sources. India-specific exposure-risk functions will provide better health benefit assessments linked to clean air targets and should be the most critical indicator in redefining its national air quality standards.

From establishing association (using a cross-sectional design) to causality (using cohorts), standard epidemiological study designs are capable of providing robust evidence of air pollution health impacts. However, it is also essential to understand the biological pathways so that early indication can be detected. Lab-based studies and biomarker analysis can help in that direction (Clark et al. 2013; Barr et al. 2020). Most importantly, a funding commitment is critical to plan and execute long-term cohorts and achieve this goal.

Acknowledgements The authors acknowledge financial support from the Clean Air Fund under 'Health and Climate Co-Benefits of Pollution Management' project and grant under SUPRA Scheme, SERB, Department of Science and Technology, GoI (SPR/). SD acknowledges financial support for the Institute Chair position and DST-FIST program (SR/FST/ESII-016/2014).

References

Afgan FR, Patidar SK. Health impacts assessment due to PM2.5, PM10 and NO2 exposure in National Capital Territory (NCT) Delhi. Pollution. 2020;6(1):115–26.

Agrawal S. Effect of indoor air pollution from biomass and solid fuel combustion on prevalence of self-reported asthma among adult men and women in India: findings from a nationwide large-scale cross-sectional survey. J Asthma. 2012;49(4):355–65.

Apte JS, Marshall JD, Cohen AJ, Brauer M. Addressing global mortality from ambient PM2.5. Environ Sci Technol. 2015;49(13):8057–66.

Balakrishnan K, Sambandam S, Ramaswamy P, Ghosh S, Venkatesan V, Thangavel G, et al. Establishing integrated rural–urban cohorts to assess air pollution-related health effects in pregnant women, children and adults in Southern India: an overview of objectives, design and methods in the Tamil Nadu Air Pollution and Health Effects (TAPHE) study. BMJ Open. 2015;5(6):e008090.

Balakrishnan K, Ghosh S, Thangavel G, Sambandam S, Mukhopadhyay K, Puttaswamy N, et al. Exposures to fine particulate matter (PM2.5) and birthweight in a rural-urban, mother-child cohort in Tamil Nadu, India. Environ Res. 2018;161:524–31.

Balakrishnan K, Dey S, Gupta T, Dhaliwal RS, Brauer M, Cohen A, Stanaway JD, Beig G, Joshi TK, Aggarwal AN, Sabde Y, Sadhu H, Frostad J, Causey K, Godwin W, Shukla DK, Kumar GA, Varghese CM, Muraleedharan P, Agrawal A, Anjana RM, Bhansali A, Bhardwaj D, Burkart K, Cercy K, Chakma JK, Chowdhury S, Christopher DJ, Dutta E, Furtado M, Ghosh S, Ghoshal A, Glenn SD, Gupta R, Jeemon P, Kant R, Kant S, Kaur T, Koul PA, Krish V, Krishna B, Larson SL, Madhipatla KK, Mahesh PA, Mohan V, Mukhopadhyay S, Mutreja P, Naik N, Nair S, Nguyen G, Odell CM, Pandian JD, Prabhakaran D, Prabhakaran P, Roy A, Salvi S, Sambandam S, Saraf DS, Sharma M, Shrivastava A, Singh V, Tandon N, Thomas NJ, Torre A, Xavier D, Yadav G, Singh S, Shekhar C, Guleria R, Vos T, Dandona R, Reddy KS, Lim SS, Murray CJL, Venkatesh S, Dandona L. The impact of air pollution on deaths, disease burden, and life expectancy across the states of India: the Global Burden of Disease Study 2017. Lancet Planet Health. 2019;3(1):26–39.

Barr DB, Puttaswamy N, Jaacks LM, Steenland K, Rajkumar S, Gupton S, Ryan PB, Balakrishnan K, Peel JL, Checkley W, Clasen T, Clark ML, (HAPIN Investigative Team). Design and rationale of the Biomarker Center of the Household Air Pollution Intervention Network (HAPIN) Trial. Environ Health Perspect. 2020;128(4):47010.

Beverland IJ, Cohen GR, Heal MR, Carder M, Yap C, Robertson C, et al. A comparison of short-term and long-term air pollution exposure associations with mortality in two cohorts in Scotland. Environ Health Perspect. 2012;120(9):1280–5.

Brabhukumr A, Malhi P, Ravindra K, Lakshmi PVM. Exposure to household air pollution during first 3 years of life and IQ level among 6–8-year-old children in India—a cross-sectional study. Sci Total Environ. 2020;709:135110.

Burnett R, Cohen A. Relative risk functions for estimating excess mortality attributable to outdoor PM2.5 air pollution: evolution and state-of-the-art. Atmosphere. 2020;11(6):589. https://doi.org/10.3390/atmos11060589.

Burnett RT, Arden Pope C, Ezzati M, Olives C, Lim SS, Mehta S, Shin HH, et al. An integrated risk function for estimating the global burden of disease attributable to ambient fine particulate matter exposure. Environ Health Perspect. 2014;122(4):397–403. https://doi.org/10.1289/ehp.1307049.

Burnett R, Chen H, Szyszkowicz M, Fann N, Bryan Hubbell C, Pope A, Apte JS, et al. Global estimates of mortality associated with long-term exposure to outdoor fine particulate matter. Proc Natl Acad Sci. 2018;115(38):9592–7. https://doi.org/10.1073/pnas.1803222115.

Chaudhary E, Dey S, Ghosh S, Sharma S, Singh N, Agarwal S, et al. Reducing the burden of anaemia in Indian women of reproductive age with clean-air targets. Nat Sustain. 2022;5(11):939–46.

Choi JY, Baumgartner J, Harnden S, Alexander BH, Town RJ, D'Souza G, Ramachandran G. Increased risk of respiratory illness associated with kerosene fuel use among women and children in urban Bangalore, India. Occup Environ Med. 2015;72(2):114–22.

Chow JC, Lowenthal DH, Chen LW, Wang X, Watson JG. Mass reconstruction methods for $PM_{2.5}$: a review. Air Qual Atmos Health. 2015;8(3):243–63.

Chowdhury S, Pozzer A, Dey S, Klingmueller K, Lelieveld J. Changing risk factors that contribute to premature mortality from ambient air pollution between 2000 and 2015. Environ Res Lett. 2020;15:074010.

Chowdhury S, Pozzer A, Haines A, Klingmüller K, Münzel T, Paasonen P, et al. Global health burden of ambient PM2.5 and the contribution of anthropogenic black carbon and organic aerosols. Environ Int. 2022;159:107020.

Clark ML, Peel JL, Balakrishnan K, Breysse PN, Chillrud SN, Naeher LP, Rodes CE, Vette AF, Balbus JM. Health and household air pollution from solid fuel use: the need for improved exposure assessment. Environ Health Perspect. 2013;121(10):1120–8.

Cohen AJ, Brauer M, Burnett R, Anderson HR, Frostad J, Estep K, et al. Estimates and 25-year trends of the global burden of disease attributable to ambient air pollution: an analysis of data from the Global Burden of Diseases Study 2015. Lancet. 2017;389(10082):1907–18.

deSouza PN, Dey S, Mwenda KM, Kim R, Subramanian SV, Kinney P. Robust relationship between ambient air pollution and infant mortality in India. Sci Total Environ. 2022;815:152755.

Dey S, Girolamo L, Donkelaar A, Tripathi S, Gupta T, Mohan M. Variability of outdoor fine particulate (PM2.5) concentration in the Indian Subcontinent: a remote sensing approach. Remote Sens Environ. 2012;127:153–61.

Dey S, Purohit B, Balyan P, Dixit K, Bali K, Kumar A, Imam F, Chowdhury S, Ganguly D, Gargava P, Shukla V. A satellite-based high-resolution (1-km) ambient $PM_{2.5}$ database for India over two decades (2000–2019): applications for air quality management. Remote Sens. 2020;12(23):3872.

Donkelaar A, Martin R, Brauer M, Kahn R, Levy R, Verduzco C, Villeneuve P. Global estimates of ambient fine particulate matter concentrations from satellite-based aerosol optical depth: development and application. Environ Health Perspect. 2010;118(6):847–55.

Donkelaar A, Martin R, Brauer M, Boys B. Use of satellite observations for long-term exposure assessment of global concentrations of fine particulate matter. Environ Health Perspect. 2015;123(2):135–43.

Gawande U, Khanvilkar A, Kadam S, Salvitthal G. Effects of ambient air pollution on respiratory health of adults: findings from a cross-sectional study in Chandrapur, Maharashtra, India. Int J Res Med Sci. 2016;4:1546–57.

GBD 2019 Risk Factors Collaborators. Global burden of 87 risk factors in 204 countries and territories, 1990–2019: a systematic analysis for the Global Burden of Disease Study 2019. Lancet. 2020;396(10258):1223–49. https://doi.org/10.1016/S0140-6736(20)30752-2.

Ghosh A, Mukherji A. Air pollution and respiratory ailments among children in urban India: exploring causality. Econ Dev Cult Chang. 2014;63(1):191–222.

Goyal N, Canning D. The association of in-utero exposure to ambient fine particulate air pollution with low birth weight in India. Environ Res Lett. 2021;16(5):054034.

Gupta U. Valuation of urban air pollution: a case study of Kanpur City in India. Environ Resour Econ. 2008;41(3):315–26.

Guttikunda SK, Goel R. Health impacts of particulate pollution in a megacity—Delhi, India. Environ Dev. 2013;6:8–20.

Guttikunda SK, Jawahar P. Application of SIM-air modeling tools to assess air quality in Indian cities. Atmos Environ. 2012;62:551–61.

Haque M, Singh RB. Air pollution and human health in Kolkata, India: a case study. Climate. 2017;5(4):77.

Islam S, Mohanty SK. Maternal exposure to cooking smoke and risk of low birth weight in India. Sci Total Environ. 2021;774:145717.

Islam S, Rana MJ, Mohanty SK. Cooking, smoking, and stunting: effects of household air pollution sources on childhood growth in India. Indoor Air. 2021;31(1):229–49.

Jain S, Sharma SK, Vijayan N, Mandal TK. Seasonal characteristics of aerosols (PM2.5 and PM10) and their source apportionment using PMF: a four year study over Delhi, India. Environ Pollut. 2020;262:114337.

James BS, Shetty RS, Kamath A, Shetty A. Household cooking fuel use and its health effects among rural women in southern India—a cross-sectional study. PLoS One. 2020;15(4):e0231757.

Janssen NA, Hoek G, Simic-Lawson M, Fischer P, Van Bree L, Ten Brink H, et al. Black carbon as an additional indicator of the adverse health effects of airborne particles compared with PM10 and PM2.5. Environ Health Perspect. 2011;119(12):1691–9.

Johnson MA, Steenland K, Piedrahita R, Clark ML, Pillarisetti A, Balakrishnan K, et al. Air pollutant exposure and stove use assessment methods for the household air pollution intervention network (HAPIN) trial. Environ Health Perspect. 2020;128(4):047009.

Joshi P, Ghosh S, Dey S, Dixit K, Choudhary RK, Salve HR, Balakrishnan K. Impact of acute exposure to ambient $PM_{2.5}$ on non-trauma all-cause mortality in the megacity Delhi. Atmos Environ. 2021;259:118548.

Kesavachandran C, Pangtey BS, Bihari V, Fareed M, Pathak MK, Srivastava AK, Mathur N. Particulate matter concentration in ambient air and its effects on lung functions among residents in the National Capital Region, India. Environ Monitor Assess. 2013;185(2):1265–72.

Kesavachandran CN, Kamal R, Bihari V, Pathak MK, Singh A. Particulate matter in ambient air and its association with alterations in lung functions and respiratory health problems among outdoor exercisers in National Capital Region, India. Atmos Pollut Res. 2015;6(4):618–25.

Kumar N, Foster A. Respiratory health effects of air pollution in Delhi and its neighboring areas, India. Environ Monit Assess. 2007;135:313–25.

Kumar N, Middey A. Prognosis of boundary layer assimilative capacity over a landlocked urban district in India: a prelude to exposure risk assessment. Urban Clim. 2022;41:101076.

Kumar R, Sharma M, Srivastva A, Thakur JS, Jindal SK, Parwana HK. Association of outdoor air pollution with chronic respiratory morbidity in an industrial town in northern India. Arch Environ Health. 2004;59(9):471–7.

Kumar R, Sharma S, Thakur JS, Lakshmi PVM, Sharma MK, Singh T. Association of air pollution and mortality in the Ludhiana city of India: A time-series study Indian Journal of Public Health. 2010;54(2):98–103.

Liao J, Liu Y, Steenland K, Pillarisetti A, Thompson LM, Dey S, et al. Child survival and early lifetime exposures to ambient fine particulate matter in india: a retrospective cohort study. Environ Health Perspect. 2022;130(1):017009.

Lippmann M, Chen LC, Gordon T, Ito K, Thurston GD. National Particle Component Toxicity (NPACT) Initiative: integrated epidemiologic and toxicologic studies of the health effects of particulate matter components. Res Rep Health Eff Inst. 2013;177:5–13.

Liu HY, Bartonova A, Schindler M, Sharma M, Behera S, Katiyar K, Dikshit O. Respiratory disease in relation to outdoor air pollution in Kanpur, India. Arch Environ Occup Health. 2013;68(4):204–17. https://doi.org/10.1080/19338244.2012.701246.

Ma Z, Dey S, Christopher S, Liu R, Bi J, Balyan P, Liu Y. A review of statistical methods used for developing large-scale and long-term PM2.5 models from satellite data. Remote Sens Environ. 2022;269:112827.

Maji S, Ahmed S, Siddiqui WA. Air quality assessment and its relation to potential health impacts in Delhi, India. Curr Sci. 2015;109(5):902–9.

Maji KM, Dikshit AK, Deshpande A. Human health risk assessment due to air pollution in 10 urban cities in Maharashtra, India. Cogent Environ Sci. 2016;2(1):1193110.

Maji S, Ahmed S, Siddiqui WA, Ghosh S. Short term effects of criteria air pollutants on daily mortality in Delhi, India. Atmos Environ. 2017a;150:210–9.

Maji K, Dikshit A, Chaudhary R. Human health risk assessment due to air pollution in the megacity Mumbai in India. Asian J Atmos Environ. 2017b;11(2):61–70.

Maji S, Ghosh S, Ahmed S. Association of air quality with respiratory and cardiovascular morbidity rate in Delhi, India. Int J Environ Health Res. 2018;28(5):471–90.

Manojkumar N, Srimuruganandam B. Health effects of particulate matter in major Indian cities. Int J Environ Health Res. 2021;31:3.

Martin R, Brauer M, Donkelaara A, Shaddick G, Narang G, Dey S. No one knows which city has the highest concentration of fine particulate matter. Atmos Environ X. 2019;3:100040.

Mehta U, Dey S, Chowdhury S, Ghosh S, Hart JE, Kurpad A. The association between ambient PM2.5 exposure and anemia outcomes among children under five years of age in India. Environ Epidemiol. 2021;5(1):e125.

Morley D, Gulliver J. A land use regression variable generation, modelling and prediction tool for air pollution exposure assessment. Environ Model Softw. 2018;105:17–23.

Mueller W, Wilkinson P, Milner J, Loh M, Vardoulakis S, Petard Z, Cherrie M, Puttaswamy N, Balakrishnan K, Arvind DK. The relationship between greenspace and personal exposure to PM2.5 during walking trips in Delhi, India. Environ Pollut. 2022;305:119294. https://doi.org/10.1016/j.envpol.2022.119294.

Odo DB, Yang IA, Dey S, Hammer MS, van Donkelaar A, Martin RV, et al. Ambient air pollution and acute respiratory infection in children aged under 5 years living in 35 developing countries. Environ Int. 2022;159:107019.

Ostro B, Feng WY, Broadwin R, Green S, Lipsett M. The effects of components of fine particulate air pollution on mortality in California: results from CALFINE. Environ Health Perspect. 2007;115(1):13–9.

Ostro B, Spadaro J, Gumy S, Mudu P, Awe Y, Forastiere F, Peters A. Assessing the recent estimates of the global burden of disease for ambient air pollution: methodological changes and implications for low- and middle-income countries. Environ Res. 2018;166:713–25.

Pandey A, Brauer M, Cropper ML, Balakrishnan K, Mathur P, Dey S, Turkgulu B, Kumar GA, Khare M, Beig G, Gupta T, Krishnankutty RP, Causey K, Cohen AJ, Bhargava S, Aggarwal AN, Agrawal A, Awasthi S, Bennitt F, Bhagwat S, Bhanumati P, Burkart K, Chakma J, Chiles TC, Chowdhury S, Christopher DJ, Dey S, Fisher S, Fraumeni B, Fuller R, Ghoshal AG, Golechha MJ, Gupta PC, Gupta R, Gupta R, Gupta S, Guttikunda S, Hanrahan D, Harikrishnan S, Jeemon P, Joshi TK, Kant R, Kant S, Kaur T, Koul PA, Kumar P, Kumar R, Larson SL, Lodha R, Madhipatla KK, Mahesh PA, Malhotra R, Managi S, Martin K, Mathai M, Mathew JL, Mehrotra R, Mohan BVM, Mohan V, Mukhopadhyay S, Mutreja P, Naik N, Nair S, Pandian JD, Pant P, Perianayagam A, Prabhakaran D, Prabhakaran P, Rath GK, Ravi S, Roy A, Sabde YD, Salvi S, Sambandam S, Sharma B, Sharma M, Sharma S, Sharma RS, Shrivastava A, Singh S, Singh V, Smith R, Stanaway JD, Taghian G, Tandon N, Thakur JS, Thomas NJ, Toteja GS, Varghese CM, Venkataraman C, Venugopal KN, Walker KD, Watson AY, Wozniak S, Xavier D, Yadama GN, Yadav G, Shukla DK, Bekedam HJ, Reddy KS, Guleria R, Vos T, Lim SS, Dandona R, Kumar S, Kumar P, Landrigan PJ, Dandona L. Health and economic impact of air pollution in the states of India: the Global Burden of Disease Study 2019. Lancet Planet Health. 2021;5(1):e25–38.

Patankar AM, Trivedi PL. Monetary burden of health impacts of air pollution in Mumbai, India: implications for public health policy. Public Health. 2011;125(3):157–64.

Prabhakaran D, Mandal S, Krishna B, Magsumbol M, Singh K, Tandon N, et al. Exposure to particulate matter is associated with elevated blood pressure and incident hypertension in urban India. Hypertension. 2020;76(4):1289–98.

Rajak R, Chattopadhyay A. Short and long term exposure to ambient air pollution and impact on health in India: a systematic review. Int J Environ Health Res. 2020;30(6):593–617.

Sacks J, Fann N, Gumy S, Kim I, Ruggeri G, Mudu P. Quantifying the public health benefits of reducing air pollution: critically assessing the features and capabilities of WHO's AirQ+ and U.S. EPA's environmental benefits mapping and analysis program—community edition (BenMAP—CE). Atmosphere. 2020;11:516.

Sapkota A, Gajalakshmi V, Jetly DH, Roychowdhury S, Dikshit RP, Brennan P, et al. Indoor air pollution from solid fuels and risk of hypopharyngeal/laryngeal and lung cancers: a multicentric case–control study from India. Int J Epidemiol. 2008;37(2):321–8.

Sasmitaa S, Kumar DB, Priyadharshini B. Assessment of sources and health impacts of PM10 in an urban environment over eastern coastal plain of India. Environ Chall. 2022;7:100457.

Sehgal M, Suresh R, Sharma VP, Gautam SK. Assessment of outdoor workers' exposure to air pollution in Delhi (India). Int J Environ Stud. 2015;72(1):99–116.

Shaddick G, Thomas ML, Amini H, Broday D, Cohen A, Frostad J, Green A, Gumy S, Liu Y, Martin RV, Pruss-Ustun A, Simpson D, Donkelaar A, Brauer M. Data integration for the assessment of population exposure to ambient air pollution for global burden of disease assessment. Environ Sci Technol. 2018;52(16):9069–78.

Shriyan P, Babu GR, Ravi D, Ana Y, van Schayck OCP, Thankachan P, Murthy GVS. Ambient and indoor air pollution in pregnancy and the risk of low birth weight and ensuing effects in infants (APPLE): a cohort study in Bangalore, South India. Wellcome Open Res. 2020;3:133. https://doi.org/10.12688/wellcomeopenres.14830.3.

Siddique S, Banerjee M, Ray MR, Lahiri T. Air pollution and its impact on lung function of children in Delhi, the capital city of India. Water Air Soil Pollut. 2010;212(1):89–100.

Singh A. Ambient air pollution and COVID-19 in Delhi, India: a time-series evidence. Int J Environ Health Res. 2021;32(11):2575–88. https://doi.org/10.1080/09603123.2021.1977258.

Singh P, Dey S. Crop burning and forest fires: long-term effect on adolescent height in India. Resour Energy Econ. 2021;65:101244.

Singh P, Roy A, Bhasin D, Kapoor M, Ravi S, Dey S. Crop fires and cardiovascular health—a study from north India. SSM Popul Health. 2021a;14:100757.

Singh N, Mhawish A, Banerjee T, Ghosh S, Singh RS, Mall RK. Association of aerosols, trace gases and black carbon with mortality in an urban pollution hotspot over central Indo-Gangetic Plain. Atmos Environ. 2021b;246:118088.

Spears D, Dey S, Chowdhury S, Scovronick N, Vyas S, Apte J. The association of early-life exposure to ambient PM 2.5 and later-childhood height-for-age in India: an observational study. Environ Health. 2019;18(1):1–10.

Thurston GD, Kipen H, Annesi-Maesano I, Balmes J, Brook RD, Cromar K, et al. A joint ERS/ATS policy statement: what constitutes an adverse health effect of air pollution? An analytical framework. Eur Respir J. 2017;49(1):1600419.

Tobollik M, Razum O, Wintermeyer D, Plass D. Burden of outdoor air pollution in Kerala, India—a first health risk assessment at state level. Int J Environ Res Public Health. 2015;12(9):10602–19.

Tonne C, Salmon M, Sanchez M, Sreekanth V, Bhogadi S, Sambandam S, et al. Integrated assessment of exposure to PM2.5 in South India and its relation with cardiovascular risk: design of the CHAI observational cohort study. Int J Hyg Environ Health. 2017;220(6):1081–8.

Wylie BJ, Coull BA, Hamer DH, Singh MP, Jack D, Yeboah-Antwi K, et al. Impact of biomass fuels on pregnancy outcomes in central East India. Environ Health. 2014;13(1):1–9.

Ye W, Thangavel G, Pillarisetti A, Steenland K, Peel JL, Balakrishnan K, et al. Association between personal exposure to household air pollution and gestational blood pressure among women using solid cooking fuels in rural Tamil Nadu, India. Environ Res. 2022;208:112756.

Air Quality Monitoring Using Geospatial Technology: A Bibliometric Analysis (1998–2021)

3

Manoranjan Mishra, Desul Sudarsan, Santosh Kumar Beja, Tamoghna Acharya, Celso Augusto Guimarães Santos, Richarde Marques da Silva, Shreerup Goswami, and Debdeep Bhattacharyya

3.1 Introduction

Air pollution is a global concern affecting millions of people around the world and creating serious health concerns due to exposure to air pollutants like particulate matter (PM), nitrogen dioxide (NO_2), sulfur dioxide (SO_2), and ozone (O_3) (Buonanno and Hänninen 2018; Cohen et al. 2017). World Health Organization (WHO) estimated that nearly 4.2 million people died due to direct or indirect impact of air pollution. The problem is going to escalate as nearly 90% of the global population live in lower air quality conditions. Consequently, a great deal of research activities have focused to understand the sources, transport mechanisms transformation processes of air pollutants at local, regional, and global scales (Intergovernmental Panel on Climate Change 2014; Kim et al. 2020). In situ air quality monitoring is critical for environmental policy and public health but is limited to small spatial and temporal scales. Hence over the years remote sensing combined with geographic information systems known as geospatial technology-based assessment of air quality has emerged as a popular tool even in developing and underdeveloped countries wherein air quality standards can be studied across large space (local, regional, and global scales) and time scales (Hadjimitsis et al. 2012; Levelt et al. 2018).

Monitoring of air pollutants by using satellite sensors like Landsat, the Advanced Very High-Resolution Radiometer (AVHRR), and GOES mission started in 1970. It was used to observe particles over the ocean and later sulfate from

M. Mishra (✉)
Department of Geography, Fakir Mohan University, Balasore, Odisha, India

D. Sudarsan
Department of Library and Information Science, Berhampur University, Berhampur, Odisha, India

S. K. Beja · D. Bhattacharyya
Department of Environmental Science, Berhampur University, Berhampur, Odisha, India

T. Acharya
Department of Marine Sciences, Berhampur University, Bhanjabihar, Odisha, India

C. A. G. Santos
Department of Civil and Environmental Engineering, Federal University of Paraíba, João Pessoa, PB, Brazil

R. M. da Silva
Department of Geosciences, Federal University of Paraíba, João Pessoa, PB, Brazil

S. Goswami
Department of Geology, Utkal University, Bhubaneswar, Odisha, India

P. K. Padhy et al. (eds.), *Air Quality and Human Health*,
https://doi.org/10.1007/978-981-97-1363-9_3

volcanoes (Fraser 1976; Mekler et al. 1977). In the late 1970s, Ozone was measured by Solar Backscatter Ultraviolet Radiometer (SBUV) and TOMS (Total Ozone Mapping Spectrometer) (Heath et al. 1975; Martin 2008). Further, with the improvement of technology, the sensors like GOME (Global Ozone Monitoring Experiment), OMI (the Ozone Monitoring Instrument), SCIAMACHY (Scanning Imaging Absorption Spectrometer for Atmospheric Cartography), TROPOMI (Tropospheric Monitoring Instrument), OMPS (Ozone Mapping Profiler Suite), Terra (MOPITT, MODIS, MISR), and CALIOP were used to measure tropospheric trace gas concentrations [NO_2, O_3, SO_2, carbon monoxide (CO)] (Bovensmann et al. 1999; Flynn et al. 2014; Levelt et al. 2018; Martin et al. 2018; Munro et al. 2016; Veefkind et al. 2012). These satellite-based imageries not only help in retrieving trace gases and aerosols but are also useful in forecasting air quality at a regional scale. However, there is no systematic review or bibliometric analysis of research works done on air quality monitoring using geospatial technology. This chapter aims to bridge this knowledge gap.

3.2 Material and Methods

Researchers are using various qualitative and quantitative methods to evaluate existing literature within a specific domain to gain a comprehensive understanding and organization of prior findings. Among these methods, bibliometrics offers a systematic, transparent, and reproducible approach, leveraging statistical measurements to assess scientific output, researchers, and scientific activities with greater objectivity and reliability (Sudarsana and Baba 2019). Figure 3.1 illustrates the research framework of this study.

The first step in the bibliometric analysis entails choosing a suitable database. In line with prior bibliometric research (Mishra et al. 2020, 2021; Shi et al. 2019; Song et al. 2019), this study employed the Web of Science (WoS) database as the primary data source for collecting literature related to air quality monitoring using geospatial technology (AQMGT). The WoS database is widely regarded as the gold standard for bibliometric analysis, primarily because it indexes only high-quality and influential publications (Modak et al. 2019).

The next step involves data extraction and filtering from the selected database. To accomplish this, the following search strategy was applied within the WoS core collection database: TS = [("air quality monitoring" AND "GIS" OR "remote sensing" OR "geographic information system" OR "geospatial technology")]. In this context, "TS" denotes the topic of a publication, indicating that the search was conducted within the title, abstract, and keywords fields. This search was carried out in August 2021, covering publications from the earliest available date up to 2021. Initially, the search generated a list of 249 publications. Following this, a manual review of the titles, abstracts, and keywords was undertaken to eliminate irrelevant publications, resulting in the exclusion of 23 articles unrelated to AQMGT. The remaining 226 records were then used for the subsequent bibliometric analysis in this study.

The third step involves conducting the actual bibliometric analysis, which can be facilitated using various tools like CiteSpace, VOSViewer, Bibexcel, SciMAT, and CitNetExplorer (Cobo et al. 2011). However, most of these tools are not suitable for all types of analysis (Aria and Cuccurullo 2017). In comparison, a more versatile option is Bibliometrix, a recently developed software package based on the R environment. Bibliometrix not only offers flexibility but also integrates the visualization capabilities of a variety of bibliometric tools (Aria and Cuccurullo 2017). Therefore, in this study, we have opted to use Bibliometrix to perform a comprehensive bibliometric analysis of the AQMGT research field. Furthermore, we used the VOSviewer tool for graphical visualization and bibliometric networks (van Eck and Waltman 2010, 2011). The network maps were created, where the size of the circle is resolute by the weight of the item and a particular parameter or cluster to which the item belongs, and the lines represent the links between the item (Van Eck and Waltman 2016).

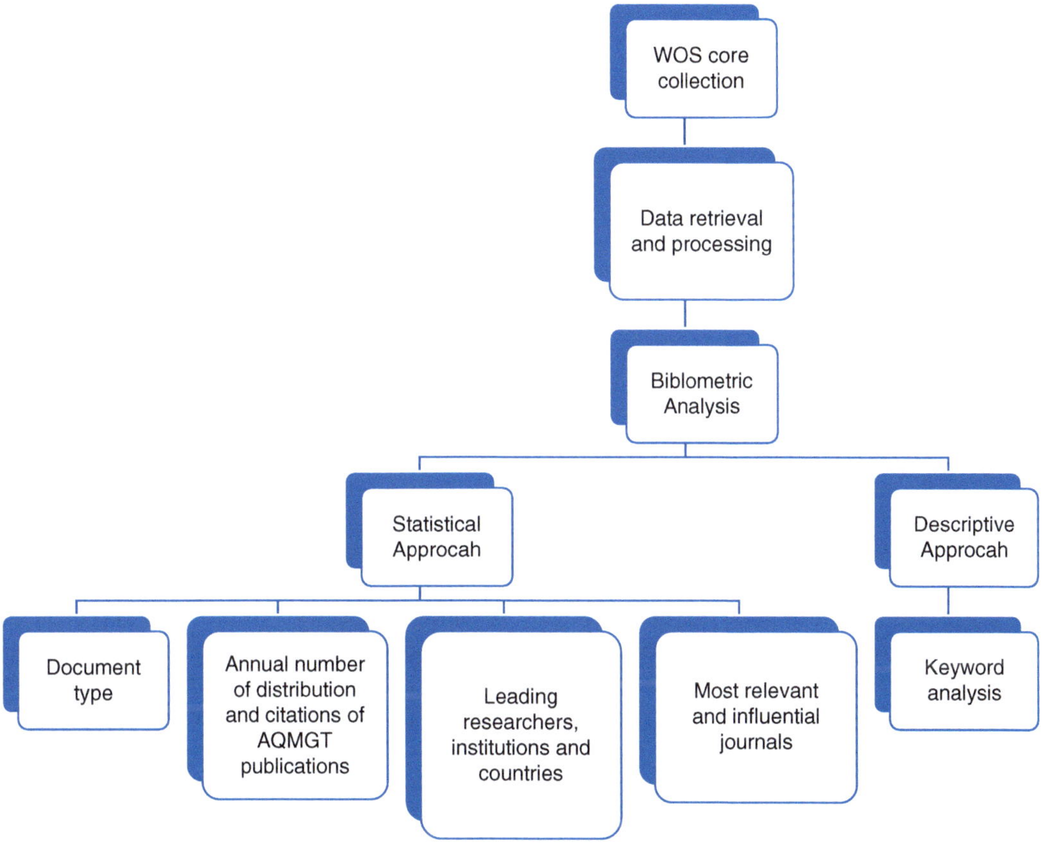

Fig. 3.1 Research framework of this study

In this study, 226 articles are analyzed to provide an overview of the AQMGT research field, including metrological analysis of AQMGT literature, researchers, institutions, countries, and journals in the AQMGT domain. Simultaneously, to delve into the intellectual structure of the AQMGT field, content analysis techniques, including keyword analysis and citation analysis, are employed to identify trending topics and research areas in the AQMGT research domain.

3.3 Results and Discussion

3.3.1 Main Information About Data

Meta information of sample data is shown in Table 3.1. There were 226 research papers related to AQMGT, published in 92 different WOS sources (e.g., journals and books) by scholars since 1998. Articles accounted for 96% of the collection, while review papers only accounted for 1.3%, as shown in Fig. 3.2. We identified

Table 3.1 Meta information about sample data

Description	Results
Publications	226
Timespan	1998:2021
Sources (Journals, Books, etc.)	92
Average citations per document	20.04
Keywords plus (ID)	677
Author's keywords (DE)	622
Authors	1162
Author appearances	1368
Authors of single-authored documents	6
Authors of multi-authored documents	1156
Single-authored documents	6
Publications per author	0.194
Authors per publication	5.14
Co-authors per publications	6.05
Collaboration index	5.25

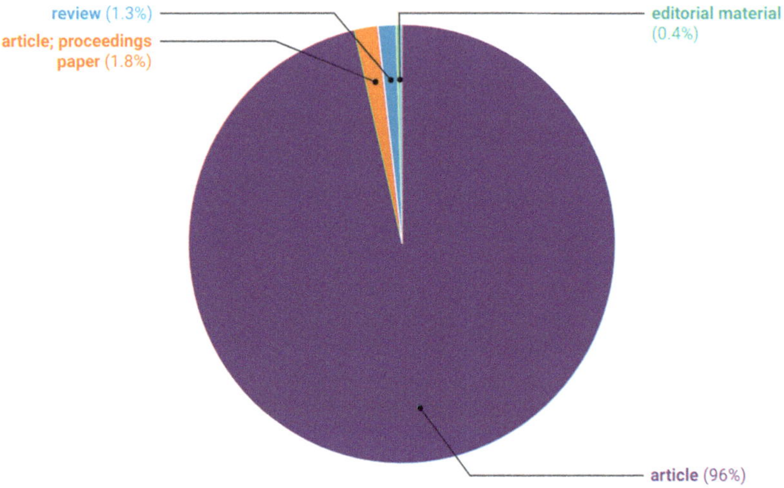

Fig. 3.2 Document type

1162 authors with 1368 author appearances in 226 papers. Single-author papers accounted for 2.65% of the collection (n = 6), and 6 single authors published them. On an average, documents per author was 0.194, authors and co-authors per document were 5.14 and 6.05, respectively, and each document has average of 20.04 citations.

3.3.2 The Annual Number of Distribution and Citations of AQMGT Publications

As depicted in Fig. 3.3, it is evident that the quantity of publications related to AQMGT exhibits a consistent upward trajectory, with a yearly growth rate of 22.6%. This trend reaches its zenith in the year 2021, underscoring the rising prominence of the AQMGT field. Notably, publications in 2005 garnered the highest number of citations, with an average of 3.55 per document. However, it is worth noting that the citation pattern displayed fluctuations from 2002 onwards.

3.3.3 Leading Researchers and Institutions

Utilizing the author data extracted from the collected dataset, it becomes possible to identify the most prominent figures within the AQMGT domain, as well as the leading institutions and countries in this field. The h-index, total citations (TC), and the number of publications (NP) of the top ten influential authors who contributed the most to the AQMGT publications are presented in Table 3.2. It can be observed from Table 3.2 that Jimmy Chi Hung Fung and Alexis Kai Hon Lau are the most prominent researchers in this domain as evidenced from their h-indices and TC. Janet E. Nichol published the highest number of publications in this domain.

To provide a comprehensive overview of the distribution of publications in the AQMGT domain, we have also analyzed the leading institutions and countries contributing to this field. The top ten institutions that have made the most significant contributions to AQMGT-related articles are presented in Table 3.3. The Hong Kong University of Science and Technology leads the pack with 26 articles, followed by the Nanjing University of Information Science and Technology with 12 articles, the University of Bremen and Wuhan University with 11 articles each, and the National Taiwan University with 9 articles. Additionally, National Central University, Hong Kong Polytechnic University, Islamic Azad University, and Peking University have all made substantial contributions to AQMGT research with 8, 7, 7, and 7 articles, respectively.

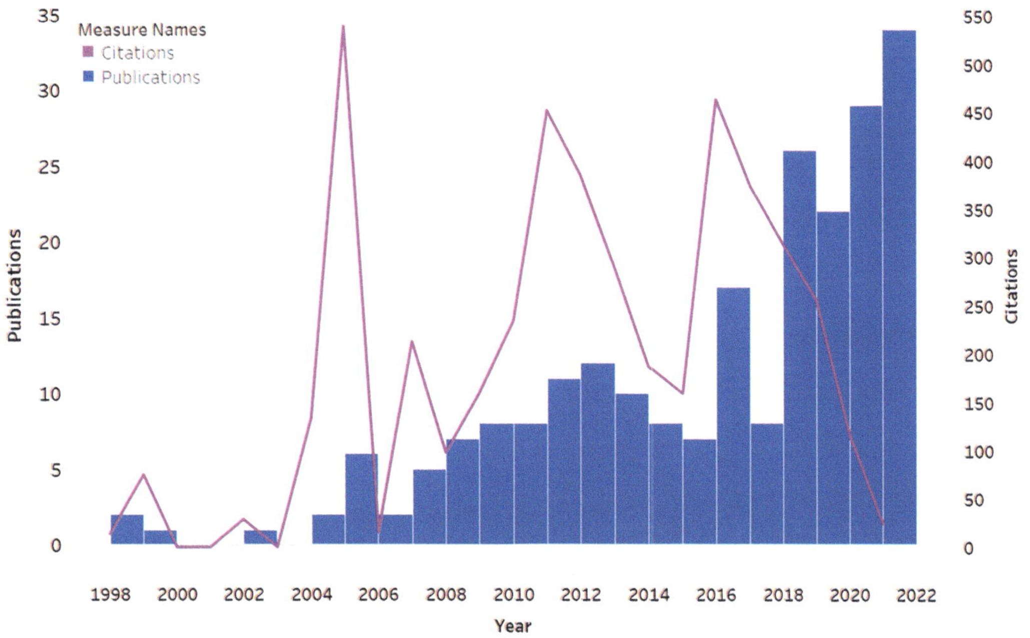

Fig. 3.3 Number of research publications and their citation pattern published between 1998 and 2021

Table 3.2 Top ten authors in the AQMGT domain

Author	Country	NP	TC	h_ index
Janet E. Nichol	Hong Kong	8	77	6
Jimmy Chi Hung Fung	Hong Kong	7	214	6
Alexis Kai Hon Lau	Hong Kong	7	214	6
Changqing Lin	Hong Kong	6	193	6
Man Sing Wong	Hong Kong	6	58	5
Chengcai Li	Hong Kong	5	187	5
Ying Li	Hong Kong	5	187	5
A. Ramachandran	India	5	506	4
Zhipeng Bai	China	3	108	3
Muhammad Bilal	China	3	24	3

Table 3.3 Top ten institutions of AQMGT research ranked by publication

Institution	Articles
Hong Kong University of Science and Technology	26
Nanjing University of Information Science and Technology	12
University of Bremen	11
Wuhan University	11
National Taiwan University	9
National Central University	8
Hong Kong Polytechnic University	7
Islamic Azad University	7
Peking University	7
Beijing Normal University	6

3.3.4 Most Widely Used Journals

The 226 AQMGT-related articles are published in 92 different journals. The number of AQMGT-related articles (NP), total citation (TC), and the *h*-index of each journal are used as measures here to identify the most relevant and influential journals in the AQMGT research domain. In Table 3.4, we present the top ten journals that have published the most AQMGT-related articles. These journals can be regarded as key sources in the AQMGT field. Table 3.4 highlights the leading journals in the AQMGT domain, showing that the top four journals, namely Atmospheric Environment, Remote Sensing, Science of the Total Environment, and Atmospheric Chemistry

Table 3.4 Top ten relevant journals on AQMGT research

Journals	NP	TC	*h*_index
Atmospheric Environment	25	833	15
Remote Sensing	13	136	6
Science of the Total Environment	10	300	8
Atmospheric Chemistry and Physics	9	263	6
Environmental Monitoring and Assessment	7	83	6
Remote Sensing of Environment	7	245	7
Atmosphere	6	32	4
Atmospheric Measurement Techniques	6	462	4
International Journal of Remote Sensing	6	64	5
Journal of the Air & Waste Management Association	6	201	6

and Physics, not only published the highest number of AQMGT-related articles but are also the most influential sources, boasting the highest *h*-index values.

3.4 Intellectual Structure of the AQMGT Research Field

In this section, we aimed to distill the essential components of knowledge base of the AQMGT research field through keywords and cluster analysis as the keyway of content analysis.

3.4.1 Keyword Analysis

Keywords are being used to describe the content of the research article (Sudarsana et al. 2018). Thus, the analysis of keywords can help identify prominent topics and themes within a research domain (Zheng et al. 2016). For our analysis, we utilized Bibliometrix to gather information about the frequency of keywords within the AQMGT domain. This statistical data was then employed to construct a word cloud, which is illustrated in Fig. 3.4. The word cloud serves as a visually informative representation of the prevailing themes in the AQMGT research field, with the size of each keyword proportional to its frequency in the dataset. Notably, the word cloud highlights key terms such as "air pollution," "air quality," "remote sensing," "GIS," "nitrogen dioxide," "pm2.5," and "modis," signifying their significance in the AQMGT research landscape.

3.4.2 Cluster Analysis

The most frequently used keywords were identified and analyzed to classify the 226 articles as part of the collected data. These keywords formed seven clusters, as shown in Fig. 3.5. Each cluster is represented by a specific keyword, with the node size indicating its frequency. The line thickness between nodes reflects co-occurrence frequency. The primary keywords per cluster are:

Remote Sensing (Red Cluster); Air Pollution (Green Cluster); Black Carbon (Blue Cluster); Satellite Remote Sensing (Yellow Cluster); Nitrogen Dioxide (Pink Cluster); Air Quality (Purple Cluster); Aerosol Optical Depth (Orange Cluster).

In addition, sustainability, tourism, management, sustainable tourism, and attitudes are identified as emerging themes. Notably, research appears to be moving toward spatiotemporal correlation, wind erosion, exposure assessment, urban air quality, and satellite remote sensing, offering new research avenues.

3.4.2.1 Cluster #1 (Red Cluster): Air Quality Monitoring Using Different Parameter

The red cluster contains a large number of keywords (12) that focus on air quality monitoring using different parameters (see Fig. 3.6). The following main keywords were identified: pm10, pm2.5, aeronet, modis, aerosol, so2, remote sensing and omi, aod.

Cluster #1, concerning remote is composed by 33 articles, from which 14 articles were analyzing areas in China, and 3 were about areas in Taiwan (Table 3.5).

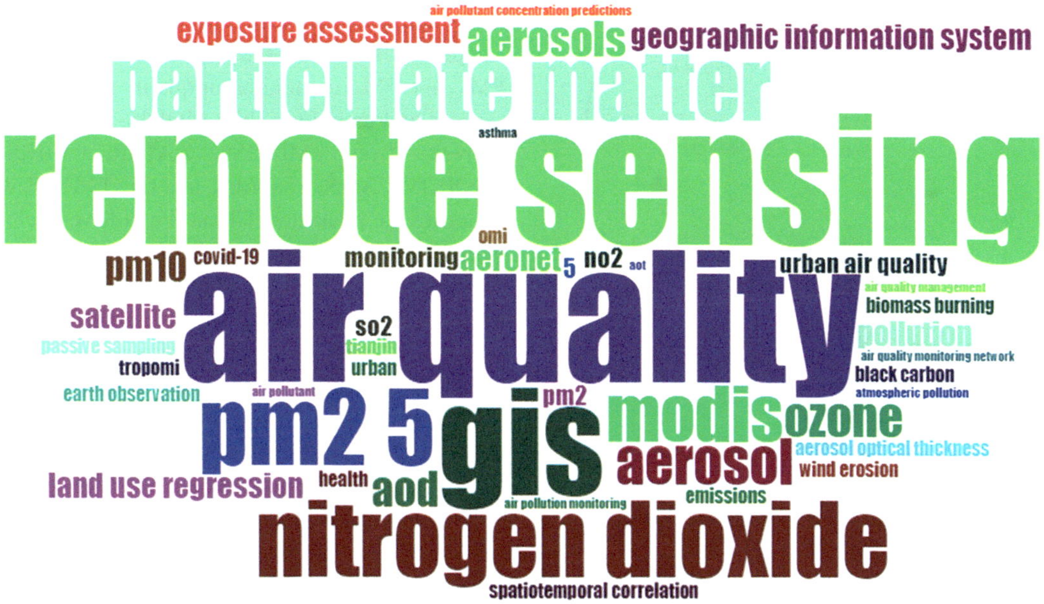

Fig. 3.4 Word cloud of keywords of the AQMGT research field

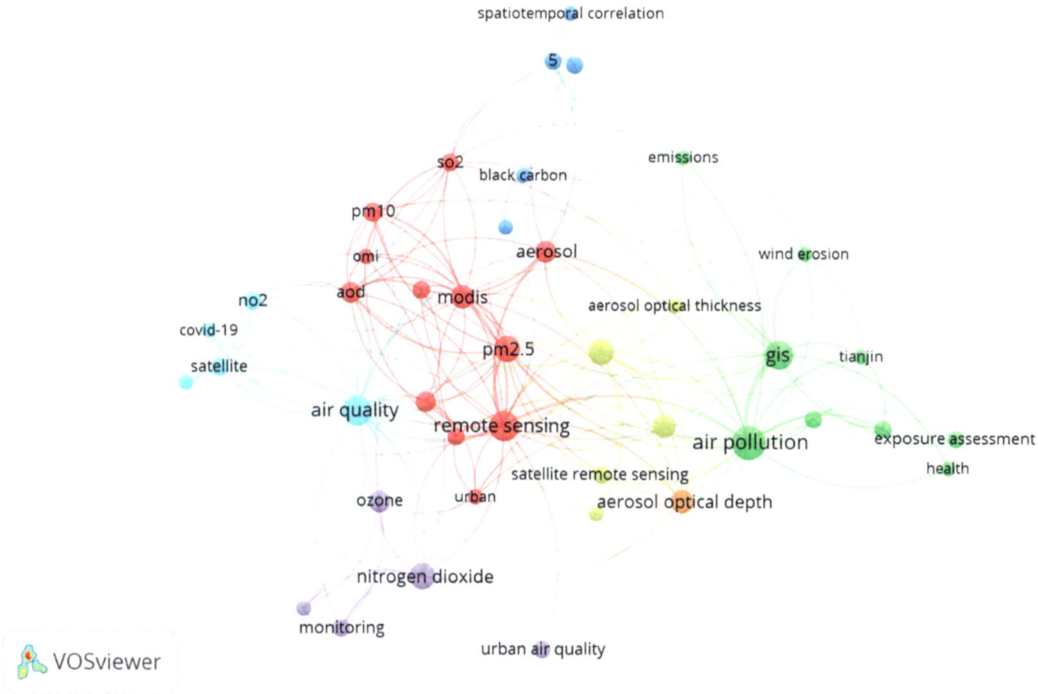

Fig. 3.5 Keyword co-occurrence network based on the publications after the detailed screening (226 publications)

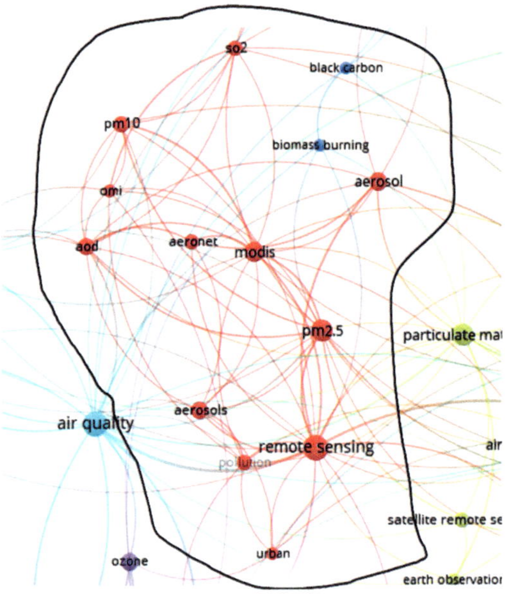

Fig. 3.6 Red cluster map

Cluster 1

The cluster 1 mostly describes air quality monitoring using different parameters like PM_1, $PM_{2.5}$, PM_{10}, aerosol variability, sulfate, carbon monoxide (CO), sulfur dioxide (SO_2), nitrogen monoxide (NO_x), nitrogen oxide (NO_2), ozone (O_3), mineral dust aerosols, dust and black carbon using different methods from different countries. Different parts of China like eastern China, Beijing, China, and Taiwan use only $PM_{2.5}$ to monitor air quality using different sensors like MODIS aerosol optical depth (AOD), SARA, Sun Photometer, and AERONET. Major findings of the region are that $PM_{2.5}$ has become a major contributor to aerosol pollution and $PM_{2.5}$ concentrations showed a very strong correlation with the terrain features, forest cover, road density, and industrial pollution whereas adults (18–65 years) and children (3–18 years) are the most vulnerable groups to the effect of $PM_{2.5}$ compared with infants and elderly people. Spain and Romania use only PM_{10} to monitor air quality using air quality monitoring network (AQMN). Many countries like China (Gobi Desert region), London (UK), Belgian, and British use both $PM_{2.5}$ and PM_{10} to monitor air quality using many sensors and techniques like MODIS-Terra, AOT derived from MERIS, AATSR, VIIRS, TROPOMI, AON, SeaWiFS, and GEOS-FP. Few countries, i.e., Pakistan, China (Pearl River Delta Area), and Republic Cyprus combine three factors, i.e., PM_1, $PM_{2.5}$, PM_{10} used (AquaTerra) MODIS, AOD, DTB, and TEOM to monitor the air quality. Using Trace gases, mineral dust aerosols, and aerosol there are many countries like Tunisia, China, Budapest (Hungary), Mongolia, Taiwan, and Czech Republic that analyze air quality using different sensors and many techniques like AQUA MODIS and AERONET, AOD, sun photometer, AOD and MISR, AOD, monthly LUR models, OMI and TROPOMI, CALIPSO, Malaysia, AERMOD model, and IVE model. Only Saudi Arabia uses Dust and Black Carbon to monitor air quality using OMI and AERONET.

3.4.2.2 Cluster #2 (Green Cluster): GIS Tools

The green cluster, which contains nine keywords, focuses on geographic information system (see Fig. 3.7). Cluster #2 is composed by 16 articles (see Table 3.6).

In cluster 2 mostly all the countries use GIS tools to analyze the different parameters. Most of the countries studied trace gases along with particulate matter only a few studies monitor using PM_{10} as well as NO_2. Colombia, Romania, Iran, Chin, Italy, the USA, Taiwan, Australia, Ireland, Madrid, South Africa, and Canada use many trace gases as well as particulate matter like $PM_{2.5}$ and PM_{10}, black carbon, CO, NO_2, SO_2, O_3, Pb and dust storm to monitor the air quality using LUR, GIS, AQCC. Spain and Madrid are the countries that use PM10 to monitor air quality using GIS and AQMN. Israel is the only country that uses both PM2.5 and PM10 to monitor air quality using AQMS.

Table 3.5 Details about the papers within Cluster #1

	Title	Study area	Year	Parameter	Sensor	Key finding	References
1	"An evaluation of MODIS-retrieved aerosol optical depth over AERONET sites in Alaska"	Alaska	2000–2014	Particulate concentrations	MODIS AOD	Products of AOD may be used in applications of quantitative measurement of air quality in Alaska during summer time	https://doi.org/10.3390/rs10091384
2	"Air Pollution in the Gobi Desert Region: Analysis of Dust Storm Events"	Gobi Desert region	2017	PM2.5 and PM10	MODIS-Terra	According to air mass trajectory, the dust cloud moved from epicenter to east and southeast which affects the central and eastern parts of country	https://doi.org/10.1002/qj.3961
3	"Aerosol uncertainty assessment: an integrated approach of remote AQUA MODIS and AERONET data"	Tunisia	2014	Aerosol variability	AQUA MODIS and AERONET	Principal source and transported path of dust can be identified using MODIS data	https://doi.org/10.1007/s12517-018-4214-y
4	"A methodological framework for improving air quality monitoring network layout. Applications to environment management"	Madrid (Spain)	2008–2017	PM 10	Air quality monitoring network (AQMN)	Fixed site within AQMN with European legislation guidelines helps to enhance the air quality management	https://doi.org/10.1016/j.jes.2020.09.009
5	"Aerosol retrievals over China with the AATSR dual view algorithm"	Eastern China	2008	AOD		Maximum and minimum AOD were observed during the summer and winter periods over the urban areas	https://doi.org/10.1016/j.rse.2011.04.041

(continued)

Table 3.5 (continued)

	Title	Study area	Year	Parameter	Sensor	Key finding	References
6	"Aerosol optical depth, aerosol composition and air pollution during summer and winter conditions in Budapest"	Budapest	2003–2004	Gaseous (CO, SO_2, NO_x, O_3), solid components (PM2.5, PM10)	Sun photometer	The nocturnal increase of the MLH coincided with a decrease in pollutant concentrations except ozone; whereas an increase in ozone concentration indicates the vertical mixing of different air layers	https://doi.org/10.1016/j.scitotenv.2007.04.037
7	"Using Multi-Angle Imaging SpectroRadiometer Aerosol Mixture Properties for Air Quality Assessment in Mongolia"	Mongolia	2011–2016	PM2.5, PM10, and SO_2	AOD and MISR	Machine regression shows good cross-validation statistics and PM2.5 estimates that capture relevant spatial and temporal patterns over the region	https://doi.org/10.3390/rs10081317
8	"Classification of aerosols over Saudi Arabia from 2004–2016"	Saudi Arabia	2004–2016	Dust and Black Carbon (BC)	OMI and AERONET	Dust aerosol was dominant in both seasonally and annually because of repeated dust storm events. Along with fine particulates, i.e., BC and Mixed (BC and Dust) were also observed, because if industrial activities	https://doi.org/10.1016/j.atmosenv.2020.117785
9	"A Geographically and Temporally Weighted Regression Model for Ground-Level PM2.5 Estimation from Satellite-Derived 500 m Resolution AOD"	Eastern China	2014–2015	PM2.5	AOD and SARA	MODIS satellite's overpass can be used for air quality monitoring and haze tracking at the urban and regional scales	https://doi.org/10.3390/rs8030262

10	"Air pollution scenario over Pakistan: Characterization and ranking of extremely polluted cities using long-term concentrations of aerosols and trace gases"	Pakistan	2003–2020	PM1, PM2.5, and PM10	(AquaTerra) MODIS, AOD, DTB	Pakistan's air quality is strongly affected by anthropogenic sources inside Pakistan, with contributions from surrounding countries	https://doi.org/10.1016/j.rse.2021.112617
11	"Analysis of Aerosol Properties in Beijing Based on Ground-Based Sun Photometer and Air Quality Monitoring Observations from 2005 to 2014"	Beijing	2005–2014	PM2.5	Sun Photometer, AERONET, and AOD	Fine mode aerosol, especially PM2.5 has become a major contributor to the aerosol pollution in Beijing	https://doi.org/10.3390/rs8020110
12	"Estimating urban PM10 and PM2.5 concentrations, based on synergistic MERIS/AATSR aerosol observations, land cover and morphology data"	London (UK)	2002–2012	PM10 and PM2.5	AOT derived from MERIS and AATSR	Methodology applied in the study can quantify spatial and temporal distribution of Black carbon, Ultra-fine particles, and different geophysical quantities	https://doi.org/10.1016/j.rse.2015.10.017
13	"MODIS AND OMI SATELLITE OBSERVATIONS SUPPORTING AIR QUALITY MONITORING"	Northern Italy	200,320,072,008	PM2.5 and NO_2	MODIS AND OMI SATELLITE	Good correlation observed between measured ground level concentrations of NO_2 with in situ data obtained from environmental agencies	https://doi.org/10.1093/rpd/ncp231
14	"Detection of aerosol pollution sources during sandstorms in Northwestern China using remote sensed and model simulated data"	Northwestern China	2014	PM10 and PM2.5	MODIS, AOD and CALIPSO	Sources of pollution are the deserted regions of Northern and Northwestern China	https://doi.org/10.1016/j.asr.2017.11.037

(continued)

Table 3.5 (continued)

	Title	Study area	Year	Parameter	Sensor	Key finding	References
15	"A Synergic Algorithm for Retrieval of Aerosol Optical Depth over Land"		2005–2006	??	MODIS, AOD, and AATSR	Synergic algorithm can be potentially used in climate change and air quality monitoring	https://doi.org/10.1007/s00376-009-7218-4
16	"Incorporating satellite-derived data with annual and monthly land use regression models for estimating spatial distribution of air pollution"	Taiwan	2015	NO_2 and PM2.5	AOD, monthly LUR models	Both monthly and annually model shows robust performance for NO_2 whereas PM2.5 models usually performed for few months, i.e., April to August	https://doi.org/10.1016/j.envsoft.2019.01.010
17	"Analysis of heavy pollution episodes in selected cities of northern China"	Northern China	2009	PM10, sulphate and nitrate		Stationary anthropogenic is the major source for pollution for NO_x/SO_2 ratios	https://doi.org/10.1016/j.atmosenv.2011.11.053
18	"Temporal Trend of PM10 and Associated Human Health Risk over the Past Decade in Cluj-Napoca City, Romania"	Cluj-Napoca City, Romania	2009–2019	PM10		Significant increase in the air quality from warm months to cold months during 2009–2019. Whereas non-carcinogenic chronic health risk was reported between 2009 and 2010, it gets its peak during the cold months of each year	https://doi.org/10.3390/app10155331
19	"High-resolution aerosol remote sensing retrieval over urban areas by synergetic use of HJ-1 CCD and MODIS data"	China	2010	Urban air quality	HJ-1 CCD and MODIS data, AERONET	Advantage of retrieval is higher resolution compared to MODIS aerosol components	https://doi.org/10.1016/j.atmosenv.2011.10.002

20	"Retrieval of High-Resolution Atmospheric Particulate Matter Concentrations from Satellite-Based Aerosol Optical Thickness over the Pearl River Delta Area, China"	Pearl River Delta Area, China	2008–2011	PM10, PM2.5 and PM1	MODIS, AOT	Given model gives spatial distribution map of annual average of PM10 of Pearl River Delta region	https://doi.org/10.3390/rs70607914
21	"Applying SPOT data to estimate the aerosol optical depth and air quality"	Northern Taiwan	1998		AOD with SPOT HRV	Variation of the atmospheric turbidity increases as the AOD value increases	https://doi.org/10.1016/S1364-8152(01)00047-0
22	"Spatial and temporal (short and long-term) variability 1 of submicron, 2 fine and sub-10 µm particulate matter (PM1, PM2.5, PM10) in Cyprus"	Cyprus	1997–2015	PM1, PM2.5, PM10	TEOM	Fine and coarse particles were significantly derived from fine mode as well as coarse mode emission from both industrial and urban areas of Cyprus	https://doi.org/10.1016/j.atmosenv.2018.07.048
23	"Full-coverage spatiotemporal mapping of ambient PM2.5 and PM10 over China from Sentinel-5P and assimilated datasets: Considering the precursors and chemical compositions"	China	2018–2020	PM2.5 and PM10	VIIRS,TROPOMI, AON and GEOS-FP	The proposed approach could efficiently present daily full-coverage results at 5-km spatial grids, which fixes the issue due to the missing AOD data	https://doi.org/10.1016/j.scitotenv.2021.148535
24	"Application of aerosol speciation data as an in situ dust proxy for validation of the Dust Regional Atmospheric Model (DREAM)"	El Paso (TX), Austin (TX), Phoenix (AZ), Salt Lake City (UT) and Bakersfield (CA)	2006	Mineral dust aerosols	Dust Regional Atmospheric Model (DREAM)	Lack of total reduction in RMSE infers model can be more improvement. Whereas for dust transport model, soil is the main component for theoretical proxy	https://doi.org/10.1016/j.atmosenv.2008.06.018

(continued)

Table 3.5 (continued)

	Title	Study area	Year	Parameter	Sensor	Key finding	References
25	"SAMIRA-Satellite Based Monitoring Initiative for Regional Air Quality"	Czech Republic	2016–2019	SO_2, NO_2 and PM	OMI and TROPOMI	Algorithm applies to OMI NO_2 data with high-resolution TROPOMI NO_2 as reference and increase in accuracy and performance after downscaling	https://doi.org/10.3390/rs13112219
26	"Spatio-temporal evolution of a severe winter dust event in the Western Mediterranean: aerosol optical and physical properties"	Iberian Peninsula	2016	African dust and PM10	Sun-photometer (CIMEL CE-318) included in AERONET	Backscattered vertical profiles show thick aerosol layer and closer to surface in both high latitude and sea surface layer	https://doi.org/10.1002/2016JD026252
27	"Atmospheric particulate matter (PM) estimation from SeaWiFS imagery"	Belgian and British	1997–2004	PM10, PM2.5	SeaWiFS	Study finds the map of PM10 SeaWiFS and low correlation between in situ and instrumental data	https://doi.org/10.1016/j.rse.2007.03.009
28	"Improved algorithm for MODIS satellite retrievals of aerosol optical thickness over land in dusty atmosphere: Implications for air quality monitoring in China"	China	2008	Surface visible (0.65 μm) and near-infrared (NIR, 2.1 μm)	MODIS, GEOS-Chem	Surface concentrations of PM10 of improved AOT show strong correlation with ground observation than GEOSCHem simulation	https://doi.org/10.1016/j.rse.2010.05.034
29	"An air pollution episode and its formation mechanism during the tropical cyclone Nuri's landfall in a coastal city of south China"	South China	2008	SO_2 and PM10	Sun photometer and satellite lidar CALIPSO	Accumulation of local aerosol and heavy emission source are important contributors in each episode. So Site differences play a major role for pollution formation investigation	https://doi.org/10.1016/j.atmosenv.2011.12.023

30	"Estimating ground-level PM2.5 over a coastal region of China using satellite AOD and a combined model"	China	2014–2016	$PM_{2.5}$	MODIS aerosol optical depth (AOD)	Spatial pattern of PM2.5 shows high concentration in urban areas compared to mountainous and sub-urban areas. Whereas concentration of PM2.5 shows high correlation with road density, industrial pollution, terrain features and forest covers	https://doi.org/10.1016/j.jclepro.2019.04.231
31	"Relationship between land use composition and PM10 concentrations in Iskandar Malaysia"	Iskandar Malaysia	2002, 2006, 2008	Land use composition and PM10	LANDSAT and MODIS aerosol optical depth	PM10 concentrations were most affected by commercial land use than industrial land use	https://doi.org/10.1007/s10098-016-1263-3
32	"GIS-based emission inventories of urban scale: A case study of Hangzhou, China"	Hangzhou, China	2004	SO_2, NO_X, and PM10	AERMOD model, IVE model	High-dense populated are similar to industrial area and are the most pollutant concentration due to combination of two zones	https://doi.org/10.1016/j.atmosenv.2008.02.012
33	"Spatial Correlation of Satellite-Derived PM2.5 with Hospital Admissions for Respiratory Diseases"	Taiwan	2008	$PM_{2.5}$	MODIS	Elderly and infant are less vulnerable as compared to adults (18–65 years) and children (3–18 years) from the effect of PM2.5	https://doi.org/10.3390/rs8110914

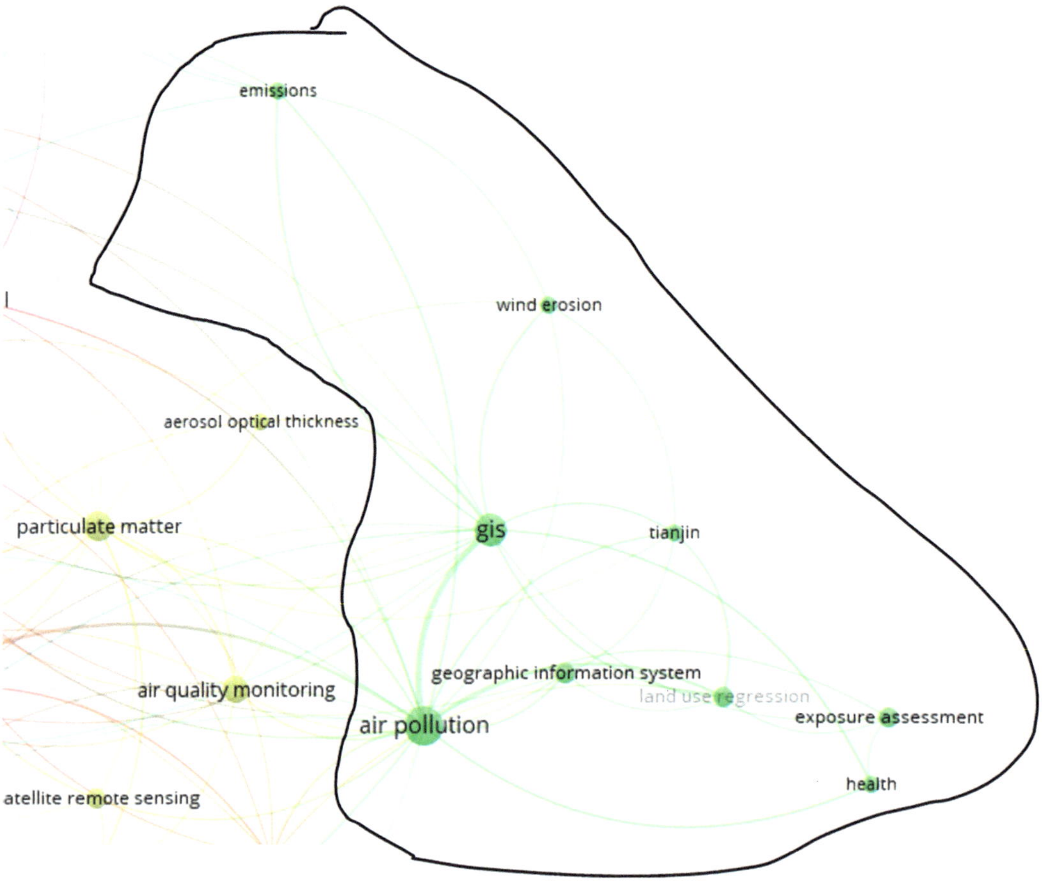

Fig. 3.7 Green cluster map

3.4.2.3 Cluster #3 (Blue Cluster): $PM_{2.5}$, PM_{10}, and Black Carbon

Cluster #3 mainly focused on mostly particulate matter, i.e., $PM_{2.5}$ and PM_{10} whereas the few studies on black carbon (see Fig. 3.8). This cluster consists of eight articles and details about these papers are presented in Table 3.7. As observed from Table 3.7, China and India use $PM_{2.5}$ to monitor the air quality using different sensors and methods like NDVI data, PBLH, AOD, and MODIS. Black carbon and few trace gases (SO_2, NO_3, K+, NH_4CO, and O_3) are used to monitor air quality in India and Finland using DMSP-OLS methods. Whereas the USA and Finland used both $PM_{2.5}$ and PM_{10} to monitor air quality using MODIS, in situ air quality measurement.

3.4.2.4 Cluster #4 (Yellow Cluster): Air Quality Monitoring

The yellow cluster contains five keywords, focusing on monitor air quality using DOAS and GIS methods (see Fig. 3.9). The most appearing keywords in this cluster were satellite remote sensing, particulate matter, aerosol optical thickness, earth observation, and air quality monitoring. Some articles from this cluster were identified and their finding was analyzed (Table 3.8). As observed from Table 3.8, Romania and Pakistan use both particulate matter ($PM_{2.5}$, PM_{10}) and trace gases (NO, NO_2, NO_x, CO, O_3, and SO_2) to monitor air quality using DOAS and GIS methods. Hong Kong and China use PM to monitor air quality using MODIS AOD and NDVI data. PM_{10}, $PM_{2.5}$, and PM_1 were used to monitor air

Table 3.6 Details about the papers within Cluster #2

	Title	Study area	Year	Parameter	Sensor	Key finding	References
1	"Monitoring the Indoor Air Quality and Microclimate for Improving Passenger Health and Comfort Level in Bucharest Subway Platforms"	Bucharest Subway Platforms	2015	CO, NO_2, SO_2, $PM_{2.5}$ and PM_{10} pollutants		Aim of study to improve comfort level and air quality in long subway	Tudor et al. (2019)
2	"Exposure levels to PM2.5 and black carbon for people with disabilities in rural homes of Colombia"	Colombia	2017–2018	PM2.5 and black carbon		Black carbon concentration in 2.5 times in rural area as compared to urban area because of high vehicle traffic and population density, control mechanism for these factors to address the indoor air pollution	https://doi.org/10.1007/s10661-020-08803-3
3	"A methodological framework for improving air quality monitoring network layout. Applications to environment management"	Madrid's (Spain)	2008–2017	PM10	GIS, AQMN	Criteria set for methodology used to delineate redundant station within AQMN	https://doi.org/10.1016/j.jes.2020.09.009
4	"Land use regression models to estimate the annual and seasonal spatial variability of sulfur dioxide and particulate matter in Tehran, Iran"	Tehran, Iran	2010	SO_2 and PM10	AQCC	According to WHO during 2010, Tehran's total population lived in area where 24-h average of SO_2 (7 ppb) and yearly mean of PM10 (20 μg/m^3) were exceeded	https://doi.org/10.1016/j.scitotenv.2014.04.106
5	"A land use regression for predicting NO_2 and PM10 concentrations in different seasons in Tianjin region, China"	Tianjin region, China	2006	NO_2 and PM10		LUR model is more effective during hot season as compared to non-heating season as well as it gives robust resolution with NO_2 compared to PM10	https://doi.org/10.1016/S1001-0742(09)60263-1
6	"A land use regression model incorporating data on industrial point source pollution"	Tianjin, China	2006	SO_2, NO_2 and PM10		Relationship between integrated and dependent variables with highly consistent with a linear relationship	
7	"A Bayesian kriging model for estimating residential exposure to air pollution of children living in a high-risk area in Italy"	Italy	2007–2008	SO_2 and NO_2	Bayesian kriging (GIS model)	Bayesian kriging is highly useful to assess the realistic exposure with ambient air pollution	
8	"Associations between maternal residential proximity to air emissions from industrial facilities and low birth weight in Texas, USA"	Texas, USA	1996–2008	PM2.5, PM10, O_3, NO_2, SO_2, CO, and Pb		Study identifies 14 pollutants as LBW risk factors except few air pollutant like ethylbenzene, zinc. Benzene and toluene	https://doi.org/10.1016/j.envint.2018.07.045

(continued)

Table 3.6 (continued)

	Title	Study area	Year	Parameter	Sensor	Key finding	References
9	"Assessment of different route choice on commuters' exposure to air pollution in Taipei, Taiwan"	Taipei, Taiwan	2014	CO, SO_2, NO_2, PM10, and PM2.5		Motorcyclists are more exposure to inhaling high levels of pollutants during commutes, which results in peaks of exposure to high concentrations for short durations	
10	"The 23rd October 2002 dust storm in eastern Australia: characteristics and meteorological conditions"	Eastern Australia	2002	Dust storm	Dust visibility index (DVI)	Records show that during storm events, Australia experiences a significantly higher dust load compare to other events in the past	https://doi.org/10.1016/j.atmosenv.2004.10.016
11	"A land use regression model for explaining spatial variation in air pollution levels using a wind sector-based approach"	Ireland	2010–2012	NO_2	LUR	The NO_2 map shows the dominant influence of traffic emissions on national (and urban) NO_2 concentrations	https://doi.org/10.1016/j.scitotenv.2018.02.317
12	"Estimating windblown PM-10 emissions from vacant urban land using GIS"	Clark County, NV	1999	PM-10		PM10 emission was mainly dependent on different parameters like surface soil erosion, increase in disturbed barren land, frequency and duration of intense erosive winds	
13	"Spatial regression analysis of NO_x and O_3 concentrations in Madrid urban area using Radial Basis Function networks"	Madrid	2002–2007	NO_x and O_3		Evolutionary program is the main mechanism for initial regression for pollutant of Madrid	https://doi.org/10.1016/j.chemolab.2009.07.012
14	"Harbor and Intra-City Drivers of Air Pollution: Findings from a Land Use Regression Model, Durban, South Africa"	Durban, South Africa	2015	PM2.5, PM10, O_3, NO_2, SO_2	LUR	Harbor variable, NO_2 which used as proxy is not derived only from emission from ship but also from heavy duty vehicle associated with port activity	https://doi.org/10.3390/ijerph17155406
15	"Temporal stability of land use regression models for traffic-related air pollution"	Metro Vancouver, Canada	2003–2010	NO_2, NO	LUR	LUR model will be invariably the same or has to improve while decreasing in variability and concentration in NO_2 and NO	
16	"Estimating multi-annual PM2.5 air pollution levels using sVOC soil tests: Ashkelon South, Israel as a case study"	Ashkelon South, Israel	2010–2012	PM10 and PM2.5	AQMS	Study confirms the feasibility of obtaining a precise estimate of the PM in air pollution using sVOC soil testing	https://doi.org/10.1016/j.atmosenv.2013.09.053

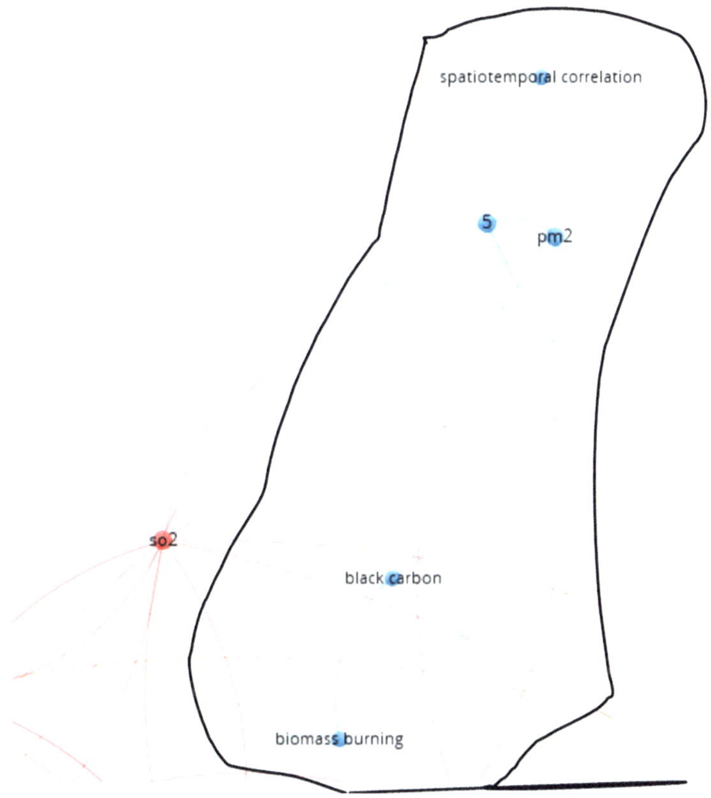

Fig. 3.8 Blue cluster map

quality in London (UK), China, and Israel using AOT derived from MERIS and AATSR. Trace gases like SO_2, CO, and NO_2 were used to monitor air quality in Iran and Belgium using ERDAS.

Cluster #5

3.4.2.5 Cluster #5 (Pink Cluster): Nitrogen Dioxide

Pink cluster refers to nitrogen dioxide, ozone and passive sampling (see Fig. 3.10) and contains 23 articles. The details of those articles are presented in Table 3.9. As observed form those articles, most of the country used trace gases to monitor air quality. Trace gases mostly NO, NO_2, HCHO, and O_3 are used to monitor air quality in the different country like Netherlands, Europe, India, China, France, Toronto, Spain, the USA, Paris, and New Zealand using different sensors, methods, and instruments like S5P/TROPOMI, GIS, Cloud and Aerosol Micro-Lidar (CAML), NASA EOS-Aura, TROPOMI, UV–VIS spectrometer, AQS, and MODIS FRP. Hungary is the only country that uses gaseous (CO, SO_2, NO_x, O_3), solid components (PM2.5, PM10) to monitor air quality using sun photometer. Saudi Arabia and India use black carbon (BC) aerosol, dust, CO, and O_3 to monitor air quality using OMI, AERONET, and DMSP-OLS.

Table 3.7 Details about the papers within Cluster #3

	Title	Study area	Year	Parameter	Sensor	Key findings	References
1	"Variations in black carbon aerosol, carbon monoxide and ozone over an urban area of Hyderabad, India, during the forest fire season"	Hyderabad, India	2004–2006	Black carbon (BC) aerosol; carbon monoxide (CO) and ozone (O_3)	DMSP-OLS	Higher incidence of forest fires on days coincides with high concentrations of CO and BC and spatial distribution of forest fires	https://doi.org/10.1016/j.atmosres.2006.10.004
2	"Long-range transport episodes of fine particles in southern Finland during 1999–2007"	Southern Finland	1999–2007	Black carbon, monosaccharide anhydrides, oxalate, succinate, malonate, SO_2, NO_3, K+ and NH_4		In every year the huge area of Europe attains peak concentration of fine particle because of the burning of open biomass in Eastern Europe	https://doi.org/10.1016/j.atmosenv.2008.11.022
3	"A Novel Recursive Model Based on a Convolutional Long Short-Term Memory Neural Network for Air Pollution Prediction"	China	2014–2019	PM2.5	NDVI data	Proposed CR-LSTM model, implementing a combination of recursive strategy, geographical rules, and deep learning, shows better performance in longer-term prediction	https://doi.org/10.3390/rs13071284
4	"Regional Air Quality Impact of Northern South America Biomass Burning Emissions"	Northern South America	2006–2016	PM2.5, PM10	MODIS	Correlation between PM10 and nos. of fires shows decreasing trends with distance increase from sources ranging from 0.6 to 0.25, like fires from Orinoco River basin shows worst air quality of area which is hundreds of km far away	https://doi.org/10.1016/j.atmosenv.2019.01.042
5	"A novel spatiotemporal convolutional long short-term neural network for air pollution prediction"	China	2016–2017	PM2.5	PBLH, AOD	Neighboring stations will provide records of PM2.5 as well as help to increase the prediction accuracy model Auxiliary data also helps to predict sudden changes in air quality	https://doi.org/10.1016/j.scitotenv.2018.11.086

6	"Long short-term memory neural network for air pollutant concentration predictions: Method development and evaluation"	Beijing City	2014–2016	PM2.5		The study shows model based on deep learning has the high precision compared to ARMA, SVR, and TDNN model. MAPE, MAE, and RMSE values indicate LSTME and traditional LSTM NN model were better prediction performance	
7	"Impact of Urban Growth on Air Quality in Indian Cities Using Hierarchical Bayesian Approach"	India	2001–2015	PM2.5	MODIS	Concentration of PM2.5 in the central part of the city is mainly derived from emission from residential area, brick kilns, and industries with 35–11%, 11.7–5.2%, and 4.2–2.8% respectively	https://doi.org/10.3390/atmos10090517
8	"Biomass burning aerosols observed in Eastern Finland during the Russian wildfires in summer 2010 e Part 1: In-situ aerosol characterization"	Eastern Finland	2010	PM2.5-PM10	Air quality measurement sites	Smoke is one of the major contributary in addition which leads to an increase of trace gas concentration, emission from traffic and local sources not affects	https://doi.org/10.1016/j.atmosenv.2011.10.067

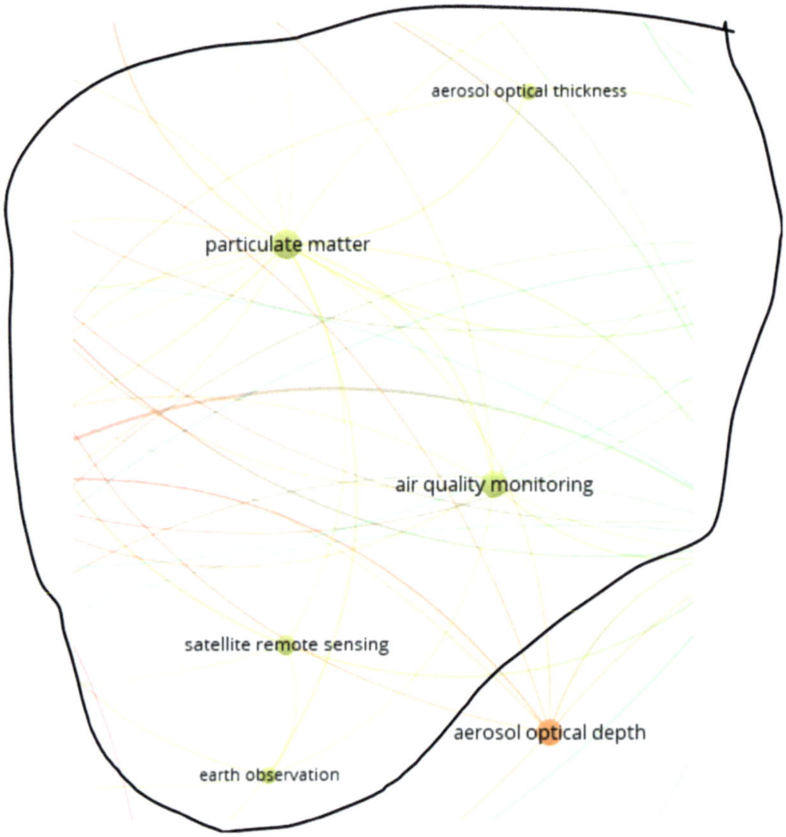

Fig. 3.9 Yellow cluster map

Cluster #6

3.4.2.6 Cluster #6 (Purple Cluster): Air Quality

The keywords in the purple cluster were satellite, air quality, NO_2, tropomi, and covid-19 (see Fig. 3.11). This cluster contains 43 articles and details of those papers are presented in Table 3.10. Cluster 6 mainly studies many trace gases as well as particulate matter. Many countries like Italy, India, Siberia, Eastern Sierra Nevada, European, and Israel use $PM_{2.5}$ as a parameter for monitoring air quality with the uses of different sensors like MODIS (Terra-Aqua/NASA platforms), MODIS, VIIRS, and SeaWiFS imagery. Only two countries in cluster-IV used PM10 parameters to monitor the air quality using GIS and AQMN sensors. Whereas both PM10 and PM2.5

Table 3.8 Details about the papers within Cluster #4

	Title	Study area	Year	Parameter	Sensor	Key finding	References
1	"Monitoring the Indoor Air Quality and Microclimate for Improving Passenger Health and Comfort Level in Bucharest Subway Platforms"	Bucharest Subway Platforms	2015	CO, NO_2, SO_2, $PM_{2.5}$ and PM_{10} pollutants		Aim of the study is to improve comfort level and air quality in long subway	Tudor et al. (2019)
2	"Measuring air pollutants in an international Romania airport with point and open path instruments"	ROMANIA AIRPORT	2008	PM2.5, PM10, NO, NO_2, NO_x, CO, O_3 and SO_2	DOAS	Maximum value recorded mainly during fueling of airplane	Ionel et al. (2011)
3	"Air pollution Monitoring in Urban Areas due to Heavy Transportation and Industries: a Case Study of Rawalpindi and Islamabad"	Rawalpindi and Islamabad		NO_2, SO_2, CO, O_3 and PM2.5		NO_2 and PM2.5 concentration was more than the maximum permissible limit given by Environmental Protection Agency of Pakistan	Hassan et al. (2013)
4	"15-Year PM2.5 Trends in the Pearl River Delta Region and Hong Kong from Satellite Observation"	Pearl River Delta Region and Hong Kong	2001–2015	PM2.5	MODIS AOD	PM2.5 concentration showed worse during 2001–2005 (tenth 5-year plan) and started recovering from 2006 to 2010 (11th FYP) in the Hong Kong and PRD region. Whereas during the 11th FYP period PM2.5 concentration was decreasing trend in urban areas of Central Hong Kong	https://doi.org/10.4209/aaqr.2017.11.0437
5	"Estimating urban PM10 and PM2.5 concentrations, based on synergistic MERIS/AATSR aerosol observations, land cover and morphology data"	London (UK)	2002–2012	PM10 and PM2.5	AOT derived from MERIS and AATSR	Methodology applied in the study can quantify spatial and temporal distribution of Black carbon, Ultra-fine particles, and different geophysical quantities	https://doi.org/10.1016/j.rse.2015.10.017
6	"Retrieval of High-Resolution Atmospheric Particulate Matter Concentrations from Satellite-Based Aerosol Optical Thickness over the Pearl River Delta Area, China"	Pearl River Delta Area, China	2008–2011	PM10, PM2.5 and PM1	MODIS, AOT	Given model gives spatial distribution map of annual average of PM10 of Pearl River Delta region	https://doi.org/10.3390/rs70607914

(continued)

Table 3.8 (continued)

	Title	Study area	Year	Parameter	Sensor	Key finding	References
7	"Assessment of contribution of SO_2, CO, and NO_2 in different urban land use in Bushehr region, Iran"	Bushehr region, Iran	2017–2018	SO_2, CO, and NO_2	ERDAS	Traffic is the main factor which still contributes SO_2, CO, and NO_2 emissions in the warm and cold season	https://doi.org/10.1007/s12517-021-07164-6
8	"The impact of street canyon morphology and traffic volume on NO_2 values in the street canyons of Antwerp"	Antwerp (Belgium)	2018	NO_2		Maximum hourly traffic volume (TVmax) and AR correlated significantly with the in situ NO_2 values, which makes an indicator for air quality of street canyons	https://doi.org/10.1016/j.buildenv.2021.107825
9	"Estimating multi-annual PM2.5 air pollution levels using sVOC soil tests: Ashkelon South, Israel as a case study"	Ashkelon South, Israel	2010–2012	PM10 and PM2.5	AQMS	Study confirms the feasibility of obtaining a precise estimate of the PM in air pollution using sVOC soil testing	https://doi.org/10.1016/j.atmosenv.2013.09.053
10	"A Novel Recursive Model Based on a Convolutional Long Short-Term Memory Neural Network for Air Pollution Prediction"	China	2014–2019	PM2.5	NDVI data	Proposed CR-LSTM model, implementing a combination of recursive strategy, geographical rules, and deep learning, shows better performance in longer-term prediction	https://doi.org/10.3390/rs13071284

Fig. 3.10 Pink cluster map

parameters were used in Istanbul using Aqua/Modis. Netherland, Madrid (Spain), Europe, North America, North Toronto, Newmarket and Egbert, French, France, India, and Belgium used NO_2 to monitor air quality using different sensor methods and instruments liAQMN, S5P/TROPOMI, UV–VIS spectrometer, and SCIAMACHY. Whereas a combination of multiple trace gases, Aerosol, and particulate matter was used to monitor air quality in Iran, Romania, China, Ukraine, low- and middle-income countries, the USA, India, Spain, Italy, Czech Republic, Norway, Iran, and Greece using various satellite sensor and methods like ANP-OWA method in GIS, MODIS, DDV, sun photometers and LIDAR's, Landsat8, UV-DCS light, RSSD, AQMN, OMI and TROPOMI, ERDAS, GPRS, GEOS-FP, and VIIRS.

Table 3.9 Details about the papers within Cluster #5

	Title	Study area	Year	Parameter	Sensor	Key finding	References
1	"The statistical evaluation and comparison of ADMS-Urban model for the prediction of nitrogen dioxide with air quality monitoring network"	Kaunas city	2011	NO_2		Model observation and values were most suitable for traffic monitoring station as compared to residential and background stations	https://doi.org/10.1007/s10661-015-4810-1
2	"Spatial estimation of surface ozone concentrations in Quito Ecuador with remote sensing data, air pollution measurements and meteorological variables"	Quito, Ecuador	2014	O_3	Landsat 8	Given model helps to infer the high-resolution spatial distribution of surface ozone in Quito	https://doi.org/10.1007/s10661-019-7286-6
3	"Concept of small satellite UV/visible imaging spectrometer optimized for tropospheric NO_2 measurements in air quality monitoring"	Beijing and Hokkaido		NO_2	SLATS and JAXA	Comprehensive analysis shows lower satellite (300 km latitude) are suitable to determine different scientific parameters of 5% for tropospheric NO_2 SCD	https://doi.org/10.1016/j.actaastro.2019.03.081
4	"Aerosol optical depth, aerosol composition and air pollution during summer and winter conditions in Budapest"	Budapest	2003–2004	Gaseous (CO, SO_2, NO_x, O_3), solid components (PM2.5, PM10)	Sun photometer	The nocturnal increase of the MLH coincided with decrease in pollutant concentrations except ozone; whereas increase in ozone concentration indicates vertical mixing of different air layers	https://doi.org/10.1016/j.scitotenv.2007.04.037
5	"Classification of aerosols over Saudi Arabia from 2004–2016"	Saudi Arabia	2004–2016	Dust and Black Carbon	OMI and AERONET	Dust aerosol types were highly concentrated seasonally and annually because of frequent events of dust storm. Fine particulates like BC and Mixed BC were observed because of industrial activities	https://doi.org/10.1016/j.atmosenv.2020.117785
6	"Calibration of Electrochemical Sensors for Nitrogen Dioxide Gas Detection Using Unmanned Aerial Vehicles"	Netherland	2018–2019	NO_2		The study finds quantification of NO_2 using electrochemical sensor boarded in UAV is highly reliable	https://doi.org/10.3390/s20247332

7	"Variations in black carbon aerosol, carbon monoxide and ozone over an urban area of Hyderabad, India, during the forest fire season"	Hyderabad, India	2004–2006	Black carbon (BC) aerosol; carbon monoxide (CO) and ozone (O_3)	DMSP-OLS	Higher incidence of forest fires on days coincides with high concentrations of CO and BC and spatial distribution of forest fires	https://doi.org/10.1016/j.atmosres.2006.10.004
8	"TROPOMI NO_2 Tropospheric Column Data: Regridding to 1 km Grid-Resolution and Assessment of their Consistency with In Situ Surface Observations"	Europe	2018–2020	NO_2	S5P/TROPOMI	Concentration of NO_2 in the winter period of 2020 was decreased than in the winter period of 2019	
9	"Intra-urban variability of ozone in a tropical city—characterization of local and regional sources and major influencing factors"	Middle Indo-Gangetic Plains of India	2013–2017	O_3, NO	GIS	O_3 level shows decreasing trend during the study period because of increase in emission of NO_x and intra-urban variability of O_3 due to traffic and local sources	https://doi.org/10.1007/s11869-018-0600-6
10	"Simulation of average monthly ozone exposure concentrations in China: A temporal and spatial estimation method"	China	2015–2017	O_3		Ozone exposure concentration primarily gathers in eastern and northern regions, and part of central mainland of China	https://doi.org/10.1016/j.envres.2021.111271
11	"Influence of vertical mixing and night time transport on surface ozone variability in the morning in Paris and the surrounding region"	Paris, France	2014	O_3, NO_2	Cloud and Aerosol Micro-Lidar (CAML)	Vertical mixing and incorporation of ozone located above the boundary layer in the morning is a main factor which control daytime surface ozone concentrations in Paris and in around	https://doi.org/10.1016/j.atmosenv.2018.10.009
12	"Biases in long-term NO_2 averages inferred from satellite observations due to cloud selection criteria"	Toronto	2005–2010	NO and NO_2	NASA EOS-Aura satellite	This investigation highlights the caution that must be applied when using satellite observations of ground-level pollution where a significant amount of data must be removed due to clouds	chttps://doi.org/10.1016/j.atmosres.2006.10.004

(continued)

Table 3.9 (continued)

	Title	Study area	Year	Parameter	Sensor	Key finding	References
13	"Incorporating satellite-derived data with annual and monthly land use regression models for estimating spatial distribution of air pollution"	Taiwan	2015	NO_2 and PM2.5	MODIS	Both monthly and annual model shows robust performance for NO_2 whereas PM2.5 models usually performed for few months, i.e., April to August	https://doi.org/10.1016/j.envsoft.2019.01.010
14	"Design of air quality monitoring networks and its application to NO_2 and O_3 in Cordova, Spain"	Cordova, Spain	2001–2002	NO_2 and O_3		Because of pollution distribution in Cordova, many diffusive samplers are benefitted by taking advantages due to low operating cost and easy use	https://doi.org/10.1016/j.microc.2009.07.007
15	"Optimization of the design of air quality monitoring networks and its application to NO_2 and O_3 in Jaen, Spain"	Jaen, Spain	2001–2002	NO_2 and O_3		Methods used to design AQMN for NO_2 and O_3 and their applications in Jaen, Southern Spain	https://doi.org/10.1016/j.microc.2010.07.002
16	"Air Quality Monitoring Network Design to Control Nitrogen Dioxide and Ozone, Applied in Granada, Spain"	Granada, Spain	1999–2001	NO_2 and O_3		Monitoring station locations are still main representatives for assessment of air quality	https://doi.org/10.1080/01919512.2011.536741
17	"Using different spatial scale measurements in a geostatistically based approach for mapping atmospheric nitrogen dioxide concentrations. Application to the French Centre region"	French	2005	NO_2		A geostatistical methodology used to examine the correlation between seasonal concentration of NO_2 and auxiliary variables	https://doi.org/10.1002/env.947
18	"Evaluation of OMI NO_2 Vertical Columns Using MAX-DOAS Observations over Mexico City"	Mexico City	2014–2017	NO_2	TROPOMI	The study shows average retrieval of OMI is 56% lower than MAX-DOAS	https://doi.org/10.3390/rs13040761
19	"OMI and Ground-Based In Situ Tropospheric Nitrogen Dioxide Observations over Several Important European Cities during 2005–2014"	Athens, Berlin, Bucharest, Lisbon, Madrid, Paris, Rome and Rotterdam	2005–2014	NO_2	UV–VIS spectrometer	Observation based on remote sensing and in-situ infers NO_2 shows negative trend for all locations of the studied area	https://doi.org/10.3390/ijerph14111415

20	"Transboundary transport of ozone pollution to a US border region: A case study of Yuma"	Yuma, USA	2005–2016	NO_2 and O_3	Yuma Supersite from the EPA Air Quality System (AQS)	Study infers slight increase in emission of nitrogen oxide from 2005 to 2016 in Mexico but decrease in Yuma	https://doi.org/10.1016/j.envpol.2020.116421
21	"Spatio-Temporal Characteristics of Tropospheric Ozone and Its Precursors in Guangxi, South China"	Guangxi, South China	2005–2016	NO_2, CO, and HCHO	MODIS FRP	Seasonal distribution of TCO was mainly derived from both long-term transportation and impact of meteorology	https://doi.org/10.3390/atmos9090355
22	"Development of a microscale land use regression model for predicting NO_2 concentrations at a heavy trafficked suburban area in Auckland, NZ"	Auckland, NZ	2013–2014	NO_2		The study has suggested micro-scale measurement and local scale model has to develop to identify small-scale variability of NO_2	
23	"Changes in ambient air quality and atmospheric composition and reactivity in the South East of the UK as a result of the COVID-19 lockdown"	South East of the UK	2015–2020	NO_2	Sentinel-5P TROPOMI	0D chemical box model simulations observed increasing in concentration of Ozone during the lockdown period in the region of hydrocarbon-limited ozone. Whereas significant decrease in NO_x than total non-methane	https://doi.org/10.1016/j.scitotenv.2020.142526

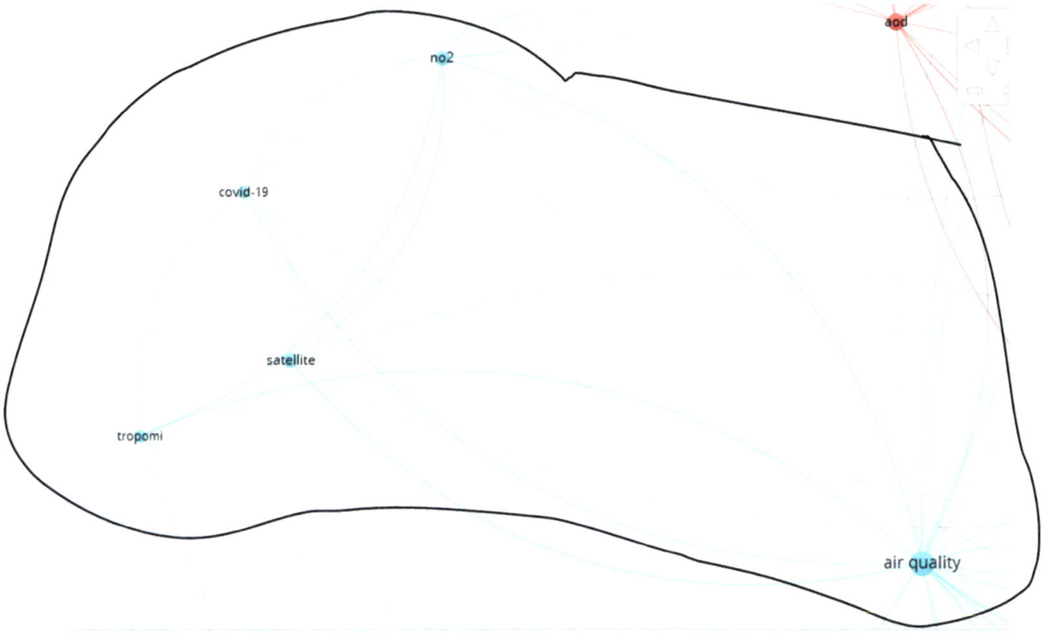

Fig. 3.11 Purple cluster map

Cluster #7

3.4.2.7 Cluster #7 (Orange Cluster): Aerosol Optical Depth

In this cluster many studies combine both particulate matter and gases to monitor air quality in the USA, Hungary, Mongolia, Taiwan, China, and India using different sensors and methods MODIS AOD, sun photometer, AOD, monthly LUR models and MISR, AERONET (see Table 3.11). Whereas Taiwan, low- and middle-income countries, China use sensors like AOD, LUR, and MODIS. Hong Kong is the only country that analyzes the aerosol to monitor the air quality using MODIS data. Whereas Finland is the only country that has analyzed the black carbon and trace metal to monitor air quality (Fig. 3.12).

Table 3.10 Details about the papers within Cluster #6

	Title	Study area	Year	Parameter	Sensor	Key finding	References
1	"Determination of the physical domain for air quality monitoring stations using the ANP-OWA method in GIS"	Iran	2007–2017	CO, NO_2, SO_2, and PM10 pollutants	ANP-OWA method in GIS	Air pollution monitoring stations were established in Tehran prominence	https://doi.org/10.1007/s10661-019-7422-3
2	"Monitoring the Indoor Air Quality and Microclimate for Improving Passenger Health and Comfort Level in Bucharest Subway Platforms"	Bucharest Subway Platforms	2015	CO, NO_2, SO_2, $PM_{2.5}$ and PM_{10} pollutants		Aim of the study is to improve comfort level and air quality in long subway	Tudor et al. (2019)
3	"Particulate matter (PM10 and PM2.5) concentrations during a Saharan dust episode in Istanbul"	Istanbul	2015	PM10 and PM2.5	Aqua/Modis	Concentration of PM10 is significantly increased whereas concentration of PM2.5 increases less variably during dust episode	https://doi.org/10.1007/s11869-020-00917-4
4	"Systematics of atmospheric environment monitoring in China via satellite remote sensing"	China	2014–2015	Aerosol, particulate matter, haze, straw burning, dust storm, gaseous pollutant (sulfur dioxide, nitrogen dioxide, and ozone), greenhouse gas (carbon dioxide and methane)	MODIS, DDV	Study finds sensors of Chinas satellite need to be improved because find difficulties in data acquisition and data quality control gap	https://doi.org/10.1007/s11869-020-00922-7
5	"A methodological framework for improving air quality monitoring network layout. Applications to environment management"	Madrid's (Spain)	2008–2017	PM10	GIS, AQMN	Fixed site within AQMN with European legislation guidelines helps to enhance the air quality management	https://doi.org/10.1016/j.jes.2020.09.009

(continued)

Table 3.10 (continued)

	Title	Study area	Year	Parameter	Sensor	Key finding	References
6	"Essential variables for air quality estimation"	Kyiv city	2005–2013	O_3, SO_2, NO_2, BaP and PM	Sun photometers and LIDAR's	PM2.5 is the most important pollutants for Ukraine's urban pollution, pollution level of Ukraine similar to some city Europe	https://doi.org/10.1080/17538947.2019.1620881
7	"An urban big data-based air quality index prediction: A case study of routes planning for outdoor activities in Beijing"	Beijing	2014–2015	PM2.5, PM10, CO, NO_2, O_3, and SO_2		Deep neural networking using different datasets along with air OP-DNN model can be predict future air quality index	https://doi.org/10.1177/2399808319862292
8	"Calibration of Electrochemical Sensors for Nitrogen Dioxide Gas Detection Using Unmanned Aerial Vehicles"	Netherlands	2018–2019	NO_2		The study finds quantification of NO_2 using electrochemical sensor boarded in UAV is highly reliable	https://doi.org/10.3390/s20247332
9	"Application of a short-term air quality action plan in Madrid (Spain) under a high-pollution episode—Part I: Diagnostic and analysis from observations"	Madrid (Spain)	2016	NO_2	AQMN	Impacts of concentration of NO_2 significantly affected by stage 3 protocol	https://doi.org/10.1016/j.scitotenv.2018.03.149
10	"TROPOMI NO_2 Tropospheric Column Data: Regridding to 1 km Grid-Resolution and Assessment of their Consistency with In Situ Surface Observations"	Europe	2018–2020	NO_2	S5P/TROPOMI	Concentration of NO_2 in the winter period of 2020 was decreased in the than winter period of 2019	

11	"A high-resolution index suitable for multi-pollutant monitoring in urban areas"	Beijing, Jinan, Shanghai and Xian		PM2.5, NO_2	MODIS, Landsat8 Operational Land Imager	DSI model is more fit for monitoring AQI, NO_2, and PM2.5 whereas it is not fit for urban areas with high brightness surface	https://doi.org/10.1016/j.scitotenv.2021.145428
12	"Particulate Matter at Surface: Northern Italy Monitoring Based on Satellite Remote Sensing, Meteorological Fields, and in situ Samplings"	Northern Italy	2007	PM2.5	MODIS (Terra-Aqua/ NASA platforms)	QUITSAT project over six validation sites of the Po valley showed good agreement	https://doi.org/10.1109/JSTARS.2009.2033948
13	"A Satellite-Based High-Resolution (1-km) Ambient PM2.5 Database for India over Two Decades (2000–2019): Applications for Air Quality Management"	India	2000–2019	PM2.5	MODIS	Poor air quality over urban-rural transect is because of regional scale problem	https://doi.org/10.3390/rs12233872
14	"Inferring ground-level nitrogen dioxide concentrations at fine spatial resolution applied to the TROPOMI satellite instrument"	North America	2018–2019	NO_2	TROPOMI satellite	Estimation by TROPOMI-inferred of exposed population to NO_2 of population exposed exceed health standard are at least three times more than for OMI-inferred estimates	
15	"Towards DCS in the UV Spectral Range for Remote Sensing of Atmospheric Trace Gases"			Trace Gases	UV-DCS light	Available sources of UV in UV-DCS have a potential to identify UV-absorbing gases	https://doi.org/10.3390/rs12203444
16	"Biases in long-term NO_2 averages inferred from satellite observations due to cloud selection criteria"	North Toronto, Newmarket, and Egbert	2005–2010	NO_2	UV–VIS spectrometer	Main source of NO_2 pollution is transported pollutant. Based on the analysis indicates significant bias in NO_2 data in long-term average	https://doi.org/10.1016/j.rse.2012.05.008

(continued)

Table 3.10 (continued)

	Title	Study area	Year	Parameter	Sensor	Key finding	References
17	"Ambient Air Pollutant Monitoring and Analysis Protocol for Low and Middle Income Countries: An Element of Comprehensive Urban Air Quality Management Framework"	Low- and middle-income countries	2016–2018	O_3, NO_2, and AOD	Remote sensing-based retrieval of satellite data; (RSSD)	Remote sensing monitoring data calibrated with in-situ measured data has a huge potential to measure air quality in a regional scale	https://doi.org/10.1016/j.atmosenv.2019.117120
18	"The use of MODIS data and aerosol products for air quality prediction"	Texas	2002	NO_x, CO, and OH	MODIS	Possibility of forecasting of air quality using remote sensing satellite data can be possible if it isolated from local pollution sources	https://doi.org/10.1016/j.atmosenv.2004.06.032
19	"Design of an Integrated System for Modeling of Functional Air Quality Index Integrated with Health-GIS Using Bayesian Neural Network"	Kolkata Municipal Corporation	2016	PM2.5, Nitrogen dioxide, Sulphur dioxide, Ozone	GIS	Model can locate the particular location particularly sensitive area of respiratory disease like asthma, bronchitis, and COPD	https://doi.org/10.1007/s12524-017-0724-4(0123456789().,-volV)(0123456789().,-volV)
20	"Temporal Trend of PM10 and Associated Human Health Risk over the Past Decade in Cluj-Napoca City, Romania"	Cluj-Napoca City, Romania	2009–2019	PM10	National Air Quality Monitoring Network	Significant increase in the air quality from warm months to cold months during 2009–2019. Whereas non-carcinogenic chronic health risk was reported between 2009 and 2010, it gets its peak during the cold months of each year	https://doi.org/10.3390/app10155331

21	"Observation of PM2.5 using a combination of satellite remote sensing and low-cost sensor network in Siberian urban areas with limited reference monitoring"	Krasnoyarsk, Siberia	2017–2018	PM2.5	MODIS	Concentration of annual PM2.5 data from satellite resulted from data restrictions that occurred during noon and data loss during winter were 20% and 30%, respectively. During summer period, 2018, study area experiences high PM2.5 concentration due to regional transport of smoke from forest wildfires	https://doi.org/10.1016/j.atmosenv.2020.117410
22	"Design of air quality monitoring networks and its application to NO_2 and O_3 in Cordova, Spain"	Cordova, Spain	2001–2002	NO_2 and O_3		Distribution in Cordova, many diffusive samplers are benefitted by taking advantages due to low operating cost and easy use	https://doi.org/10.1016/j.microc.2009.07.007
23	"Optimization of the design of air quality monitoring networks and its application to NO_2 and O_3 in Jaen, Spain"	Jaen, Spain	2001–2002	NO_2 and O_3		Methods used to design AQMN for NO_2 and O_3 and their applications in Jaen, Southern Spain	https://doi.org/10.1016/j.microc.2010.07.002
24	"Air Quality Monitoring Network Design to Control Nitrogen Dioxide and Ozone, Applied in Granada, Spain"	Granada, Spain	1999–2001	NO_2 and O_3		Monitoring station locations are still main representatives for assessment of air quality	https://doi.org/10.1080/01919512.2011.536741
25	"Using different spatial scale measurements in a geostatistically based approach for mapping atmospheric nitrogen dioxide concentrations. Application to the French Centre region"	French	2005	NO_2		A geostatistical methodology used to examine the correlation between seasonal concentration of NO_2 and auxiliary variables	https://doi.org/10.1002/env.947

(continued)

Table 3.10 (continued)

	Title	Study area	Year	Parameter	Sensor	Key finding	References
26	"Apportioning Smoke Impacts of 2018 Wildfires on Eastern Sierra Nevada Sites"	Eastern Sierra Nevada	2018	PM2.5	MODIS and VIIRS	Strategy for large fires was not feasible which leads to worse air quality as compared to slower-growing fires	https://doi.org/10.3390/atmos11090970
27	"OMI and Ground-Based In Situ Tropospheric Nitrogen Dioxide Observations over Several Important European Cities during 2005–2014"	Athens, Berlin, Bucharest, Lisbon, Madrid, Paris, Rome and Rotterdam	2005–2014	NO_2	UV–VIS spectrometer	Observation based on remote sensing and in situ infers NO2 shows negative trend for all locations of the studied area	https://doi.org/10.3390/ijerph14111415
28	"Recent trends in tropospheric NO_2 over India observed by SCIAMACHY: Identification of hot spots"	India	2002–2012	NO_2	SCIAMACHY	Study identifies 12 different hotspots across the country which shows drastic increasing trend	https://doi.org/10.5094/APR.2013.040
29	"Multi-objective optimization of air quality monitoring"	Brescia, Italy	1995–1998	SO_2, NO_2 and O_3	AQMN	Study area has significant potential to increase cost-effectiveness of air quality monitoring networks mainly in the urban area and regional scales	https://doi.org/10.1007/s10661-007-9725-z
30	"SAMIRA-Satellite Based Monitoring Initiative for Regional Air Quality"	Poland to Romania, the Czech Republic, and Southern Norway	2016–2019	NO_2, SO_2, AOD, PM2.5 and PM10	OMI and TROPOMI	Algorithm applies to OMI NO_2 data with high-resolution TROPOMI NO_2 as reference and increase in accuracy and performance after downscaling	https://doi.org/10.3390/rs13112219
31	"Assessment of contribution of SO_2, CO, and NO_2 in different urban land use in Bushehr region, Iran"	Bushehr region, Iran	2017–2018	SO_2, CO, and NO_2	ERDAS	Traffic is the main factor which still contributes SO_2, CO, and NO_2 emissions in the warm and cold season	https://doi.org/10.1007/s12517-021-07164-6

32	"Geographical information systems and air pollution simulation for Megalopolis' electric power plant in Peloponnese, Greece"	Megalopolis' electric power plant in Peloponnese, Greece	2002	^{226}Ra, ^{232}Th, ^{234}Th, ^{40}K and ^{238}U	GIS	Both vegetation and air quality are mainly affected by industrial waste	https://doi.org/10.1080/10934529.2014.895557
33	"An Internet of Things Platform for Air Station Remote Sensing and Smart Monitoring"	Andalusian, South of Spain		PM2.5, PM10, CO, O_3, NO_2 and SO_2	GPRS	Combined action of Cloud Computing and Big Data can process and correlate different remote sensed data from station which compute the automatic and personalized alert for health risk assessment	https://doi.org/10.32604/csse.2020.35.005
34	"Atmospheric particulate matter (PM) estimation from SeaWiFS imagery"	European	1997–2004	PM2.5	SeaWiFS imagery	Study finds the map of PM10 SeaWiFS and low correlation between in situ and instrumental data	https://doi.org/10.1016/j.rse.2007.03.009
35	"The impact of street canyon morphology and traffic volume on NO_2 values in the street canyons of Antwerp"	Antwerp (Belgium)	2018	NO_2		Maximum hourly traffic volume (TVmax) and AR correlated significantly with the in situ NO_2 values, which makes an indicator for air quality of street canyons	https://doi.org/10.1016/j.buildenv.2021.107825
36	"Improved algorithm for MODIS satellite retrievals of aerosol optical thickness over land in dusty atmosphere: Implications for air quality monitoring in China"	China	2008	surface visible (0.65 μm) and near-infrared (NIR, 2.1 μm)	MODIS, GEOS-Chem	Surface concentrations of PM10 of improved AOT show strong correlation with ground observation than GEOSCHem simulation	https://doi.org/10.1016/j.rse.2010.05.034

(continued)

Table 3.10 (continued)

	Title	Study area	Year	Parameter	Sensor	Key finding	References
37	"Estimating daily full-coverage near surface O_3, CO, and NO_2 concentrations at a high spatial resolution over China based on S5P-TROPOMI and GEOS-FP"	China	2020	O_3, CO, and NO_2	S5P-TROPOMI and GEOS-FP	Due to restriction of anthropogenic activity China experiences distribution of NSC of NO_2, CO, and O_3 during lockdown periods	https://doi.org/10.1016/j.isprsjprs.2021.03.018
38	"A GIS-based multi-source and multi-box modelling approach (GMSMB) for air pollution assessment—A North American case study"	State of California	2008	CO, NO_2, SO_2 and PM2.5	GIS-based multi-source and multi-box modeling approach (GMSMB)	Assessment of air pollution both in local and regional scale can be controlled and managed by GIS-based model	https://doi.org/10.1080/10934529.2012.707597
39	"Fine-Grained Air Quality Inference with Remote Sensing Data and Ubiquitous Urban Data"	Beijing, China		CO_2 and SO_2	Microsoft Research Asia (MSRA)	Besides air quality (currently PM2.5, PM10, and NO_2), our approach can be extended to infer other types of urban dynamics, such as CO_2 and SO_2	https://doi.org/10.1145/3340847
40	"Environmental effects of prohibiting urban fireworks and firecrackers in Jinan, China"	Jinan, China	2014–2018	SO_2, CO, and NO_2	AQI	AQI recorded highest in west and lowest in eastern region during 2017, western region shows highest because of high built-up activity and lowest recorded in developed area	

41	"The Impacts of the COVID-19 Lockdown on Air Quality in the Guanzhong Basin, China"	The Guanzhong Basin, China		PM2.5, PM10, SO_2, NO_2, CO and O_3	VIIRS	PM concentrations were decreased significantly in Guanzhong Basin during the lockdown period, whereas drastically decreased in the initial period and slower thereafter	https://doi.org/10.3390/rs12183042
42	"Spatial-temporal variations in regional ambient sulfur dioxide concentration and source-contribution analysis: A dispersion modeling approach"	Dallas County, Texas	1996–2002	SO_2		Results show very minimal decrease in annual SO_2 concentration in small areas of Dallas country Analysis results whereas other regions experience increasing trends of SO_2	https://doi.org/10.1016/j.atmosenv.2011.05.073
43	"Estimating multi-annual PM2.5 air pollution levels using sVOC soil tests: Ashkelon South, Israel as a case study"	Ashkelon South, Israel	2010–2012	PM10 and PM2.5	AQMS	Study confirms the feasibility of obtaining a precise estimate of the PM in air pollution using sVOC soil testing	https://doi.org/10.1016/j.atmosenv.2013.09.053

Table 3.11 Details about the papers within Cluster #7

	Title	Study area	Year	Parameter	Sensor	Key finding	References
1	"An evaluation of MODIS-retrieved aerosol optical depth over AERONET sites in Alaska"	Alaska	2000–2014	particulate pollution concentrations	MODIS AOD	Products of AOD may be used in applications of quantitative measurement of air quality in Alaska during summer time	https://doi.org/10.3390/rs10091384
2	"Aerosol optical depth, aerosol composition and air pollution during summer and winter conditions in Budapest"	Budapest	2003–2004	Gaseous (CO, SO_2, NO_x, O_3), solid components (PM2.5, PM10)	Sun photometer	The nocturnal increase of the MLH coincided with a decrease in pollutant concentrations except ozone; whereas an increase in ozone concentration indicates vertical mixing of different air layers	https://doi.org/10.1016/j.scitotenv.2007.04.037
3	"Using Multi-Angle Imaging Spectro Radiometer Aerosol Mixture Properties for Air Quality Assessment in Mongolia"	Mongolia	2011–2016	PM2.5, PM10, and SO_2	AOD and MISR	Machine regression shows good cross-validation statistics and PM2.5 estimates that capture relevant spatial and temporal patterns over the region	https://doi.org/10.3390/rs10081317
4	"Incorporating satellite-derived data with annual and monthly land use regression models for estimating spatial distribution of air pollution"	Taiwan	2015	NO_2 and PM2.5	AOD, monthly LUR models	Both monthly and annual model shows robust performance for NO_2 whereas PM2.5 models usually performed for few months, i.e., April to August	https://doi.org/10.1016/j.envsoft.2019.01.010
5	"High-resolution aerosol remote sensing retrieval over urban areas by synergetic use of HJ-1 CCD and MODIS data"	China	2010	Urban air quality	HJ-1 CCD and MODIS data, AERONET	Advantage of retrieval is higher resolution compared to MODIS aerosol components	https://doi.org/10.1016/j.atmosenv.2011.10.002
6	"Applying SPOT data to estimate the aerosol optical depth and air quality"	Northern Taiwan	1998		AOD with SPOT HRV	Variation of the atmospheric turbidity increases as the AOD value increases	
7	"Estimating ground-level PM2.5 over a coastal region of China using satellite AOD and a combined model"	China	2014–2016	$PM_{2.5}$	MODIS aerosol optical depth (AOD)	Spatial pattern of PM2.5 shows high concentration in urban areas compared to mountainous and sub-urban areas. Whereas concentration of PM2.5 shows high correlation with road density, industrial pollution, terrain features and forest covers	https://doi.org/10.1016/j.jclepro.2019.04.231

8	"Spatial Correlation of Satellite-Derived PM2.5 with Hospital Admissions for Respiratory Diseases"	Taiwan	2008	$PM_{2.5}$	MODIS	Elderly and infant are less vulnerable as compared to adults (18–65 years) and children (3–18 years) from the effect of PM2.5	https://doi.org/10.3390/rs8110914
9	"Evaluation of Using Satellite-Derived Aerosol Optical Depth in Land Use Regression Models for Fine Particulate Matter and Its Elemental Composition"	Taiwan	2014	Aluminum, calcium, chromium, iron, potassium, manganese, sulfur, silicon, titanium, vanadium, and zinc	AOD, LUR	Concentration of PM2.5 shows performance of model should be improved during high pollution season	https://doi.org/10.3390/atmos12081018
10	"A 3D aerosol and visibility information system for urban areas using remote sensing and GIS"	Hong Kong	2007	Aerosol	MODIS	The study shows source and precision of input AOT data along with their storage and retrieval in GIS platform which provides information on basis of 3D geographical location	
11	"Long-range transport episodes of fine particles in southern Finland during 1999–2007"	Southern Finland	1999–2007	Black carbon, monosaccharide anhydrides, oxalate, succinate, malonate, SO_2, NO_3, K+ and NH_4		In every year a huge area of Europe attains peak concentration of fine particle because of burning of open biomass in Eastern Europe	https://doi.org/10.1016/j.atmosenv.2008.11.022
12	"Opportunities and challenges for filling the air quality data gap in low- and middle-income countries"	Low- and middle-income countries		PM2.5		Air quality monitoring can be enhanced by the combination of emerging technologies, adaptation to country-specific conditions, and building capacity	https://doi.org/10.1016/j.atmosenv.2019.06.032
13	"Sensors and systems for air quality assessment monitoring and management: A review"	India	1990–2017	PM, CO, SO_2 and NO_2		A Satellite-based system will provide global half-hourly coverage for air quality by 2023 infers, their adaptability in future AQM. Whereas integrated systems are future solution to AQM	https://doi.org/10.1016/j.jenvman.2021.112510

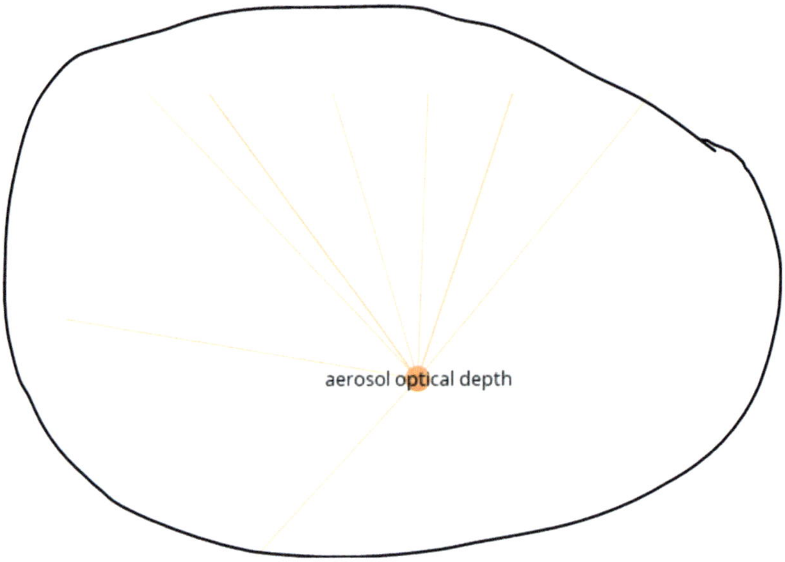

Fig. 3.12 Orange cluster map

3.5 Conclusions

This study carried out a bibliometric analysis and recent advances on air quality pollution, mainly studies between exposure of the population to air pollutants and its adverse health impact. This extensive review of air quality articles shows the widespread interest and concern with the problem of poor air quality, particularly in large urban centers. We can conclude that most of the studies were carried out in China, Hong Kong (China), and India, which have severe air pollution problems with rapid industrialization. In addition, the researched articles highlight an increase in the use of Remote Sensing products and Geographic Information Systems techniques, which are used to spatialize and understand the behavior of changes in air quality across the planet. However, this review showed a lack of proposals or recommendations to improve the state of the air in large cities. Thus, more review studies and articles that propose measures to reduce air pollution in large urban centers are essential to improve the quality of life of urban populations.

References

Aria M, Cuccurullo C. Bibliometrix: an R-tool for comprehensive science mapping analysis. J Informetr. 2017;11(4):959–75.

Bovensmann H, Burrows JP, Buchwitz M, Frerick J, Noël S, Rozanov VV, Chance KV, et al. SCIAMACHY: mission objectives and measurement modes. J Atmos Sci. 1999;56(2):127–50.

Buonanno G, Hänninen O. Air quality and health. Int J Environ Res Public Health. 2018;15(11):1–9.

Cobo MJ, López-Herrera AG, Herrera-Viedma E, Herrera F. Science mapping software tools: review, analysis, and cooperative study among tools. J Am Soc Inform Sci Technol. 2011;62(7):1382–402.

Cohen AJ, Brauer M, Burnett R, Anderson HR, Frostad J, Estep K, Balakrishnan K, et al. Estimates and 25-year trends of the global burden of disease attributable to ambient air pollution: an analysis of data from the Global Burden of Diseases Study 2015. Lancet. 2017;389(10082):1907–18.

Flynn L, Long C, Wu X, Evans R, Beck CT, Petropavlovskikh I, McConville G, et al. Performance of the ozone mapping and profiler suite (OMPS) products. J Geophys Res Atmos. 2014;119(10):6181–95.

Fraser RS. Satellite measurement of mass of Sahara dust in the atmosphere. Appl Opt. 1976;15(10):2471–9.

Hadjimitsis DG, Themistocleous K, Nisantzi A. Air pollution monitoring using earth observation & GIS. In: Air pollution: monitoring, modelling and health, vol. 117. Norderstedt: BoD–Books on Demand; 2012.

Hassan M, Malik AH, Waseem A, Abbas M. Air pollution monitoring in urban areas due to heavy transportation and industries: a case study of Rawalpindi and Islamabad. J Chem Soc Pak. 2013;35(6):1623.

Heath DF, Krueger AJ, Roeder HA, Henderson BD. The Solar backscatter ultraviolet and total ozone mapping spectrometer (SBUV/TOMS) for NIMBUS G. Opt Eng. 1975;14(4):323–31. https://doi.org/10.1117/12.7971839.

Intergovernmental Panel on Climate Change. Climate change 2013—the physical science basis. Cambridge: Cambridge University Press; 2014. https://doi.org/10.1017/CBO9781107415324.

Ionel I, Nicolae D, Popescu F, Talianu C, Belegante L, Apostol G. Measuring air pollutants in an international Romania airport with point and open path instruments. Rom J Phys. 2011;56(3–4):507–19.

Kim J, Jeong U, Ahn M-H, Kim JH, Park RJ, Lee H, Song CH, et al. New era of air quality monitoring from space: geostationary environment monitoring spectrometer (GEMS). Bull Am Meteorol Soc. 2020;101(1):E1–22.

Levelt PF, Joiner J, Tamminen J, Veefkind JP, Bhartia PK, Stein Zweers DC, Duncan BN, et al. The Ozone Monitoring Instrument: overview of 14 years in space. Atmos Chem Phys. 2018;18(8):5699–745.

Martin RV. Satellite remote sensing of surface air quality. Atmos Environ. 2008;42(34):7823–43.

Martin NI, Kelly N, Terry PC. A framework for self-determination in massive open online courses: design for autonomy, competence, and relatedness. Australas J Educ Technol. 2018;34(2):35–55.

Mekler Y, Quenzel H, Ohring G, Marcus I. Relative atmospheric aerosol content from ERTS observations. J Geophys Res. 1977;82(6):967–70.

Mishra M, Sudarsan D, Kar D, Naik AK, Das PP, Santos CAG, da Silva RM. The development and research trend of using dsas tool for shoreline change analysis: a scientometric analysis. J Urban Environ Eng. 2020;14(1):69–77.

Mishra M, Sudarsan D, Santos CAG, Mishra SK, Kar D, Baral K, Pattnaik N. An overview of research on natural resources and indigenous communities: a bibliometric analysis based on Scopus database (1979–2020). Environ Monit Assess. 2021;193(2):59.

Modak NM, Merigó JM, Weber R, Manzor F, de Ortúzar JD. Fifty years of transportation research journals: a bibliometric overview. Transp Res A Policy Pract. 2019;120:188–223.

Munro R, Lang R, Klaes D, Poli G, Retscher C, Lindstrot R, Huckle R, et al. The GOME-2 instrument on the Metop series of satellites: instrument design, calibration, and level 1 data processing—an overview. Atmos Meas Tech. 2016;9(3):1279–301.

Shi J, Miao W, Si H. Visualization and analysis of mapping knowledge domain of urban vitality research. Sustainability. 2019;11(4):988.

Song Y, Chen X, Hao T, Liu Z, Lan Z. Exploring two decades of research on classroom dialogue by using bibliometric analysis. Comput Educ. 2019;137:12–31.

Sudarsana D, Baba MS. Global nuclear fuel research during 2000 to 2017: a scientometric analysis. Ann Libr Inf Stud. 2019;66:85–93.

Sudarsana D, Meenachi NM, Venkatesh T, Gunta V, Gowtham R, Sai Baba M. Method for automatic key concepts extraction: application to documents in the domain of nuclear reactors. Electron Libr. 2018;37(1):2–15.

Tudor G, Bontos MD, Vasiliu OM, Vasiliu CG. Monitoring the indoor air quality and microclimate for improving passenger health and comfort level in Bucharest subway platforms. Rev Chim. 2019;70(10):3495–501.

van Eck NJ, Waltman L. Software survey: VOSviewer, a computer program for bibliometric mapping. Scientometrics. 2010;84(2):523–38.

van Eck NJ, Waltman L. Text mining and visualization using VOSviewer. 2011. http://arxiv.org/abs/1109.2058.

Van Eck NJ, Waltman L. Text mining and visualization. 2016. https://doi.org/10.1201/b19007.

Veefkind JP, Aben I, McMullan K, Förster H, de Vries J, Otter G, Claas J, et al. TROPOMI on the ESA Sentinel-5 Precursor: a GMES mission for global observations of the atmospheric composition for climate, air quality and ozone layer applications. Remote Sens Environ. 2012;120:70–83.

Zheng T, Wang J, Wang Q, Nie C, Shi Z, Wang X, Gao Z. A bibliometric analysis of micro/nano-bubble related research: current trends, present application, and future prospects. Scientometrics. 2016;109(1):53–71.

Air Pollution and Human Health: A Medical Geology Perspective

4

Amina Khatun, Trisha Barman, and Pulak Kumar Patra

4.1 Introduction

Air, water, and soil, all of which are influenced by regional and local geology, directly impact human health and well-being. A growing number of people worldwide are becoming aware of the adverse effects of natural conditions on human health. Governments are increasingly investing in efforts to reduce these risks. Many multidisciplinary health-related research projects have been initiated involving earth scientists as integral members of the research team. This has led to a remarkable growth in the field of medical geology (Hasan 2021). Medical geology can be defined as the study impact of earth processes and materials on human and ecosystem health (Bunnell 2004). Geological health studies encompass both natural and manmade sources of potential health problems and their regional distribution. Local geology has a direct impact on the nature and amount of mineral nutrients present in a region's water and soil. In addition, it impacts whether they are abundant or scarce. In addition to impacting human health, geological materials and processes also have an impact on plants and animals.

A. Khatun · T. Barman · P. K. Patra (✉)
Department of Environmental Studies, Visva-Bharati, Santiniketan, West Bengal, India
e-mail: pulakpatra@visva-bharati.ac.in

All chemical components originate from the rocks of the earth. The crust, or the topmost layer of the earth, provides all nutrients. Through weathering, rock gradually decomposes into its component minerals and eventually into different elements which are released into the environment to provide both essential nutrients and hazardous compounds. This intricate and gradual geologic process is influenced by the regional climate, the chemistry of the rock, and the characteristics of the biota. Furthermore, several harmful chemical components and compounds, coarse particles, and mineral-laden dust can be released into the atmosphere by natural geological events including dust storms, volcanic eruptions, and earthquakes, leading to poor and frequently dangerous air quality (Radhakrishna 2005).

A good number of natural and artificial elements and chemicals have been attributed to adversely affecting human health. The term "earth materials" refers to the solid materials that are produced, released, and transported by geological processes and human activities from rocks of the earth's crust. These particulate and dust materials can enter the air, water, and soil medium and pollute them. Plants also uptake and translocate these contaminants, eventually leading to contaminated food from both plant-based and animal-based sources, and water sources, leading to health issues. Thus, our health is affected by the geology of the area we live in.

P. K. Padhy et al. (eds.), *Air Quality and Human Health*,
https://doi.org/10.1007/978-981-97-1363-9_4

Hippocrates, the father of medicine, was the first person to recognize the relationship between geology and human health in the fifth century BC (Duzgoren-Aydin 2008). However, the concept has been gradually unheeded by medical professionals.

Medical geology is a multi-disciplinary field comprising medical or health sciences, earth sciences, and life sciences. Its research areas involve different scales. On one hand, it tries to understand the role of earth materials on our health system through microscopic studies of individual microbes or chemical compounds within tissues. On the other hand, it also examines the global or planetary scale movement/transport of particulates and pathogens to find out their source and origin. Modern medical geology also studies the effects of natural processes on human health, such as earthquakes, volcanic eruptions, and climate change. These studies can help identify potential health risks and develop new strategies for disease prevention. The standardization of procedures to link the regional water and soil quality with the health status of the inhabitants is in progress. Our understanding of the endemic health hazards due to groundwater fluoride and arsenic contamination has been vastly improved within recent decades. However, the medical geology of air pollution, and more particularly, atmospheric particulates, is still in its infancy. This is mainly due to the highly dispersive nature of the air pollutants, which can travel hundreds of kilometres from their source.

In the field of medical geology, atmospheric particulates are looked at as a mixture of natural and man-made earth materials that float through the air. Even though many experiments and epidemiological studies have shown that links exist between airborne particulates and human health, we still don't know much about the exact mechanisms that cause diseases that are caused by particulates.

This paper presents a discussion on the atmospheric particle cycles, their complex pathways, and health risks. It also discusses the challenges and emerging research areas related to the medical geology of air particulates.

4.2 Atmospheric Particle Cycles

The availability of major and minor chemical elements (nutrients) is influenced by the chemical elements and compounds found in geological materials and processes. Hazardous geological events, including earthquakes, landslides, and volcanic eruptions, release chemical components and substances into the environment, which can have simultaneous positive as well as negative effects on all living organisms. Rocks are the source of all chemical components that occur naturally. Weathering is the process of complicated physical, chemical, and metabolic interactions that crustal rocks are continuously subjected to, depending on their position and environment at the Earth's surface. As a result of this contact, rocks break down into individual mineral grains, releasing several substances and elements into the environment in the process. During the weathering process, elements and compounds are released, which are then carried from one area to another by geologic processes, including the action of wind, water, and ice. Geologic materials accumulate in a new area as a result of erosion, movement, and deposition. Additionally, it causes an imbalance or overabundance of certain chemical substances at different points on the earth, both of which have a significant bearing on human health (Radhakrishna 2005).

Natural geological processes such as weathering, erosion, and deposition can cause the lithosphere to directly produce atmospheric particles, particularly mineral dusts. The lithology of the bedrock, the thickness of the regolith (soil blanket), the predominant sediment types, and the climatic conditions (temperature, atmospheric deposit types, relative humidity, wind directions and speeds, etc.). It is important to note that earth materials and atmospheric particulates are influenced by many factors that influence their physicochemical properties. The main producers of air particulates in dry areas are unconsolidated surface deposits. These deposits include periodically deposited fine water-lake sediments, actively developing alluvial fan deposits, and fine lake sediments. The particles formed in such environments make up a sizable share of the global atmospheric particle budget. The creation of new sources of atmospheric particulates is exacerbated by anthropogenic activities such as urbanization, deforestation, and excessive agricultural activities, in addition to natural processes. Due to the influence of anthropogenic activities both directly and indirectly on biogeochemical cycles, it is increasingly difficult to distinguish between natural and anthropogenic components (Duzgoren-Aydin 2008).

The first step in establishing the geo-availability of consolidated, lithospheric materials among surface processes is weathering. The word "geo-availability" refers to the percentage of a compound in an earth material that can be released into the environment or biosphere. Weathering is a process where rocks and other surface materials are broken down by physical, chemical, and biological processes. This releases the minerals they contain into the environment, making them available for further processing. The mineralogy and chemical composition of the surface environment is influenced by weathering processes (i.e., the type, mode of occurrence, and composition of the parent rock) and climatic conditions (temperature, humidity, and rainfall) (Duzgoren-Aydin 2008). Other natural occurrences, particularly spontaneous natural disasters like landslides, earthquakes, wildfires, and volcanic activity, can contribute a sizable number of particles to the atmosphere. Volcanic ashes may be the most important because they are abundant in supply, move around the world, and stay bioactive (Duzgoren-Aydin 2008).

Dust detachment from the surface, as well as entrainment and subsequent transport by the wind, play an important role in governing the global atmospheric particle cycles (Derbyshire et al. 1998; Derbyshire 2005) concluding that these processes depend on several variables, including: (a) the wind speed; (b) the instability of the atmosphere; (c) the particle size; (d) the roughness of the land surface and moisture content; and (e) the degree of particle exposure. Moreover, he noted that the process by which atmospheric dust is entrapped and transported (especially solid particles) varies from a local to a global perspective, and that particle size decreases with distance travelled, thereby decreasing the proportion of particles that are respirable.

Atmospheric particles are formed from natural elements because of geological processes or manmade activity. They are classified as primary or secondary based on whether they are directly discharged into the atmosphere or produced in the atmosphere due to chemical interactions between primary components. Mineral dusts like asbestos, silica, and kaolinite, as well as re-suspended surface dusts like urban soils, road dusts, and concrete/cement dusts, are examples of atmospheric particulates (Duzgoren-Aydin 2008). So are volcanic ashes and gases and smog, which is polluted air with particulate matter from burning fossil fuels and industrial activity.

Natural sources of air particulates include windblown dust and sea sprays. Besides, wetlands emit CH_4, and plants emit ethene (C_2H_4), isoprene (C_5H_8), methyl butanol, terpenes ($C_{10}H_{16}$), and sesquiterpenes ($C_{15}H_{24}$) in different seasons; animals emit NH_3 and CH_4; lightning emits NO; and volcanoes emit sulphur compounds and ashes; oceanic biota emit dimethyl sulphide (DMS), and farmlands emit N_2O.

Particulate matter is directly emitted into the atmosphere as a result of numerous human activities. These activities include: (a) mining and processing of minerals; (b) smelting and refining of mineral ores and concentrates; (c) burning of fossil fuels (oil and gas) for energy production; (d) other industrial and manufacturing activities, such as those related to the metallurgical and chemical industries, cement production, and the manufacture of ceramics and glass; (e) agricultural practices; (f) transportation; and (g) construction and demolition. Depending on location, timing, and chemicals, some sources may dominate others. Urban environments are where these behaviours' detrimental effects on human health are most pronounced (Duzgoren-Aydin 2008).

At the surface environment, "atmospheric particle cycles" can happen in a matter of weeks, which is faster than many biogeochemical cycles. Particles in the air can move between continents, which shows that they stick around for a long time. This depends on their individual (particle) and group (bulk) properties as well as the conditions in the air. Surface environments can be both "sinks" and "sources" of atmospheric particulates. This is because particulates eventually settle on the surface, and depending on their size and the conditions in the air, they can be released back into the atmosphere. Both wet and dry precipitation in the atmosphere is part of the cycles of particles in the atmosphere. Surface runoff can carry particles from the air into water bodies. The quality of the water is getting worse because of surface runoff. In fact, urban surface runoff is known to be the second-most important source of trace metal pollution (such as Cu, Cr, Pb, and Zn) in aquatic environments. Atmospheric particles can also fall directly into water, which can be harmful for the ecosystem because they carry bacteria and toxic (like As, Hg, and Pb) or potentially toxic (like Cr, Cu, and Zn) trace metals and metalloids (Duzgoren-Aydin 2008). Figure 4.1 shows how atmospheric particle disperse through various pathways.

Weathering contributes to the release of mineral dusts including potentially toxic elements into the environment. This is how the geo-availability of earth materials and trace elements is determined. The intensity and type of weathering depend on many factors and vary from place to place. This influences the geo-availability of a given rock (the parent rock) at a particular place. In the same way, the type and intensity of human activities and their contributions to the global budget of airborne particles depend on the location. The presence of toxic elements in the environment only may not be considered as a health risk. The access of particulate matter and associated chemical elements for organisms depend on their chemical and mineralogy, and solubility (Reeder et al. 2006).

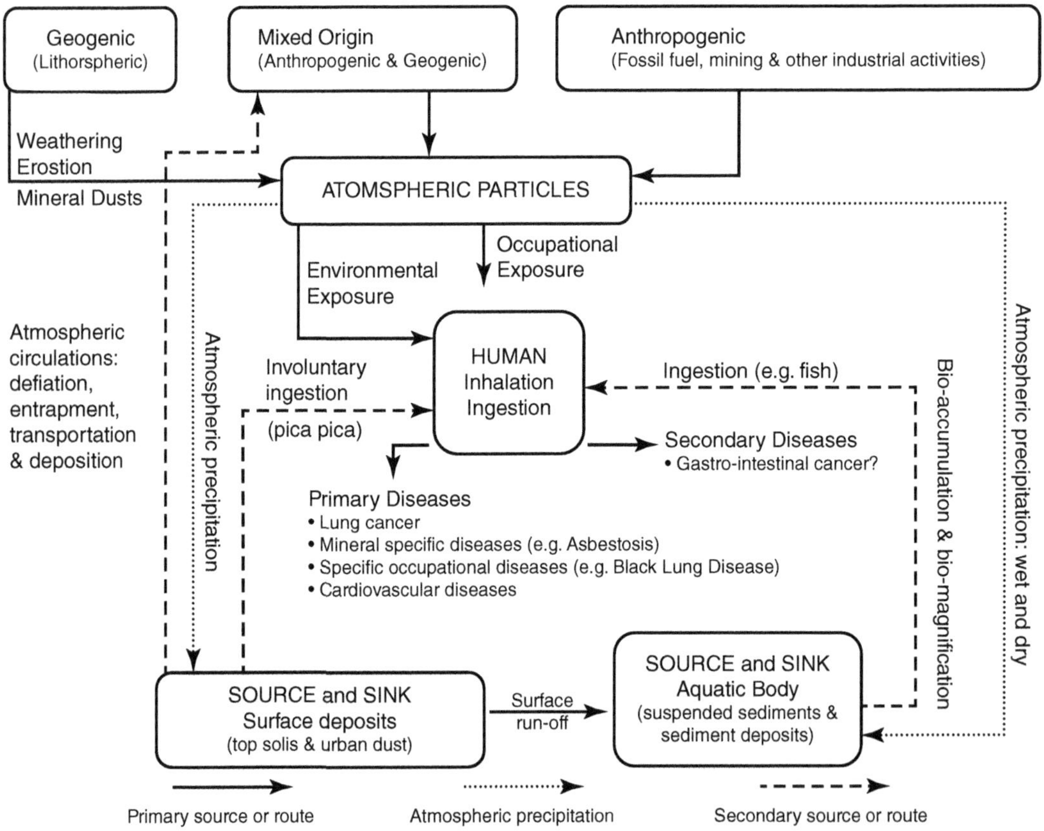

Fig. 4.1 Atmospheric particle dispersion pathways (Source: Duzgoren-Aydin 2008). (Reprinted with permission from Taylor and Francis Group)

Once the materials are available and get into the body, the following terms explain how they might cause diseases. Plumlee and Ziegler (2005) found that the bio-accessibility of a substance's abundance is usually given by: total concentration > geo-available fraction > bio-accessible fraction > bioavailable fraction. The terms "bio-durability" and "bio-persistence" are used to describe substances that can stay in the body for a long time after exposure. Bio-durability refers to the time duration a substance stays in the body. On the other hand, Bio-persistence measures how long something stays in the body despite chemical, physical, and biological ways to get rid of it. "Bio-solubility" relates to the quantity of earth material that can be dissolved in the fluid of the human body. The term "Bio-reactivity" is used to quantify the capability of the earth material to change the key body fluid parameters such as pH, redox species, and concentrations of major electrolytes (Plumlee and Ziegler 2005; Plumlee et al. 2006). Bio-solubility is caused by two different things: (a) thermodynamics and (b) kinetics. The bio-durability of several minerals can be attributed to the absence of a thermodynamic force that would dissolve them in the medium. As a result, certain minerals, including calcium phosphate, calcium carbonate, and quartz, are at a high level of saturation in the body fluids. In contrast, the plasma, interstitial fluid, and intercellular fluid compositions are all significantly under-saturated in a wide range of bio-durable aluminosilicate minerals, such as chrysotile, various amphiboles, and other minerals. This suggests that these minerals are bio-durable because of kinetic controls on dissolution (Sahai et al. 2006). An element's or mineral's bioactivity can be short-term or long-term. Acutely bio-reactive materials are those that can cause big changes in the chemistry of a fluid quickly and could hurt tissues along the route of exposure. Chronically bio-reactive materials, on the other hand, are bio-durable. This means that they react with the fluids and tissues around them mostly at their surface and slowly release chemicals into them (Duzgoren-Aydin 2008).

4.3 Atmospheric Geochemistry and Chemical Transport Model

Atmospheric geochemistry is a branch of geochemistry concerned with chemical processes occurring within the Earth's atmosphere. This research is critical to improving our understanding of climate forcings, air quality forcing, and the reciprocal interactions of the atmosphere and biosphere. This multidisciplinary approach to research draws on environmental chemistry, physics, meteorology, computer modelling, oceanography, geology, volcanology, and other disciplines. The composition and chemistry of the atmosphere require critical understanding. For the definition of the interaction between living organisms and the atmosphere. Knowing more about the gases and aerosols in the atmosphere, where they are, where they come from, and their concentrations will help us manage the health of all living organisms and the planet. The study of the chemical composition of the Earth's atmosphere seeks to understand the causes of these problems, and, by gaining a theoretical understanding of them, can be used to test possible solutions and evaluate the effects of policy changes.

A chemical transport model (CTM) is a type of computer-based numerical model that typically simulates atmospheric chemistry and may provide air pollution forecasting. CTMs may be classified according to their methodology and their species of interest, as well as more generic characteristics (e.g., dimensionality, degree of resolution) (Baklanov et al. 2011). To construct a CTM, a complete interactive set of equations for atmospheric dynamics, chemistry, and radiative processes is required. In practice, there is a need to make compromises between accuracy and computational efficiency as well as to parameterize processes that cannot be resolved by the model. Based on the modelling goals (i.e., the specific situation to be studied), these choices must be made.

From simple methods of transporting passive tracers to more sophisticated models that consider chemical processes on a diurnal to decadal scale, chemical tracer models have evolved over time. The advancement of these technologies is due to a deeper understanding of atmospheric processes as well as faster computers. According to spatial dimension, there are three levels of model hierarchy. The assumption is that all quantities are well mixed in zero spatial dimensions. One-dimensional models (I-D) describe variations in the vertical direction. Consequently, it represents an atmospheric average that is measured zonally and meridianally. Tracer experiments with 1-D gradients provide an expression for vertical transport. Tracers with smooth source distributions and lifetimes of several years (e.g., CFCs) are in general well-represented by this type of model. Consequently, 2-D models describe zonally averaged quantities since they include the meridian-vertical plane. In addition to zonal-mean transport, 2-D models can also describe chemical distributions (McIntyre 1980; Garcia and Solomon 1983). On a decadal scale, these models are used for the study of chemical processes and chemical effects. Climate studies typically involve lengthy integrations. These models can be used to model chemical species with half-lives ranging from a few days to several years.

4.4 Particulate Characters

There are several properties to consider when assessing the fate of solid particulates within the body, including composition, mineralogy, crystal structure, density, size, shape (morphology), reactivity, solubility, and surface charges. Material that is crystallized has a fixed structure and composition that fluctuates within defined limits. Compositional changes might alter solubility, reactivity, and surface charges (Plumlee et al. 2006; Guthrie Jr 1997; Plumlee and Ziegler 2005; Reeder et al. 2006). Mineral compositions may vary from their optimum stoichiometry. Most silicate mineral families, such as amphibole, serpentine, and zeolite, show a broad variety of compositional changes, called solid solution. A solid solution is a mineral structure in which particular sites are occupied by two or more distinct chemical elements (or ionic groups). Compositional differences may affect material characteristics, bioreactivity, and toxicity. Mineral dusts' toxicity is determined by their mineral concentration, individual mineral composition, and/or mineral structure.

Size and shape of a particle play a big role in its fate and thus particle size and shape are the most examined mineralogical properties when studying mineral-induced pathogenesis (Heaney and Banfield 1993). They influence the nature of the particle displacement/entrainment, physical movement, and cellular diffusion process. Size and shape also affect the particle surface area, types of reaction sites, interactions between particles and cells, and catalysis (Hochella et al. 1993; Guthrie Jr 1995; Guthrie Jr 1997; Reeder et al. 2006; Plumlee et al. 2006). For example, fine particles, botryoidal crystals, and acicular crystals are more easily dissolved than coarser, more uniform particles of the same mineral. This is because the smaller particles have more surface area per unit mass to react with body fluids (Plumlee and Ziegler 2005). Besides, however, appearance, the mineral's surface properties also influence its negative effects on human health. Earth material-induced pathogenesis can occur through several possible reaction pathways. However, mineral's surface properties, such as its chemical and electrical reactivity and its mechanical/dimensional–microtopographic features play the key role in the interactions between a mineral and a human (cell or body fluid) (Plumlee et al. 2006; Reeder et al. 2006; Hochella et al. 1993).

Most minerals have a limited ability to exchange cations because sorption (like adsorption) only happens on the surface. This means that their ability to exchange cations and their reactivity are limited to their surface area. Some groups of minerals, such as zeolites, have very high cation exchange capacities. This is because ions can move quickly from the surface of the mineral to its interior, so the whole particle can act as a buffer (Guthrie Jr and Mossman 1993; Gottardi and Galli 2012). Cation exchange may have a big impact on both cytotoxicity and gene response (Guthrie Jr and Heaney 1995). Also, some minerals, such as zeolites and some clay minerals (like kaolinite), can work in a way that is similar to how enzymes do. This is mostly because these minerals can give and take electrons or protons to create a stable surface for reacting parts and keep molecules of a certain shape or size from getting to the catalytic sites. For example, many important geochemical processes that take place between minerals and the human body (like dissolution, binding/sorption, cation exchange, and electron conduction) happen at the surface of the minerals. The chemical, electrical, and structural properties of the surfaces of individual mineral particles are often very different when they are looked at as a whole. This is especially true for minerals in the amphibole group, which include several fibrous (asbestiform) minerals that are known to cause serious health problems like mesothelioma, lung cancer, and asbestosis (Duzgoren-Aydin 2008).

4.5 Air Particulates from Natural Hazards

Toxins and pollutants are released into the atmosphere through natural cycles that occur across air, land, and water, and eventually end up in oceanic or lake sediments buried under waves. Volcanic eruptions, earthquakes, and storms, however, can release huge amounts of pollutants in a single surge of activity. Below are some aspects of these pollutants.

Volcanic Eruption: Large-scale geologic events, like volcanic eruptions, may release copious amounts of hazardous solids, toxic acids, gases, and aerosols into the atmosphere without prior notice, exposing people to potentially damaging and frequently fatal substances. The chemical and physical properties of volcanic ash depend a lot on the tectonic setting and the type of explosion. However, the speed and direction of the wind at the surface affect how quickly and where the volcanic ashes disperse. Glass or aggregate glass particles, as well as a variety of typical silicates like feldspar, pyroxene, and amphibole, make up most of these materials. Volcanic ashes can travel great distances across the world and ascend to extremely high heights in the stratosphere, changing the earth's radiation balance (Duzgoren-Aydin 2008). Studies that observed high ambient SO_2 concentrations downwind from volcano plumes have demonstrated the effect of volcanoes on air quality. Volcanoes can send a lot of particles into the air. These particles can be as small as 50 nanometres (nm) or as big as PM10 (Trejos et al. 2021).

Due to the generation procedures, mineral dust and volcanic eruptions' particle matter differ greatly. The first is the dust that is deposited on the pavement's surface and is released into the atmosphere by outside forces (air, vehicular passage, among others). The second is composed of magma and debris that were created by an eruption of a volcano (Malek et al. 2019). Several different types of minerals, including silica, clay, micas, feldspar, carbonates, sulphates, and phosphates, make up the re-suspended dust. In contrast, silica (including vitreous glass, quartz, and other polymorphs) and plagioclase feldspar make up most of the particulate material from volcanic emissions (Malek et al. 2019).

Earthquake and Landslides: Strong earthquakes as well as landslides cause buildings and other structures to collapse, releasing a fine dust-like material that is packed with a range of dangerous chemicals. Fine particles can float in the air for an extended period before entering living things, like people, through inhalation or ingestion. Although many cases of physical harm and fatalities linked to earthquakes have been documented, breathing in dust can also have negative health effects (Radhakrishna 2005). Earthquakes release a lot of dust and muck, which can temporarily harm respiratory health and cause other illnesses like diabetes and cardiovascular disease (Uprety et al. 2019).

Dust Strom: Most atmospheric dusts come from terrestrial sources, with significant land-atmosphere interaction (Saxena and Naik 2018). Dust storms have influential effects on ambient air quality besides their role on climate variability, nutrient dynamics, biogeochemical cycles of oceans, and ground and air transportation systems (Sonwani and Kulshrestha 2018). According to satellite and surface observation studies, arid and semi-arid desert regions contribute to the long-distance transport of dust particles carried out by strong winds (Saxena and Sonwani 2019). Among the naturally occurring aerosols on the ground, desert dust makes up a significant fraction. Compared to desertification processes, natural weathering has a greater impact on Asian dust emissions and dust storm occurrences (Sonwani and Kulshrestha 2018). Every dust storm is followed by net radiative cooling at the surface. Additionally, NOx concentrations increase as a result of the dust storm. The mineral dust coming from the Sahara and Thar deserts is found to be rich in nitrates, which convert NOx by a "re-noxification" process in the presence of sunlight (Ndour et al. 2009). Many epidemiological studies have found a link between particulate matter concentrations and the risk of respiratory and cardiovascular diseases (Loyo-Berrios et al. 2007; Pope III 2000). The effects of dust storms also cause several asthma cases all over the world, including Australia (Rutherford et al. 1999), the Middle East (Thalib and Al-Taiar 2012), and areas impacted by long-distance transport, such as northeast Asia (Lee and Lee 2014).

4.6 Exposure and Health Risks

The main factors that influence how air particles and the human body interact are as follows:

1. Source: the types, quantities, and physical, chemical, mineralogical, and surface characteristics of earth materials.
2. Humans: an individual's physiological and socioeconomic circumstances, as well as their age, gender, genetic make-up, and personal behaviours (such as smoking and eating a certain diet), as well as occupational exposure and a medical history.
3. Exposure and pathways: dose, route of exposure (inhalation and, as a result, the respiratory system), chemical conditions along the route of exposure, intensity and length of exposure.
4. Interactions: types of interactions encompassing defence mechanism types, circumstances like pH and bodily fluid types, and dislocation mechanisms.
5. Other: things that make it hard to tell, like the presence of more viruses or harmful substances, as well as how they interact with each other and what happens as a result (Duzgoren-Aydin 2008).

When earth materials, especially particles in the air, impact people's health, the effects are often not severe, unless changes add up over time and cause chronic effects. There are, however, responses that only last a short time after acute exposures. For chronic or repeated exposures, the dose of interest can be determined by the cumulative uptake if the effects on the target tissues are cumulative and clearance or recovery is slow. On the other hand, when the agent is quickly removed from the body or when its effects are quickly and completely reversed when the person is no longer exposed to it, the rate of delivery may be the most important dose parameter (Lippmann 2000).

In general, the type, amount, and route of exposure are important factors that influence how people and earth materials interact. The relative importance of these factors, on the other hand, changes a lot. The main way that people get exposed to particles in the air is by breathing them in. It is a way to be exposed all the time. The size of the dose from inhaled materials depends on: (a) their volume; (b) the amount that is transferred across epithelial barriers, which is the respiratory tract; (c) the amount that is carried by circulating fluids to target tissues; and (d) the amount that is taken up by the target tissues (Lippmann 2000; Alloway et al. 2005; Lindh 2005). Several physical, chemical, and physiological processes in the respiratory tract affect how things end up in the body and how they affect health. In other words, the effects of atmospheric particulates on health depend on how well they can get into the lungs despite the body's defences, which include deposition, clearance, retention, and absorption (Plumlee et al. 2006; Plumlee and Ziegler 2005; Lippmann 2000; Alloway et al. 2005; Lindh 2005).

Particulate clearance mechanisms are usually effective enough to remove inhaled particles larger than 10 μm from the air that is breathed in. Particles smaller than this are called "inhalable," which means they can get into the lungs and settle near where the airways split (branches). As there are a lot of nerve endings at these sites, the mechanical stimulation of deposited particulates often causes reflex coughing and bronchoconstriction, which leads to faster breathing rates and less pulmonary compliance (the lung's ability to give in to pressure increases without being damaged) (Plumlee et al. 2006; Plumlee and Ziegler 2005; Sullivan Jr et al. 2001; McClellan 1995).

Particulate deposition includes: (a) sedimentation onto the airway walls due to gravity settling; (b) electrostatic deposition of freshly generated particles with high surface charges; (c) interception of elongated particulates by the airway walls; and (d) deposition of fine particulates (2.5 μm) due to Brownian diffusion, which is when air molecules hit the particulates, increasing the chance of random motion, which increases the chance of deposition on airway walls. It is very important to realize that the same universal physical and chemical principles that control the complex interactions of large-scale geochemical processes in the geological environment also control the interactions of earth materials in the human body (Sahai et al. 2006; Williams 1997).

After particles have been dropped, their staying power may depend on how well they are cleared. In general, the time it takes to clear a site increases as it gets farther away from where the waste was dumped. It could take anywhere from a week to a few months for particles to leave the alveolar region. People generally think that slow clearance of particles from the respiratory system is bad because toxic substances stay in contact with sensitive tissue for longer (Sullivan Jr et al. 2001; McClellan 1995). From nasal hairs in the upper respiratory system to sneezing and coughing reflexes that get rid of mucus in which particles have been trapped or removed from deeper lung tissue, the ways in which the respiratory tract clears itself out change a lot along the path of exposure. Mucus, dust, and carbon dioxide are also moved out of the airways by the cilia (Plumlee et al. 2006). Particles with sizes ranging from 0.1 to 2.5 μm are best deposited in the lungs. As the concentration of toxic substances like Pb, Zn, Cr, Hg, and S increases as particle size decreases, the actual toxicity of these small particles may be higher than that of larger particles. Also, the huge surface area of small particles makes it possible for toxic chemical species to react and dissolve quickly. Since they stay in the alveolar region for a long time, substances like lead can be taken out and moved to other parts of the body (Sullivan Jr et al. 2001).

Other factors could influence how earth materials interact with the human body. Those can be (a) intrinsic factors, like gender, age, race, or a health problem that was there before, or (b) extrinsic factors, like the profile of exposures to pollutants, exposure to other toxic materials at the same time, nutritional status and lifestyle, socio-economic status, and other demographic factors (Godish and Fu 2019; Carbone et al. 2002; Levy 2006). Other confusing factors include: (a) interactions between two or more pollutants; (b) interactions between pollutants and meteorological variables like temperature and relative humidity; and (c) interactions between the health effects of atmospheric particulate pollutants and infectious diseases.

4.7 Conclusions

Cardiopulmonary diseases are closely associated with particulate matter air pollution exposure. There are both long-term and short-term health risks associated with particulate matter (including PM10 and PM2.5). The role that geochemical and mineralogical properties of particulates play in health effects caused by air pollution has only recently been recognized. From the perspective of medical geology, the characteristics of particulate matter air pollution, especially in urban settings, represent an immediate challenge for gaining a deeper understanding of the effects of particulate matter on human health. At present, metal-bearing mine waste, volcanic ash, volcanic gas condensates, coal, coal fly ash, wildfire smoke, pathogen-bearing dusts and soils, dust generated during earthquakes, and sediments left behind after storm events are among the earth materials that have been the subject of health concerns. The type and diversity of earth materials capable of causing or exacerbating health and environmental problems illustrates the vulnerability of the biogeochemical cycle. Over the past few decades, the role played by environmental geochemists and mineralogists has remarkably diversified and reinforced our understanding of the nature of interfaces and mechanisms that govern human interactions with the surface environment.

Even in the same location, the association between particulate matter air pollution and the group of people being investigated can vary significantly, depending on the type of data collection and processing techniques used, as well as the season and duration of sample collections. People who have no genetic susceptibility, such as young children and the elderly, are at a higher risk. It is essential to further investigate both particulate matter's individual and source characteristics, including their chemical, physical, and mineralogical characteristics, as well as their ability to carry a variety of potentially toxic elements. These characteristics ultimately determine particulate matter's bioreactivity. Research on the mechanisms and levels of regional atmospheric loading on the global budget and their effects on human health is still in its infancy.

The bioreactivity of mineral dust is determined not only by their mineral content, but also by their individual mineral composition, structure, and surface properties. To facilitate future comparisons between studies on the aetiology of mineral-induced diseases, it is necessary to present the level of chemical, mineralogical, and physical (e.g., morphology, size, and surface) heterogeneities of mineral dusts in detail. There is an urgent need to standardize the identification, characterization, and speciation processes, especially chemically, physically, and mineralogically for effective characterization of earth materials. To determine the distribution of health problems across different natural and anthropogenic environments, it is essential to examine the relationship between surface characteristics (such as geology and geomorphology) and processes (such as weathering and erosion) that determine geochemical backgrounds. As a sampling strategy, it would be ideal to target a comprehensive data set with all relevant variables. Confounding factors include variations in earth material source characteristics, distribution, and dispersion. Evidence indicates that genetic factors may have a far greater effect on mineral-induced pathogenesis than previously thought, requiring them to be included in epidemiological studies. Our understanding of the interactions between earth materials and the human body will be improved using new technologies such as spectroscopy, molecular modelling, and high-resolution image analysis. For materials such as atmospheric particulates, this is especially important, because they are nanoscale composites, whose physical, chemical, and biological characteristics, as well as their stability, reactivity, and bio-accessibility, differ significantly from those of their bulk, macroscopic counterparts.

Interfacial sciences tools will probably be used in future studies to assess quantitatively in situ surface, chemical, and mineralogical characteristics of atmospheric particles (and earth materials in general). A complementary research area would be to determine how trace elements function at the level of cells in living organisms.

References

Alloway B, Centeno JA, Finkelman RB, Fuge R, Lindh U, Smedley P. In: Selinus O, editor. Essentials of medical geology. Cambridge: Academic Press; 2005.

Baklanov A, Mahura A, Sokhi R. Integrated Systems of meso-meteorological and chemical transport models. Springer-Verlag Berlin Heidelberg. 2011;186p.

Bunnell JE. Medical geology: emerging discipline on the ecosystem-human health interface. Eco Health. 2004;1:15–8. https://doi.org/10.1007/s10393-004-0068-8.

Carbone M, Kratzke RA, Testa JR. The pathogenesis of mesothelioma. Semin Oncol. 2002;29(1):2–17.

Derbyshire E. Natural aerosolic mineral dusts and human health. In: Selinus O, Alloway B, Centeno JA, Finkelman RB, Fuge R, Lindh U, Smedley P, editors. Essentials of medical geology: impacts of the

natural environment on public health, vol. 459–80. Amsterdam: Elsevier; 2005. p. 40.
Derbyshire E, Meng XM, Kemp RA. Provenance, transport and characteristics of modern aeolian dust in Western Gansu Province, China, and interpretation of the quaternary loess record. J Arid Environ. 1998;39(497–516):31.
Duzgoren-Aydin NS. Health effects of atmospheric particulates: a medical geology perspective. J Environ Sci Health Part C. 2008;26(1):1. https://doi.org/10.1080/10590500801907340.
Garcia RR, Solomon S. A numerical model of the zonally averaged dynamical and chemical structure of the middle atmosphere. J. Geophys. Res. 1983;88:1379.
Godish T, Fu JS. Air quality. Boca Raton: CRC; 2019.
Gottardi G, Galli E. Natural zeolites, vol. 18. Cham: Springer; 2012.
Guthrie GD Jr. Mineralogical factors affect the biological activity of crystalline silica. Appl Occup Environ Hyg. 1995;10(12):1126–31.
Guthrie GD Jr. Mineral properties and their contributions to particle toxicity. Environ Health Perspect. 1997;105(suppl 5):1003–11.
Guthrie GD Jr, Heaney PJ. Mineralogical characteristics of silica polymorphs in relation to their biological activities. Scand J Work Environ Health. 1995;21(Suppl 2):5–8.
Guthrie GD Jr, Mossman BT. Health effects of mineral dusts. In: Proceedings (No. CONF-9310406-Vol. 28), vol. 28. Washington, DC: Mineralogical Society of America; 1993.
Hasan SE. Medical geology. In: Encyclopedia of geology, vol. 684. Cambridge: Academic; 2021.
Heaney PJ, Banfield JA. Health effects of mineral dusts. Washington, DC: Mineralogical Society of America; 1993.
Hochella MF, Guthrie JD, Mossman BT. Health effects of mineral dusts. Rev Mineral Geochem. 1993;28:275–308.
Lee JW, Lee KK. Effects of Asian dust events on daily asthma patients in Seoul, Korea. Meteorol Appl. 2014;21(2):202–9. https://doi.org/10.1002/met.1351.
Levy BS, editor. Occupational and environmental health: recognizing and preventing disease and injury. New York: Lippincott Williams & Wilkins; 2006.
Lindh U. In: Selinus O, Alloway B, Centeno RB, Finkelman R, Fuge R, Lindh U, Smedley P, editors. Essentials of medical geology impacts of the natural environment on public health. Biological functions of the elements; 2005. p. 115–60.
Lippmann M. Environmental toxicants: human exposures and their health effects, vol. 2000. New York: Wiley; 2000.
Loyo-Berrios NI, Irizarry R, Hennessey JG, Tao XG, Matanoski G. Air pollution sources and childhood asthma attacks in Catano, Puerto Rico. Am J Epidemiol. 2007;165(8):927–35.
Malek MA, Eom HJ, Hwang H, Do Hur S, Hong S, Hou S, Ro CU. Single particle mineralogy of microparticles from Himalayan ice-cores using SEM/EDX and ATR-FTIR imaging techniques for identification of volcanic ash signatures. Chem Geol. 2019;504:205–15. https://doi.org/10.1016/j.chemgeo.2018.11.010.
McClellan RO. Concepts in inhalation toxicology. Boca Raton: CRC; 1995.
McIntyre ME. Towards a Lagrangian-mean description of stratospheric circulations and chemical transports. Philos. Trans. R. Soc. London.1980;A296:120–148
Ndour M, Conchon P, D'Anna B, Ka O, George C. Photochemistry of mineral dust surface as a potential atmospheric renoxification process. Geophy. Res. Let. 2009;36:L05816. https://doi.org/10.1029/2008GL036662.
Plumlee GS, Ziegler TL. The medical geochemistry of dusts. Soils, and other earth materials/environmental geochemistry. Amsterdam: Elsevier; 2005. p. 263–310.
Plumlee GS, Morman SA, Ziegler TL. The toxicological geochemistry of earth materials: an overview of processes and the interdisciplinary methods used to understand them. Rev Mineral Geochem. 2006;64(1):5–57.
Pope CA III. Epidemiological basis for particulate air pollution health standards. Aerosol Sci Technol. 2000;32(1):4–14. https://doi.org/10.1080/027868200303885.
Radhakrishna BP. Medical geology. J Geol Soc India. 2005;66(4):395. https://doi.org/10.1016/b978-0-12-409548-9.12523-0.
Reeder RJ, Schoonen MA, Lanzirotti A. Metal speciation and its role in bioaccessibility and bioavailability. Rev Mineral Geochem. 2006;64(1):59–113.
Rutherford S, Clark E, McTainsh G, Simpson R, Mitchell C. Characteristics of rural dust events shown to impact on asthma severity in Brisbane. Aust Int J Biometeorol. 1999;42(4):217–25. https://doi.org/10.1007/s004840050108.
Sahai N, Schoonen MA, Skinner HCW. The emergent field of medical mineralogy and geochemistry. Rev Mineral Geochem. 2006;64(1):1–4.
Saxena, P., & Naik, V. (2018). Air pollution: sources, impacts and controls. Oxford: CABI.
Saxena P, Sonwani S. Criteria air pollutants: chemistry, sources and sinks. In: In criteria air pollutants and their impact on environmental health. Singapore: Springer; 2019. p. 7–48. https://doi.org/10.1007/978-981-13-9992-3_2.
Sullivan JB Jr, Levine RJ, Bangert JL, Maibach H, Hewitt P, Sullivan JB, Krieger G. Clinical dermatoxicology. In: Sullivan Jr JB, Krieger G, editors. Clinical environmental health and exposures. 2nd ed. Philadelphia: Lippincott Williams and Wilkins; 2001. p. 182–206.
Sonwani S, Kulshrestha U. Morphology, elemental composition and source identification of airborne particles in Delhi. India. J. Indian Geophys. Union. 2018;22(6):607–620.
Thalib L, Al-Taiar A. Dust storms and the risk of asthma admissions to hospitals in Kuwait. Sci Total Environ. 2012;433:347–51. https://doi.org/10.1016/j.scitotenv.2012.06.082.

Trejos EM, Silva LFO, Hower JC, Flores EMM, González CM, Pachón JE, Aristizábal BH. Volcanic emissions and atmospheric pollution: a study of nanoparticles. Geosci Front. 2021;12(2):746–55. https://doi.org/10.1016/j.gsf.2020.08.013.

Uprety A, Ozaki A, Higuchi A, Leppold C, Tanimoto T. The 2015 Nepal earthquake and worsening air pollution in Kathmandu. Lancet Planet Health. 2019;3(1):e8–9. https://doi.org/10.1016/S2542-5196(18)30247-X.

Williams RJP. The natural selection of the chemical elements. Cell Mol Life Sci CMLS. 1997;53(10):816–29.

Application of Land Use Regression (LUR) Models in Air Pollution Assessment

5

Bijaya Kumar Padhi and Pawan Kumar

5.1 Introduction

Pollution, defined as unwanted waste thrown into the air, land, and water bodies without regard for cost or consequence, poses an existential threat to health and jeopardizes the viability of modern societies (Fuller et al. 2022). One of the most serious and immediate environmental risks to human health is pollution which causes millions of deaths yearly. Rising urbanization, expanding industrialization, and associated anthropogenic activities are the primary contributors to air pollution emissions and declining air quality. Pollution is linked to various adverse health effects, including respiratory problems, cancer, cardiovascular diseases, neurological issues, and birth outcomes (The Lancet Planetary Health 2022). Rapid industrialization and modernization have contributed to the global increase in pollution, which is now becoming a significant public health risk (Fuller et al. 2022). A key contributor to early death and disease is air pollution. The Sustainable Development Goals (SDGs) agenda calls for a reduction in air pollution-related illnesses and deaths. Air pollution not only endangers health and shortens life but also reduces economic productivity. Generally, the air is contaminated by several pollutants, including particulate matter (PM), ozone, nitrogen oxide, and sulfur (Pandey et al. 2021).

Exposure assessment remains a challenge for epidemiologists evaluating long-term exposure to air pollution (Hoek et al. 2008). Exposure assessment, which involves analysing various exposure pathways (such as air, soil, and water) and routes (such as ingestion, inhalation, and dermal contact) across several levels, can assess the details, the degree, and linkage between sources of contamination, hazard concentrations as well as receptors (such as humans) in the environment (Habermann et al. 2015). Interpolation, dispersion models, exposure indicator variables, and land-use regression (LUR) models are techniques used to measure intra-urban air pollution contrasts. LUR modelling, in particular, has gained popularity in recent years as one such technique (Hoek et al. 2008). This chapter provides a detailed overview of the application of the LUR models in air pollution assessment.

5.2 Land Use Regression Models (LUR)

LUR, or land-use regression, is a statistical method that uses geospatial data to build prediction models in the health and environmental sciences. It is most commonly used in air pollution assessment and epidemiological research to accurately forecast pollutant concentrations at

B. K. Padhi (✉) · P. Kumar
Department of Community Medicine and School of Public Health, Postgraduate Institute of Medical Education and Research, Chandigarh, India

P. K. Padhy et al. (eds.), *Air Quality and Human Health*,
https://doi.org/10.1007/978-981-97-1363-9_5

specific locations based on past monitoring data. The development of LUR models requires the aggregation of measured data from numerous monitoring stations that sample the variable of interest. The modelling was performed using multiple linear regression analysis, with the dependent variable serves as the variable of interest (pollution concentrations) and potential predictor variables were extracted from the data on the surrounding environment (land use and road-traffic characteristics). The modelling was typically done in a Geographic Information System (GIS) framework. Potential predictor variables typically includes land cover, traffic intensity, road type, road length, population density, and meteorological conditions. These are treated as independent variables and are extracted from the monitoring site. The amounts of particulate matter and potential predictors are input into a supervised machine learning (ML) process, which attempts to develop a simple model. The model with the best-identified predictor variables can subsequently be used to forecast pollutant concentrations in other areas (Mölter and Lindley 2021).

Annual mean levels of NO_2, NO, PM2.5, and their soot content, and volatile organic compounds (VOCs) have been effectively modelled using LUR techniques in a variety of environments, including cities in North America and Europe (Abera et al. 2021). Using LUR models to calculate human exposure to air pollution is gaining popularity. Studies using LUR models have significantly increased in recent years with applications in epidemiological studies, burden of disease studies, and health impact assessment studies. This rise in popularity may be attributable to the effectiveness of LUR models.

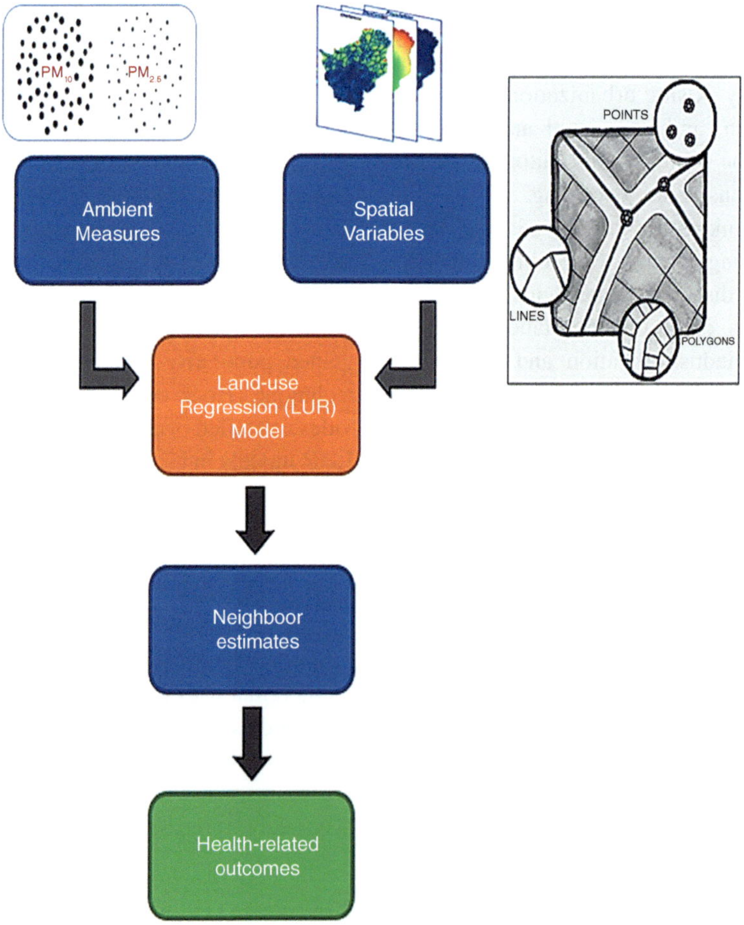

Fig 1. Land-use Regression (LUR) Model

- Relatively fast establishment and minimal start-up expenses.
- As GIS-based data sets and technologies become more widely available, researchers may quickly examine a large number (100+ in some instances) of potential predictor factors.
- Easy to obtain input data (pollutant concentration measurements and variables related to geography, land use, and transportation).
- Show alterations in spatial pollution patterns with a comparatively high spatial resolution.
- Models have demonstrated temporal stability and can be used for up to 10 years.
- LUR models are quite flexible, as they can be developed using data from variety of sources including satellites, mobile devices, and stationary measurements (Khreis et al. 2019).

5.3 Role of Geographical Information System (GIS) in LUR Modelling

A GIS is software that gathers, stores, verifies, and presents information related to locations of the Earth's surface. GIS can be used to map features such as vegetation, buildings and streets and utilizing location information like coordinates (latitude and longitude), addresses, or ZIP codes (GIS (Geographic Information System) n.d.. A wide range of academic disciplines benefit from GIS, including climatology, which investigates the causes and effects of global warming; epidemiology, which seeks the source of an outbreak of an infectious disease; political consulting, which creates election campaign strategies; and many others. Although, GIS has been used for many years to assess healthcare systems, its scope has grown considerably in recent years (McLafferty 2003).

Geographic information systems (GISs) can play a valuable role in LUR modelling for air pollution assessment. GIS organizes and visualizes data on land use, traffic, and other relevant-factors related to air pollution, and allowing for spatially analysis essential in developing a LUR model. For instance, GIS can create maps that show the distribution of different land use types in an area, and to identify areas where there is a high concentration of traffic or other sources of pollution. This information can then be used to develop a LUR model that incorporates these factors, allowing for the assessment of air pollution exposure levels at various locations within the study area (Atari and Luginaah 2009).

In LUR modelling, GIS is frequently used to gather data for multiple regression equations, characterizing the relationship between sample locations and environment variables. As a result, an algorithm is created that can forecast pollution concentrations at unmeasured locations using information about predictor variables at specific sites. To create the exposure distribution, a raster graphic data image of the site is generated and intersected with population-related data at the area level ("Land use regression model—Wikipedia' n.d.).

ArcGIS, featuring a dedicated LUR toolbox, is one of the most popular GIS-based package for creating predictor variables in LUR models. It is one of the most widely used tools for exposure assessment to calculate concentrations of air pollution in unmonitored areas. The LUR model analyses the association between predictor variables such as land use and traffic patterns near monitoring sites and observed air pollution concentrations in the multivariate regression model. The Geographic Information System creates predictor variables that depict land use and traffic conditions at specific locations (GIS). However, many researchers find it difficult and time-consuming to construct these variables. Calculates the predictor variables frequently used in the LUR model at each monitoring station ("LUR Tools|ArcGIS Hub' n.d.). GIS can be a valuable tool for supporting LUR modelling for air pollution exposure assessment and can help to improve both the accuracy and the usefulness of model results.

5.4 Applications of LUR Modelling in Air Pollution Assessment

LUR models were initially designed to assess exposure to air pollution from automobile traffic, but have lately been expanded to include the epidemiology of air pollution. These assessments are often funded by the Environmental Protection Agency (EPA), which collects hourly data from several American cities. The data examines how pollution levels vary over time and monitor the local health consequences ("U.S. Environmental Protection Agency I US EPA' n.d.).

The amounts of benzene, ethylbenzene, toluene, and M/P/0-xylene (BTEX) in Ontario, Canada, were predicted using LUR. To enhance LUR models, predictor factors and accurate estimate of BTEX concentrations, were gathered from 39 locations over a period of two weeks (Atari and Luginaah 2009).

In a 20-year study conducted in Gothenburg, Sweden, the NO2 content was evaluated to assess urban pollution. The results provided improved predictions on how the effect of altitude and traffic volume affect pollution levels in a particular area. The model was designed to assess outdoor concentrations in metropolitan areas, but had demonstrated lower accuracy in sparsely populated regions, such as islands or rural areas (Habermann et al. 2015).

Traditional fixed-site measurement methods are replaced with mobile monitoring. Even with a small number of monitoring units, mobile monitoring offers sufficient spatial coverage. Due to financial constraints, investigations are made possible through mobile surveillance (Hankey and Marshall 2015).

5.5 Advantages of LUR Modelling

There are several advantages to using LUR models in assessing air pollution. These include:

- LUR models can provide highly accurate estimates of air pollution levels at a given location, because they take into account multiple factors that can influence air pollution levels, such as land use, traffic, and other sources of pollution (Hoek et al. 2008).
- LUR models are relatively easy to create and use, even for individuals or organizations with limited technical expertise (Huang et al. 2017).
- LUR models can be used to predict air pollution levels in a wide range of locations, including both urban and rural areas.
- LUR models can be updated and refined over time, as new data becomes available or as factors that influence air pollution levels change (Mölter and Lindley 2021).

The use of LUR models in air pollution assessment enables policymakers, researchers, and other stakeholders to better understand and address air pollution, thereby taking more effective steps to improve air quality in their communities (Atari and Luginaah 2009).

5.6 Limitations of LUR Modelling

Although land-use regression (LUR) models are useful tools for air pollution assessment, they have several limitations. These include:

- LUR models are based on statistical relationships, rather than direct measurements of air pollution levels. As a result, there is some inherent uncertainty in the predictions made by these models (de Hoogh et al. 2014).
- LUR models require data on land use, traffic, and other factors that can influence air pollution levels. In some cases, these data may not be available or may be of poor quality, which can affect the accuracy of the predictions made by the model (Huang et al. 2017).
- LUR models may not be able to accurately predict air pollution levels at very fine spatial scales, such as at the level of a single building or street (Jin et al. 2019).
- LUR models can only predict air pollution levels at a given location based on the factors included in the model. If there are other factors that are not included in the model, these may not be taken into account in the predictions (Huang et al. 2017).

Therefore, it is important to carefully acknowledge the limitations of LUR models when using them for air pollution assessment and to take steps to mitigate these issues wherever possible (Table 5.1).

Table 5.1 Literature highlighting the importance of the Land-use Regression (LUR) models in assessing air pollution assessment

Author's name	Country & Year	Key findings
Eeftens M. et al.	Switzerland (2016)	The study showed that the results of the current LUR model were found to be greater for PM2.5 absorbance, PNC levels, LDSA and moderate for PM2.5, PM10, and PM coarse. In terms of cross-validation and independent external validation, LUR models (study area specific) for nitrogen dioxide exceeded combined-area alpine & non-alpine models, and were better able to account for between-area variability. Traffic predictor variables and national dispersion model estimates were significant predictors in the study (Eeftens et al. 2016)
Habermann M. et al.	Sweden (2015)	In this study, the LUR model was used to simulate the geographical distribution of pollutant concentrations in the urban region. During the period 7–20 May 2001, 25 samplers were deployed throughout the city to collect NO2 concentrations. GIS was used to estimate predictive variables such as 'altitude, density, road types, road traffic, and land use in buffers ranging from 50 to 500 m'. Linear regression came out around 5% and was conducted between NO2 and the predictive predictor, and the most robust variables with no collinearity were selected for the multivariate regression model. The average of the measurements was 23.5% with 180 predictive variables found. As predictor factors, the final model revealed 59% of the variance in NO2 levels with the 'inclusion of altitude and the sum of traffic'. The correlation between measured NO_2 and projected NO_2 was $r = 0.77$(Habermann et al. 2015)
Son Y. et al.	Mexico (2018)	The authors used mixed effect models using the Selection Operator and the least absolute shrinkage to form finer spatial-temporal LUR models for "PM2.5, PM10, O_3, NO_2, SO_2, CO in this research work (LASSO)'. Traffic density was added as a temporal variable, along with climatic and holiday data, hourly. Based on the standardized indices, the majority of the LUR models met their performance targets. LUR models with temporal scales >1 h were effectively built using LASSO with mixed effect models. The model performance was better than that of former LUR models, particularly for a time period of more than a day. To improve personal exposure estimations, the new models will be updated in conjunction with current air pollution sampling efforts in Mexico City (Son et al. 2018)
Knibbs D. L. et al.	Australia (2018)	In this research, the authors attempted to construct a model to predict NO_2 levels covering Australia using satellite measures of NO_2 (tropospheric) columns along with other predictor variables. The annual LUR model explained spatial variation in NO_2 was 81% while the top monthly model revealed 76%. We forecast NO_2 concentrations for 3.5 lacs census mesh blocks across the nation using our algorithms. The simple approach based on NO_2 column data models with little better than other approaches. Models captured within-urban NO_2 variability and can capture ambient NO_2 levels monthly, annual time scales across the country from 2006 to 2011 (Knibbs et al. 2014)

(continued)

Table 5.1 (continued)

Author's name	Country & Year	Key findings
Meng X. et al.	China (2015)	The author used the LUR model to estimate NO_2 levels in Shanghai, China, from 2008–11. The final LUR model retained four predictor variables: Road length around buffer monitoring sites, industrial sources (except power plants), agricultural land, and population counts. The surface of NO_2 conc prediction based on the LUR had a high spatial resolution based on ratio and difference techniques after 1 year. The NO_2 model outweighs the 'Kriging & inverse distance weighted interpolation approaches'. The study findings indicate that in developing countries the LUR model can provide a cost-effectiveness in measuring air pollution concentrations (Meng et al. 2015)
Lee M. et al.	Hong Kong (2017)	LUR is a typical method for estimating exposure by forecasting the geographical variability of air pollution. The quantities of NO_2, NO, BC and Fine PM 2 were detected during 2 sampling rounds in Hong Kong along with 365 relevant geographical predictor factors to form 2D LUR models. The majority of the models included traditional LUR predictors such as 'Land use, road length and car park density'. The findings corresponded to the current literature on geographical variation in air pollutant concentrations and their relationship to significant emission sources in Hong Kong. The success of these models implies that LUR is acceptable for higher-density and high-rise cities (Lee et al. 2017)
Kashima S. et al.	Japan (2009)	The author effectively used the LUR model to anticipate traffic-related pollution in Shizuoka, Japan. We used NO_2 and SPM levels between 2000–2006 in the study area. Road type, land use, traffic intensity, and physical component are the four geographic characteristics believed to predict traffic-related pollution. We then built a model using geographical variables to forecast the NO_2 and SPM values. 'The mean concentrations of NO_2 and SPM were 35.75 g/m^3 and 28.67 g/m^3, respectively'. The final NO_2 regression model included 5 independent variables. The regression model for the SPM level had a single independent variable. The SPM model had a low R2 value compared to NO_2 model. we could predict NO_2 levels even with regulatory monitoring air quality data effectively. This finding may stimulate the widespread usage of LUR models in the Asian continent (Kashima et al. 2009)
Sanchez M. et al.	India (2018)	The performance of the LUR model is uncertain in less urbanized settings in LMICs due to complicated sources of ambient air pollution and insufficient land use data. To address these concerns, 'LUR models have been formed for a peri-urban area outside of Hyderabad, India, using satellite images (e.g., vegetation, urbanicity) and manually obtained data (e.g., roads, industries, non-residential sites) from a survey of the environment. Fine PM 2.5 and BC levels were measured at 23 sites as part of the CHAI project'. The annual mean (SD) for PM2.5 was 34.1 & 2.7 for BC. The LUR model for annual BC indicated 78% of total variance and included predictors on both the local (energy supply points) and regional (road) scales. LUR models projected significant variability (within the village) for both pollutants, which is a crucial characteristic for future epidemiological investigations (Sanchez et al. 2018)

Table 5.1 (continued)

Author's name	Country & Year	Key findings
Tularam H. et al.	South Africa (2021)	The LUR technique is used to forecast ambient NO_2, SO_2, and Particulate Matter (PM10). The research was carried out in Durban where one area had a heavy industrial characteristics and the other a commercial and residential profile. Using the 'European Study of Cohorts for Air Pollution Effects (ESCAPE) technique', multiple hybrid models were created in the city including air pollution dispersion modelling for 'source-specific NO_2, SO_2 and PM10 levels' into LUR models. The higher amounts were projected in the southern part of Durban compared to northern part of Durban, so Multiple hybrid models were created in Durban by including air pollution dispersion modelling projections for 'source-specific levels of NO_2, SO_2, and PM10 levels' into LUR models North Durban is expected to have somewhat higher levels of PM10 than south Durban, which can be ascribed to 'traffic, bush burning, or home fuel burning' (Tularam et al. 2021)
Weichenthal et al.	Canada (2016)	The author suggested that, according to the available evidence, acute cardiorespiratory morbidity may be exacerbated by ambient ultrafine particles (UFPs). In this study, we used mobile monitoring data gathered from approximately 400 segments of road throughout 2011–12, we created an LUR for UFPs in Montreal to meet this demand. 'Standard multivariate linear regression and kernel-based regularized least squares (KRLS)', a machine learning (ML) method that deduces the functional form of covariate impacts on ambient UFP concentrations from the data, were both investigated as methods for model creation. 'Population density, ambient temperature, wind speed, land use factors (park and open space), local road and rail length, and anticipated yearly average NOx emissions' from transportation were all incorporated in the final models. The KRLS model described around 80% of the variance, while the final multivariate linear regression model explained around 60% of the regional variation in ambient UFP concentrations. When tested against an external data set or through cross-validation, the KRLS model outperformed the linear regression model by a small margin (Weichenthal et al. 2016)

References

Abera A, Malmqvist E, Mandakh Y, Flanagan E, Jerrett M, Gebrie GS, et al. Measurements of NOx and development of land use regression models in an east-African City. Atmos. 2021;12(4):519. https://doi.org/10.3390/atmos12040519.

Atari DO, Luginaah IN. Assessing the distribution of volatile organic compounds using land use regression in Sarnia, "Chemical Valley", Ontario, Canada. Environ Health. 2009;8(1):16. https://doi.org/10.1186/1476-069X-8-16.

de Hoogh K, Korek M, Vienneau D, Keuken M, Kukkonen J, Nieuwenhuijsen MJ, et al. Comparing land use regression and dispersion modelling to assess residential exposure to ambient air pollution for epidemiological studies. Environ Int. 2014;73:382–92. https://doi.org/10.1016/j.envint.2014.08.011.

Eeftens M, Meier R, Schindler C, Aguilera I, Phuleria H, Ineichen A, et al. Development of land use regression models for nitrogen dioxide, ultrafine particles, lung deposited surface area, and four other markers of particulate matter pollution in the Swiss SAPALDIA regions. Environ Health. 2016;15(1):1–14. https://doi.org/10.1186/S12940-016-0137-9/TABLES/6.

Fuller R, Landrigan PJ, Balakrishnan K, Bathan G, Bose-O'Reilly S, Brauer M, et al. Pollution and health: a progress update. Lancet Planet Health. 2022;6(6):e535–47. https://doi.org/10.1016/S2542-5196(22)00090-0.

GIS (Geographic Information System). n.d.. https://education.nationalgeographic.org/resource/geographic-information-system-gis.

Habermann M, Billger M, Haeger-Eugensson M. Land use regression as method to model air pollution. previous results for Gothenburg/Sweden. Proc Eng. 2015;115:21–8. https://doi.org/10.1016/J.PROENG.2015.07.350.

Hankey S, Marshall JD. Land use regression models of on-road particulate air pollution (Particle Number, Black Carbon, PM 2.5 , Particle Size) using mobile monitoring. Environ Sci Technol. 2015;49(15):9194–202. https://doi.org/10.1021/acs.est.5b01209.

Hoek G, Beelen R, de Hoogh K, Vienneau D, Gulliver J, Fischer P, Briggs D. A review of land-use regression models to assess spatial variation of outdoor air pollution. Atmos Environ. 2008;42(33):7561–78. https://doi.org/10.1016/j.atmosenv.2008.05.057.

Huang L, Zhang C, Bi J. Development of land use regression models for PM2.5, SO2, NO2 and O3 in Nanjing, China. Environ Res. 2017;158:542–52. https://doi.org/10.1016/j.envres.2017.07.010.

Jin L, Berman JD, Warren JL, Levy JI, Thurston G, Zhang Y, et al. A land use regression model of nitrogen dioxide and fine particulate matter in a complex urban core in Lanzhou, China. Environ Res. 2019;177:108597. https://doi.org/10.1016/j.envres.2019.108597.

Kashima S, Yorifuji T, Tsuda T, Doi H. Application of land use regression to regulatory air quality data in Japan. Sci Total Environ. 2009;407(8):3055–62. https://doi.org/10.1016/j.scitotenv.2008.12.038.

Khreis H, Cirach M, Mueller N, de Hoogh K, Hoek G, Nieuwenhuijsen MJ, Rojas-Rueda D. Outdoor air pollution and the burden of childhood asthma across Europe. Eur Respir J. 2019;54(4):1802194. https://doi.org/10.1183/13993003.02194-2018.

Knibbs LD, Hewson MG, Bechle MJ, Marshall JD, Barnett AG. A national satellite-based land-use regression model for air pollution exposure assessment in Australia. Environ Res. 2014;135:204–11. https://doi.org/10.1016/j.envres.2014.09.011.

Land use regression model—Wikipedia. n.d.. https://en.wikipedia.org/wiki/Land_use_regression_model. Accessed 6 Dec 2022.

Lee M, Brauer M, Wong P, Tang R, Tsui TH, Choi C, et al. Land use regression modelling of air pollution in high density high rise cities: a case study in Hong Kong. Sci Total Environ. 2017;592:306–15. https://doi.org/10.1016/j.scitotenv.2017.03.094.

LUR Tools|ArcGIS Hub. n.d.. https://hub.arcgis.com/content/de058ff6b6d44ac98a180fa1b7bcbf82/about. Accessed 30 Nov 2022.

McLafferty SL. GIS and Health care. Annu Rev Public Health. 2003;24(1):25–42. https://doi.org/10.1146/annurev.publhealth.24.012902.141012.

Meng X, Chen L, Cai J, Zou B, Wu C-F, Fu Q, et al. A land use regression model for estimating the NO2 concentration in shanghai, China. Environ Res. 2015;137:308–15. https://doi.org/10.1016/j.envres.2015.01.003.

Mölter A, Lindley S. Developing land use regression models for environmental science research using the XLUR tool—more than a one-trick pony. Environ Model Softw. 2021;143:105108. https://doi.org/10.1016/J.ENVSOFT.2021.105108.

Pandey A, Brauer M, Cropper ML, Balakrishnan K, Mathur P, Dey S, et al. Health and economic impact of air pollution in the states of India: the global burden of disease study 2019. Lancet Planet Health. 2021;5(1):e25–38. https://doi.org/10.1016/S2542-5196(20)30298-9.

Sanchez M, Ambros A, Milà C, Salmon M, Balakrishnan K, Sambandam S, et al. Development of land-use regression models for fine particles and black carbon in peri-urban South India. Sci Total Environ. 2018;634:77–86. https://doi.org/10.1016/j.scitotenv.2018.03.308.

Son Y, Osornio-Vargas ÁR, O'Neill MS, Hystad P, Texcalac-Sangrador JL, Ohman-Strickland P, et al. Land use regression models to assess air pollution exposure in Mexico City using finer spatial and temporal input parameters. Sci Total Environ. 2018;639:40–8. https://doi.org/10.1016/j.scitotenv.2018.05.144.

The Lancet Planetary Health. The air that we breathe. Lancet Planet Health. 2022;6(1):e1. https://doi.org/10.1016/S2542-5196(21)00357-0.

Tularam H, Ramsay LF, Muttoo S, Brunekreef B, Meliefste K, de Hoogh K, Naidoo RN. A hybrid air pollution/land use regression model for predicting air pollution concentrations in Durban. Environ Poll. 2021;274:116513. https://doi.org/10.1016/j.envpol.2021.116513.

U.S. Environmental Protection Agency|US EPA. n.d.. https://www.epa.gov/. Accessed 8 Dec 2022.

Weichenthal S, Van Ryswyk K, Goldstein A, Bagg S, Shekkarizfard M, Hatzopoulou M. A land use regression model for ambient ultrafine particles in Montreal, Canada: a comparison of linear regression and a machine learning approach. Environ Res. 2016;146:65–72. https://doi.org/10.1016/j.envres.2015.12.016.

Particulate Matters, Microplastics, and Mitochondrial Health

6

Reshmi Ravindran and Dhyan Chandra

6.1 Introduction

Environmental pollutants affecting human health have been a grave concern. Particulate matter is a most ubiquitous air pollutant with serious consequences to health. With their diversity in chemical composition, these minute pollutants are difficult to track and minimize. Microplastics are an emerging group of pollutants to be intensively studied owing to their small size and widespread use of non-biodegradable plastics. Both particulate matters and microplastics are implicated in various disease conditions. Not surprisingly, both these pollutants are able to positively affect the hallmarks of cancer and aid in cancer progression. Mitochondria are crucial to cells' energy metabolism and normal homeostasis. Particulate matter and microplastics have an impact on mitochondrial health. This chapter aims to understand the nature of particulate pollutants and microplastics, their toxicity and their effects on various pathological conditions, and the impact of these potent pollutants on mitochondrial health.

R. Ravindran · D. Chandra (✉)
Department of Pharmacology and Therapeutics, Roswell Park Comprehensive Cancer Center, Buffalo, NY, USA
e-mail: dhyan.chandra@roswellpark.org

6.2 Particulate Matter

Human civilization has long been victims of pollutants that surround them, causing serious threats to survival. Air pollution is one prominent health disrupter, and the World Health Organization (WHO) states that air pollution is responsible for causing 4.2 million deaths annually. Particulate matter (PM) contributes significantly to polluting the air and deteriorating human health. Particulate matter as a potent threat to human life was first evident from increased mortality due to smog in early 1900 in Belgium, Pennsylvania, and the UK (Stanek et al. 2011). Particulate matter has been associated with causing cardiovascular and respiratory distress. In addition, it blocks solar radiation, modulates climate, and seriously affects air visibility apart from influencing ozone production and impacting precipitation (Liu et al. 2020; Zeb et al. 2018). Particulate matter is a heterogenous mixture of minute differently sized particles with varying physio-chemical properties and suspension of liquid droplets in the air. The chemical composition of PM mainly consists of carbonaceous particles with various chemicals and metals attached to them. The inorganic matters of concern including silicates, nitrates, sulfates, and organic components like polyaromatic hydrocarbons also pose a threat. The presence of heavy metals adsorbed to their surface (Fe, Cu, Ni, Zn) enhances the toxicity of PM (Hamanaka and

P. K. Padhy et al. (eds.), *Air Quality and Human Health*,
https://doi.org/10.1007/978-981-97-1363-9_6

Mutlu 2018; Zhang et al. 2018). PM is also reported to harbor spores and pathogenic cell fragments which can cause opportunistic infections (Miri et al. 2018).

Based on their aerodynamic diameter, particulate matter is characterized as coarse particles, having diameter greater than 2.5 μm; fine particles with diameter 0.1–2.5 μm and ultra-fine particulate matter with diameter less than 0.1 μm (Xiao et al. 2018). Coarse particulate matters are formed from mechanical or physical processes such as grinding. $PM_{2.5}$ are byproducts of combustion reactions either from anthropogenic sources like emissions from traffic, fossil fuel burning, industrial effluents, or natural sources like forest fires, dust storms, and volcanic eruptions (Miri et al. 2018; Wu et al. 2018a, b). PM has been reported to sustain in the air for a substantial time frame and is also reported to cause serious health effects to human being, in addition to causing modulations in climate (Deng et al. 2019; Miri et al. 2018; Zhang et al. 2018).

6.3 Particulate Matter: Impact on Human Health

Particulate matters are a serious threat to human health. Coarse particulate matter is reported to deposit in upper respiratory tracts. Fine particulate matter gets deposited in the lungs and is a potent respiratory irritant. $PM_{2.5-10}$ are associated with respiratory diseases. Asthma, chronic obstructive pulmonary disease (COPD), pneumonia, and allergic rhinitis are linked to short-term exposure to particulate matters, and emphysema, bronchitis, and pleuritis are associated with long-term exposures (Cho and Choi 2021; Wu et al. 2018a, b). Respiratory distress is an aftermath of inflammatory reactions caused due to oxidative stress or direct toxicity brought about by particulate matters as evidence points out exposure to PM leads to the release of inflammatory mediator (Delfino et al. 2010). $PM_{2.5}$ are inducers of neutrophils and eosinophils, which account for allergic rhinitis in addition to promoting TH2-and TH13-mediated allergic responses (Wu et al. 2018a, b). Oxidative stress has been reported in lung epithelial cells upon exposure to PM which in turn results in apoptosis of the cells (Huang et al. 2014).

Ultra-fine particulate matter could diffuse out of alveoli and even penetrate to circulation through blood vessels (Mannucci et al. 2015). Particulate matters from wildfires are reported to be linked to pulmonary edema and cardiovascular diseases (Ghio et al. 2012). Cardiovascular diseases associated with particulate matter exposure include ischemic heart disease, myocardial infarction, congestive heart failure, and carotid artery thrombosis (Wu et al. 2018a, b). Endothelial inflammation due to particulate matters in circulation results in increase in incidences of atherosclerosis (Dutta et al. 2012). Another cause of concern is evidence of premature death incidents due to particulate matter exposure.

Ultra-fine particulate matter sustains for a longer time in the air and is reported to be a carrier of pathogenic microbes and viruses (Setti et al. 2020). PM is reported to be the carriers of viruses acting as vectors of pathogenicity (Alonso et al. 2015; Yang et al. 2011). A report from Italy hypothesizes the presence of SARs-CoV2 on particulate matter (Setti et al. 2020). It is also found that chronic exposure to PM could make the population susceptible to SARs-CoV2 (Yao et al. 2020). Reports also suggest that the mechanism of action of particulate matter by recruiting inflammatory mediators and affecting the airways can impact the prognosis of COVID patients (Kelly and Fussell 2011; Yao et al. 2020). These findings suggest that in the present scenario of rising COVID cases, particulate matter pollution will further worsen pandemic-associated morbidity.

Particulate matters are also associated with cancer progression. It acquired the status of being a carcinogen in 2013 as proclaimed by the WHO and International Agency for Research on Cancer (IARC). Increased particulate matter in the atmosphere has risked human life to lung cancers predominantly (Turner et al. 2011). $PM_{2.5}$ is also associated with cancers of the breast, colorectal, bladder, kidney, stomach, liver, pancreas, and cervix (Deng et al. 2019; Turner et al. 2020; Wu et al. 2019). Hodgkin lymphoma and leukemia are also reported to be linked to particulate matter exposure (Coleman et al. 2020). Particulate matters are implicated to affect all major hallmarks of cancer (Santibanez-Andrade et al. 2019). They are reported to activate proliferation in lung cancer lines (Yang et al. 2016). $PM_{2.5}$ prevents bronchial cell apoptosis (Ferecatu et al. 2010). Components associated with particulate matters like PAHs and heavy metals activate anti-apoptotic signaling pathways (Lovera-Leroux et al. 2015; Shang et al. 2017). $PM_{2.5}$ is reported to inhibit the mTOR pathway in lung cell lines and hence induce autophagy (Liu et al. 2015). Particulate matters accelerate angiogenesis and help in tumor growth (Yang et al. 2018). They aid in epithelial-mesenchymal transition, which further increases the invasiveness of cancer and by means of reactive oxygen species (ROS) cause DNA damage (Wang et al. 2020a, b, c; Wei et al. 2017). Particulate matters as epigenetic modifiers are yet to be studied conclusively. However, reports suggest that exposure to particulate matter led to changes in methylation status and caused alterations in the expression of micro-RNAs and long non-coding RNAs inducing malignant genes and pathways sustaining the tumor growth and progression (Hesselbach et al. 2017; Liu et al. 2015). Mitochondrial health is also compensated when exposed to ultra-fine particulate matters leading to changes in mitochondrial membrane potential, mtDNA content, and energy metabolism (Bhargava et al. 2018).

6.4 Microplastics

Plastics are one of the most ubiquitous commodities that humans use. The fact that they are non-biodegradable is overlooked without any check on their production or proper disposal. Plastics are so prevalent that they are now used as indicator of Anthropocene, evident from deposits which remarkedly act as indicators of stratigraphy (Zalasiewicz et al. 2016). Basically, polymers of various chemicals like polyethylene, polystyrene, and polyvinyl chloride are made from non-renewable resources like fossil fuels or are bio-based. Plastics pose a threat to life and nature alike. A recent UN report points to the fact that there is an increasing trend in the number of publications on plastic debris. Not surprisingly, WEF also revealed the observation that plastic particles in oceans will surpass the marine fish population by 2050.

Microplastics are considered as a particulate pollutant owing to their minute sizes and have earned ample interest in terms of threatening human health. Marine ecologist Richard Thomas coined the term in 2004 although the earliest literature on these minute contaminants dates to the 1990s. There isn't any conclusive definition of microplastics, but their signature is of particles having sizes in the range of 1 nm—<5 mm. Microplastics may be artificially synthesized or are formed from natural or chemical modification of plastics. Microplastics are in market, incorporated in scrubs, cosmetic products, and detergents. These are classified as primary microplastics. Secondary microplastics are formed from plastic by means of their degradation or modification by exposure to UV radiations, and weathering of physio-chemical treatments. Microplastics are widespread in oceans, and dumping of industrial sewage into water bodies can be a major cause. Combustion of plastics on large scale can increase the microplastic content in the atmosphere. The main sources of microplastics in the atmosphere are synthetic cloth fibers, tire industry, industrial effluents, and combustion of chemicals. Microplastics are also found in various food products like honey, for example (Liebezeit and Liebezeit 2013). Use of canned foods and storing drinking water in plastic cans (El-Ziney 2016) are major cause of concern for the presence of microplastics in water and food. The accumulations of microplastics in soil alter their biochemical properties which can affect flora and fauna dependent upon the soil (de Souza Machado et al. 2018). The leaching of such particles into the ocean or water bodies again harm aquatic life and is even reported to enter the food chain endangering the lives of animals as well (Waring et al. 2018).

Microplastic toxicity is being studied in terrestrial and marine life. Elucidating the harmful effects of microplastics is still a challenge, owing to the fact they are of varied shapes and sizes with no proper detection methods, studies are inconclusive with regard to how to tackle the burden of microplastics, how to determine their age and biochemistry, routes of exposure to marine and terrestrial life.

6.5 Microplastics and Human Health

Microplastics are a silent threat to human health, minuscule but with immense potential to cause harm. Aptly put by Rachal Carson in his book *Silent spring*, "If we are going to live so intimately with these chemicals - eating and drinking them, taken them into the very marrow of our bones, we had better know something about their nature and power" (Carson 2002). Microplastics are omnipresent, recorded in the very air we breathe in, the soil, in the oceans, associated with microbes and even in unimaginable conditions as the Polar Regions (Campanale et al. 2020b; Pabortsava and Lampitt 2020; Peeken et al. 2018). The very same reason without doubt validates the exposure of humans to microplastics.

The major ways in which microplastics pave their way into human system include inhalation of these minute particles from the atmosphere (Prata 2018), ingestion of some microparticles through food (Bernatsky et al. 2016) or drinking water (Koelmans et al. 2019), absorption of microparticles via skin (De-la-Torre 2020). Though very few reports point to the absorption of microparticles via skin, it cannot be overlooked. Microparticles find their way to our system through glands of skin or skin follicles and even act as opportunistic attackers through cuts or injuries in skin. It is reported that microparticles are inhaled in concentrations of 0.4–56.5 ppm indoors and 0.3–1.2 ppm concentrations outdoors (Dris et al. 2017). Microplastics are found to enter the food chain, especially through marine ecosystem and they tend to accumulate with each passing hierarchy of the food chain to finally culminate into human system (Rist et al. 2018). Marine fauna such as the mollusks and other sea foods are reported sources of entry to human through their dining. The organism accumulates microplastics in their digestive tracks and the consumption of fish as whole or even dry fishes pose a threat of exposure to these chemicals (Bernatsky et al. 2016).

6.6 Respiratory System

Airway exposure of microparticles results in accumulation in the respiratory pathway leading to irritation, inflammation mainly in bronchioles, alveoli, and lungs. This in turn can cause serious consequences in the form of respiratory diseases like asthma, bronchitis, and lung cancer (Valavanidis et al. 2013). Additives used in microplastics like phthalates are a major concern with respect to asthma and allergic reactions (Ait Bamai et al. 2014). Although un-noticed even cloth fibers cause respiratory distress or difficulty in breathing, dyspnea (Prata 2018). Polyvinyl chloride polymers are found to cause cytotoxicity in human lung. In the COVID era with the increasing use of PPE made of polyvinyl polymers, there is a greater risk of respiratory distress due to microplastics (Fadare and Okoffo 2020).

6.7 Digestive System

Through drinking water or routine food intake, especially sea foods (Bernatsky et al. 2016), we ingest microparticles. Microparticles get caught in mucous layers but many of these get transported and lodged in the intestine (Prata 2018). They are reported to be found in human stool (Schwabl et al. 2019). Microparticles are said to even penetrate the barriers of the intestinal epithelium and get translocated to other organs via the lymphatic system or the circulatory system (Rahman et al. 2021).

6.8 Immunology

Microplastics have the ability to affect the immune system by perturbing the immune cells. They reach the lymphatic and circulatory system crossing epithelial and lung barriers (Campanale

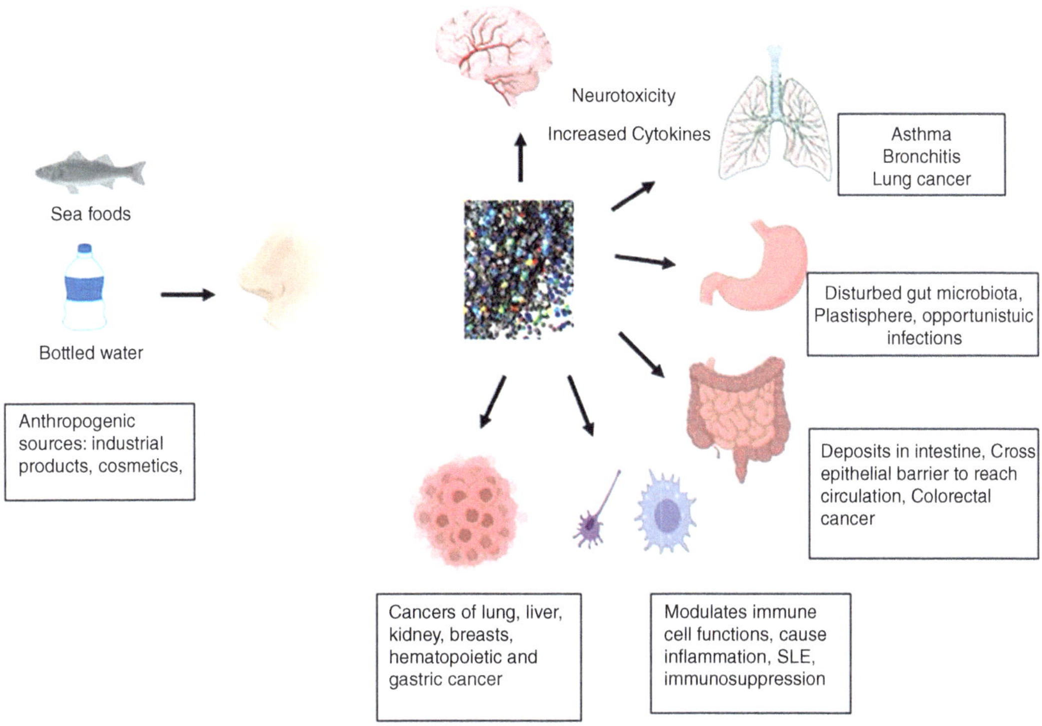

Fig. 6.1 Impact of microplastics on human health

et al. 2020a). They trigger allergic reaction by up-regulating histamines (Hwang et al. 2019). Macrophages are reported to accumulate toxic polymers from microplastic like polystyrene and result in autophagy causing cytotoxicity in lung epithelium and macrophages. The cytotoxic effect of microplastics is an aftermath of excessive ROS production caused by oxidative stress (Geiser et al. 2005). They are also reported to get into secondary immune cells like M cells and dendritic cells. By stimulating the overproduction of cytokines, microplastics play a major role in causing chronic inflammatory reactions which lead to cytotoxicity and are also related to cancer (Prata 2018). Other less-studied potential health hazards from microplastic exposure include autoimmune diseases like SLE (Fernandes et al. 2015), Rheumatoid arthritis (Bernatsky et al. 2016), and immune suppression (Prata et al. 2020) (Fig. 6.1).

6.9 Reproductive Disorders

Additives used in the production of microplastics like bisphenol (BPA) are reported to be toxic to the reproductive system (Campanale et al. 2020a). Polybrominated diphenyl ether (PBDE), a class of additive used as flame retardant, is a known endocrine modulator (Linares et al. 2015). Microplastics can be considered as a new class of teratogen. Recently, microplastics are found to cross the human placenta causing retardation of fetal growth (Ragusa et al. 2021).

6.10 Neurotoxicity

There are very few conclusive studies on how microplastics affect the nervous system. Studies of microplastics affecting cerebral epithelial cells indicate some level of oxidative stress

(Barboza et al. 2018). Neurotoxicity, resulting from increased cytokine production by microplastics, is a cause of concern (Mohan Kumar et al. 2008).

6.11 Pathogenicity

Microplastics harbor harmful microorganisms on their surface forming a biofilm. These biofilms are pathogenic (Lobelle and Cunliffe 2011). The presence of antibiotic-resistant bacteria in microplastics is a reason to worry for sure and must be studied extensively (Zhang et al. 2020). Association of microplastic with gut microbiome referred as plastisphere can alter the microbiome of the gut. This has been reported to have endotoxic effects in addition to that it results in various opportunistic diseases (Zettler et al. 2013).

6.12 Cancer

The ability of microplastics to cause chronic inflammation is the major cause of cancers due to microplastics (Prata 2018). DNA damage by microplastics is also a reported strategy (Prata 2018). Heavy metals used as additives in microplastics production are reported to be potent carcinogens (Tchounwou et al. 2012). Arsenic toxicity has been linked to cancers of lung, liver, kidney, and bladder (Hahladakis et al. 2018; Jan et al. 2015). Metals, like antimony and aluminum, are associated with breast cancer (Byrne et al. 2013; Hahladakis et al. 2018; Byrne et al. 2013). Additives like phthalate (Böckers et al. 2021) and tri-cresyl phosphate (Bockers et al. 2020) have been reported to be tumorigenic in MCF-7 lines. Various chemicals used as additives in microplastics have endocrine-modulating function and are implicated in breast cancer, prostate, and testicular cancers (Böckers et al. 2021). Evidence indicates occupational exposure of microplastics with the development of lung cancer, hematopoietic cancer, colorectal cancer, esophageal cancer, cancers of cervix, and gastric cancer (Cho and Choi 2021; Ibrahim et al. 2021; Prata 2018; Yan et al. 2020). Report suggests possible gene mutation due to long-term exposure of microplastics as a cue for developing cancers (Hu et al. 2021).

6.13 Particulate Matters, Microplastics, and Mitochondrial Health

Mitochondrial health can be a classic marker for exposure to harmful pollutants like particulate matter and microplastics. Most studies point to oxidative stress due to excessive ROS production resulting from exposure to pollutants as a major pathway of cytotoxicity (Wang et al. 2020a, b, c). Reports suggest perturbations in mitochondrial morphology because of particulate matters and microplastics. Changes in morphology are particularly vivid in the form of disruption of mitochondrial membrane, mitochondrial edema, and disappearance of mitochondrial cristae as reported in studies of polystyrene microplastics on rat hepatocytes (Li et al. 2021), particulate exposure on cardiomyocytes (Yang et al. 2018), and macrophages (Wei et al. 2021). Damage to mitochondrial membrane in turn is reported to open mitochondrial permeability transition pore which results in the swelling of mitochondria (Miao et al. 2019). Because of increased ROS, the decline in mitochondrial membrane potential has been observed in various studies (Chew et al. 2020). Exposure to microplastics causes lysosome-induced mitochondrial membrane depolarization (Wang et al. 2020a, b, c). Mitochondrial depolarization has been reported to occur because of microplastics in size-dependent manner in caco2 cells (Wang et al. 2020a, b, c).

Imbalance in mitochondrial membrane potential has a serious impact on mitochondrial respiration as it uncouples the mitochondrial electron transport chain and hence decreases ATP significantly. This is validated in murine macrophages exposed to microplastics (Merkley et al. 2021), and polystyrene exposed to caco2 cells (Wang et al. 2020a, b, c). Particulate matter has the ability to alter mitochondrial electron transport chain (ETC) enzyme complexes and hence deteriorate ATP (Myers et al. 2010). Metabolic alteration due to exposure to microplastics has been implicated in intestinal epithelial cells (Myers et al. 2010). Elevated oxidative stress has a marked impact on decreasing the mtDNA copy number in leucocytes exposed to particulate matters (Wang et al. 2020a, b, c) and is also implicated in studies on exposure of poly aromatic hydrocarbons (Pardo et al. 2020). mtDNA is required for crucial processes and depletion of mtDNA results in a decrease in the biogenesis of mitochondria. Disturbed mitochondrial membrane potential and reduction in mtDNA contribute to excessive ROS production by mitochondria and hence induce apoptosis, as evident from studies on exposure of kidney epithelial cells to microplastics (Wang et al. 2021). Increased ROS is also reported to induce lipid peroxidation of the mitochondrial membrane which aids in the accumulation of toxic particulate matter in mitochondria, further increasing mitochondrial dysfunction (Yang et al. 2018). Particulate matters are reported to modulate mitochondrial dynamics by increasing mitochondrial fission (Wang et al. 2019). Imbalance in the dynamics of mitochondria results in poor mitochondrial quality control. Mitochondria-mediated apoptosis is initiated in most cases of particulate matter exposure. Polystyrene additives in microplastics have been reported to trigger mitophagy (Pan et al. 2021). Till date various studies have elucidated the impact of microplastics and particulate matter on mitochondrial health; however, a conclusive modality to minimize the impact and to what extent these pollutants affect human health is yet to be explored (Fig. 6.2).

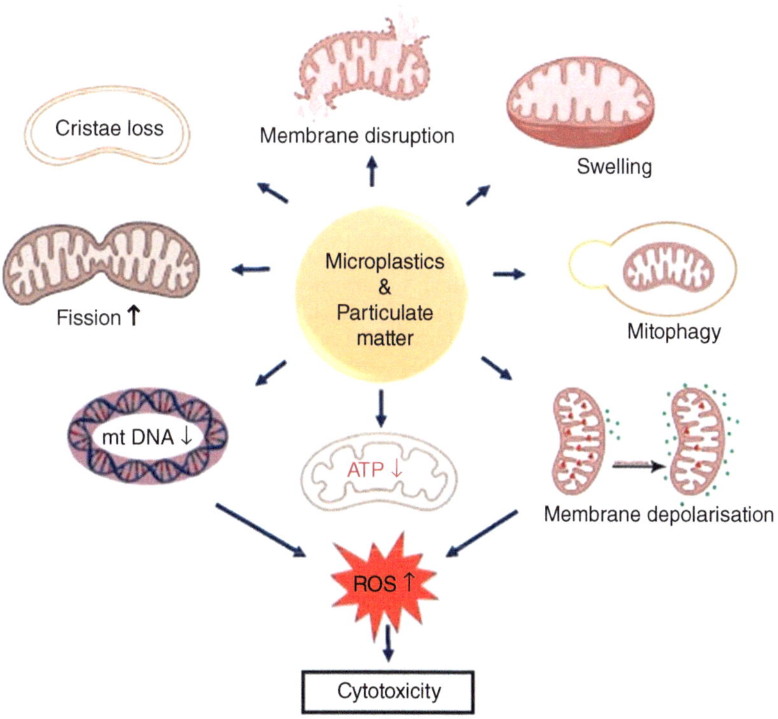

Fig. 6.2 Effect of microplastics and particulate matter on mitochondrial health

6.14 Conclusions and Future Perspectives

Microplastics and particulate matters are potent pollutants capable of causing harm to human health. The ubiquitous presence of these particles continues to pose a threat. They are also increasing on a day-to-day basis owing to various anthropogenic activities. It is a matter of concern that despite various studies on the impact of microplastics and particulate matter on human health has been conducted, preventive and protective strategies are still underexplored. The heterogeneity of these pollutants makes them difficult to be mapped for studies. Real-time analysis of the presence of these pollutants must be elucidated to avoid a potential threat. Minimizing sources can be a way forward to limiting their release. Reducing and reusing plastics, using proper protective gears to avoid on-site exposure, and proper treatment of drinking water can be few steps toward a greater goal of minimizing human exposure to these toxic chemicals. Keeping the environment free of these pollutants should be a way forward by using sustainable resources and avoiding the unchecked release of plastics and other particulate pollutants.

References

Ait Bamai Y, Shibata E, Saito I, Araki A, Kanazawa A, Morimoto K, Kishi R. Exposure to house dust phthalates in relation to asthma and allergies in both children and adults. Sci Total Environ. 2014;485–486:153–63. https://doi.org/10.1016/j.scitotenv.2014.03.059.

Alonso C, Raynor PC, Davies PR, Torremorell M. Concentration, size distribution, and infectivity of airborne particles carrying swine viruses. PLoS One. 2015;10(8):e0135675. https://doi.org/10.1371/journal.pone.0135675.

Barboza LGA, Vieira LR, Branco V, Figueiredo N, Carvalho F, Carvalho C, Guilhermino L. Microplastics cause neurotoxicity, oxidative damage and energy-related changes and interact with the bioaccumulation of mercury in the European seabass, Dicentrarchus labrax (Linnaeus, 1758). Aquat Toxicol. 2018;195:49–57. https://doi.org/10.1016/j.aquatox.2017.12.008.

Bernatsky S, Smargiassi A, Barnabe C, Svenson LW, Brand A, Martin RV, Joseph L. Fine particulate air pollution and systemic autoimmune rheumatic disease in two Canadian provinces. Environ Res. 2016;146:85–91. https://doi.org/10.1016/j.envres.2015.12.021.

Bhargava A, Tamrakar S, Aglawe A, Lad H, Srivastava RK, Mishra DK, Mishra PK. Ultrafine particulate matter impairs mitochondrial redox homeostasis and activates phosphatidylinositol 3-kinase mediated DNA damage responses in lymphocytes. Environ Pollut. 2018;234:406–19. https://doi.org/10.1016/j.envpol.2017.11.093.

Bockers M, Paul NW, Efferth T. Organophosphate ester tri-o-cresyl phosphate interacts with estrogen receptor alpha in MCF-7 breast cancer cells promoting cancer growth. Toxicol Appl Pharmacol. 2020;395:114977. https://doi.org/10.1016/j.taap.2020.114977.

Böckers M, Paul N, Efferth T. Butyl octyl phthalate interacts with estrogen receptor α in MCF-7 breast cancer cells to promote cancer development. World Academy Sci J. 2021;3(2):1. https://doi.org/10.3892/wasj.2021.92.

Byrne C, Divekar SD, Storchan GB, Parodi DA, Martin MB. Metals and breast cancer. J Mammary Gland Biol Neoplasia. 2013;18(1):63–73. https://doi.org/10.1007/s10911-013-9273-9.

Campanale C, Massarelli C, Savino I, Locaputo V, Uricchio VF. A detailed review study on potential effects of microplastics and additives of concern on human health. Int J Environ Res Public Health. 2020a;17(4):1212. https://doi.org/10.3390/ijerph17041212.

Campanale C, Stock F, Massarelli C, Kochleus C, Bagnuolo G, Reifferscheid G, Uricchio VF. Microplastics and their possible sources: the example of Ofanto river in Southeast Italy. Environ Pollut. 2020b;258:113284. https://doi.org/10.1016/j.envpol.2019.113284.

Carson R. Silent spring. Boston: Houghton Mifflin Harcourt; 2002. p. 1962.

Chew S, Kolosowska N, Saveleva L, Malm T, Kanninen KM. Corrigendum to "impairment of mitochondrial function by particulate matter: implications for the brain" [Neurochem. Int. 135 (May 2020) 104694]. Neurochem Int. 2020;139:104776. https://doi.org/10.1016/j.neuint.2020.104776.

Cho YM, Choi KH. The current status of studies of human exposure assessment of microplastics and their health effects: a rapid systematic review. Environ Anal Health Toxicol. 2021;36(1):e2021004–0. https://doi.org/10.5620/eaht.2021004.

Coleman NC, Burnett RT, Higbee JD, Lefler JS, Merrill RM, Ezzati M, Pope CA 3rd. Cancer mortality risk, fine particulate air pollution, and smoking in a large, representative cohort of US adults. Cancer Causes Control. 2020;31(8):767–76. https://doi.org/10.1007/s10552-020-01317-w.

de Souza Machado AA, Kloas W, Zarfl C, Hempel S, Rillig MC. Microplastics as an emerging threat to terrestrial ecosystems. Glob Chang Biol. 2018;24(4):1405–16. https://doi.org/10.1111/gcb.14020.

De-la-Torre GE. Microplastics: an emerging threat to food security and human health. J Food Sci

Technol. 2020;57(5):1601–8. https://doi.org/10.1007/s13197-019-04138-1.

Delfino RJ, Staimer N, Tjoa T, Arhami M, Polidori A, Gillen DL, Sioutas C. Associations of primary and secondary organic aerosols with airway and systemic inflammation in an elderly panel cohort. Epidemiology. 2010;21(6):892–902. https://doi.org/10.1097/EDE.0b013e3181f20e6c.

Deng Q, Deng L, Miao Y, Guo X, Li Y. Particle deposition in the human lung: health implications of particulate matter from different sources. Environ Res. 2019;169:237–45. https://doi.org/10.1016/j.envres.2018.11.014.

Dris R, Gasperi J, Mirande C, Mandin C, Guerrouache M, Langlois V, Tassin B. A first overview of textile fibers, including microplastics, in indoor and outdoor environments. Environ Pollut. 2017;221:453–8. https://doi.org/10.1016/j.envpol.2016.12.013.

Dutta A, Ray MR, Banerjee A. Systemic inflammatory changes and increased oxidative stress in rural Indian women cooking with biomass fuels. Toxicol Appl Pharmacol. 2012;261(3):255–62. https://doi.org/10.1016/j.taap.2012.04.004.

El-Ziney MG. Migration levels of monostyrene from polystyrene containers to dairy products. MOJ Food Process Technol. 2016;3(2):0–8. https://doi.org/10.15406/mojfpt.2016.03.00066.

Fadare OO, Okoffo ED. Covid-19 face masks: a potential source of microplastic fibers in the environment. Sci Total Environ. 2020;737:140279. https://doi.org/10.1016/j.scitotenv.2020.140279.

Ferecatu I, Borot MC, Bossard C, Leroux M, Boggetto N, Marano F, Andreau K. Polycyclic aromatic hydrocarbon components contribute to the mitochondria-antiapoptotic effect of fine particulate matter on human bronchial epithelial cells via the aryl hydrocarbon receptor. Part Fibre Toxicol. 2010;7:18. https://doi.org/10.1186/1743-8977-7-18.

Fernandes EC, Silva CA, Braga AL, Sallum AM, Campos LM, Farhat SC. Exposure to air pollutants and disease activity in juvenile-onset systemic lupus erythematosus patients. Arthritis Care Res (Hoboken). 2015;67(11):1609–14. https://doi.org/10.1002/acr.22603.

Geiser M, Rothen-Rutishauser B, Kapp N, Schurch S, Kreyling W, Schulz H, Gehr P. Ultrafine particles cross cellular membranes by nonphagocytic mechanisms in lungs and in cultured cells. Environ Health Perspect. 2005;113(11):1555–60. https://doi.org/10.1289/ehp.8006.

Ghio AJ, Soukup JM, Case M, Dailey LA, Richards J, Berntsen J, Rappold A. Exposure to wood smoke particles produces inflammation in healthy volunteers. Occup Environ Med. 2012;69(3):170–5. https://doi.org/10.1136/oem.2011.065276.

Hahladakis JN, Velis CA, Weber R, Iacovidou E, Purnell P. An overview of chemical additives present in plastics: migration, release, fate and environmental impact during their use, disposal and recycling. J Hazard Mater. 2018;344:179–99. https://doi.org/10.1016/j.jhazmat.2017.10.014.

Hamanaka RB, Mutlu GM. Particulate matter air pollution: effects on the cardiovascular system. Front Endocrinol (Lausanne). 2018;9:680. https://doi.org/10.3389/fendo.2018.00680.

Hesselbach K, Kim GJ, Flemming S, Haupl T, Bonin M, Dornhof R, et al. Disease relevant modifications of the methylome and transcriptome by particulate matter (PM2.5) from biomass combustion. Epigenetics. 2017;12(9):779–92. https://doi.org/10.1080/15592294.2017.1356555.

Hu X, Biswas A, Sharma A, Sarkodie H, Tran I, Pal I, De S. Mutational signatures associated with exposure to carcinogenic microplastic compounds bisphenol a and styrene oxide. NAR Cancer. 2021;3(1):zcab004. https://doi.org/10.1093/narcan/zcab004.

Huang Q, Zhang J, Peng S, Tian M, Chen J, Shen H. Effects of water soluble PM2.5 extracts exposure on human lung epithelial cells (A549): a proteomic study. J Appl Toxicol. 2014;34(6):675–87. https://doi.org/10.1002/jat.2910.

Hwang J, Choi D, Han S, Choi J, Hong J. An assessment of the toxicity of polypropylene microplastics in human derived cells. Sci Total Environ. 2019;684:657–69. https://doi.org/10.1016/j.scitotenv.2019.05.071.

Ibrahim YS, Tuan Anuar S, Azmi AA, Wan Mohd Khalik WMA, Lehata S, Hamzah SR, et al. Detection of microplastics in human colectomy specimens. JGH Open. 2021;5(1):116–21. https://doi.org/10.1002/jgh3.12457.

Jan AT, Azam M, Siddiqui K, Ali A, Choi I, Haq QM. Heavy metals and human health: mechanistic insight into toxicity and counter defense system of antioxidants. Int J Mol Sci. 2015;16(12):29592–630. https://doi.org/10.3390/ijms161226183.

Kelly FJ, Fussell JC. Air pollution and airway disease. Clin Exp Allergy. 2011;41(8):1059–71. https://doi.org/10.1111/j.1365-2222.2011.03776.x.

Koelmans AA, Mohamed Nor NH, Hermsen E, Kooi M, Mintenig SM, De France J. Microplastics in freshwaters and drinking water: critical review and assessment of data quality. Water Res. 2019;155:410–22. https://doi.org/10.1016/j.watres.2019.02.054.

Li S, Shi M, Wang Y, Xiao Y, Cai D, Xiao F. Keap1-Nrf2 pathway up-regulation via hydrogen sulfide mitigates polystyrene microplastics induced-hepatotoxic effects. J Hazard Mater. 2021;402:123933. https://doi.org/10.1016/j.jhazmat.2020.123933.

Liebezeit G, Liebezeit E. Non-pollen particulates in honey and sugar. Food Addit Contam Part A Chem Anal Control Expo Risk Assess. 2013;30(12):2136–40. https://doi.org/10.1080/19440049.2013.843025.

Linares V, Belles M, Domingo JL. Human exposure to PBDE and critical evaluation of health hazards. Arch Toxicol. 2015;89(3):335–56. https://doi.org/10.1007/s00204-015-1457-1.

Liu T, Wu B, Wang Y, He H, Lin Z, Tan J, et al. Particulate matter 2.5 induces autophagy via inhibition of the phosphatidylinositol 3-kinase/Akt/mammalian target of rapamycin kinase signaling pathway in human bronchial epithelial cells. Mol Med Rep. 2015;12(2):1914–22. https://doi.org/10.3892/mmr.2015.3577.

Liu H, Cao C, Huang J, Chen Z, Chen G, Lai Y. Progress on particulate matter filtration technology: basic concepts, advanced materials, and performances. Nanoscale. 2020;12(2):437–53. https://doi.org/10.1039/c9nr08851b.

Lobelle D, Cunliffe M. Early microbial biofilm formation on marine plastic debris. Mar Pollut Bull. 2011;62(1):197–200. https://doi.org/10.1016/j.marpolbul.2010.10.013.

Lovera-Leroux M, Crobeddu B, Kassis N, Petit PX, Janel N, Baeza-Squiban A, Andreau K. The iron component of particulate matter is antiapoptotic: a clue to the development of lung cancer after exposure to atmospheric pollutants? Biochimie. 2015;118:195–206. https://doi.org/10.1016/j.biochi.2015.09.030.

Mannucci PM, Harari S, Martinelli I, Franchini M. Effects on health of air pollution: a narrative review. Intern Emerg Med. 2015;10(6):657–62. https://doi.org/10.1007/s11739-015-1276-7.

Merkley S, Moss HC, Goodfellow SM, Ling CL, Meyer-Hagen JL, Weaver J, Castillo EF. Polystyrene microplastics induce an immunometabolic active state in macrophages. Cell Biol Toxicol. 2021;38(1):31–41. https://doi.org/10.21203/rs.3.rs-259689/v1.

Miao X, Li W, Niu B, Li J, Sun J, Qin M, Zhou Z. Mitochondrial dysfunction in endothelial cells induced by airborne fine particulate matter (<2.5 mum). J Appl Toxicol. 2019;39(10):1424–32. https://doi.org/10.1002/jat.3828.

Miri M, Alahabadi A, Ehrampush MH, Rad A, Lotfi MH, Sheikhha MH, Sakhvidi MJZ. Mortality and morbidity due to exposure to ambient particulate matter. Ecotoxicol Environ Saf. 2018;165:307–13. https://doi.org/10.1016/j.ecoenv.2018.09.012.

Mohan Kumar SM, Campbell A, Block M, Veronesi B. Particulate matter, oxidative stress and neurotoxicity. Neurotoxicology. 2008;29(3):479–88. https://doi.org/10.1016/j.neuro.2007.12.004.

Myers CR, Antholine WE, Myers JM. The pro-oxidant chromium(VI) inhibits mitochondrial complex I, complex II, and aconitase in the bronchial epithelium: EPR markers for Fe-S proteins. Free Radic Biol Med. 2010;49(12):1903–15. https://doi.org/10.1016/j.freeradbiomed.2010.09.020.

Pabortsava K, Lampitt RS. High concentrations of plastic hidden beneath the surface of the Atlantic Ocean. Nat Commun. 2020;11(1):4073. https://doi.org/10.1038/s41467-020-17932-9.

Pan L, Yu D, Zhang Y, Zhu C, Yin Q, Hu Y, Xiong X. Polystyrene microplastics-triggered mitophagy and oxidative burst via activation of PERK pathway. Sci Total Environ. 2021;781:146753. https://doi.org/10.1016/j.scitotenv.2021.146753.

Pardo M, Qiu X, Zimmermann R, Rudich Y. Particulate matter toxicity is Nrf2 and mitochondria dependent: the roles of metals and polycyclic aromatic hydrocarbons. Chem Res Toxicol. 2020;33(5):1110–20. https://doi.org/10.1021/acs.chemrestox.0c00007.

Peeken I, Primpke S, Beyer B, Gutermann J, Katlein C, Krumpen T, Gerdts G. Arctic Sea ice is an important temporal sink and means of transport for microplastic. Nat Commun. 2018;9(1):1505. https://doi.org/10.1038/s41467-018-03825-5.

Prata JC. Airborne microplastics: consequences to human health? Environ Pollut. 2018;234:115–26. https://doi.org/10.1016/j.envpol.2017.11.043.

Prata JC, da Costa JP, Lopes I, Duarte AC, Rocha-Santos T. Environmental exposure to microplastics: an overview on possible human health effects. Sci Total Environ. 2020;702:134455. https://doi.org/10.1016/j.scitotenv.2019.134455.

Ragusa A, Svelato A, Santacroce C, Catalano P, Notarstefano V, Carnevali O, Giorgini E. Plasticenta: first evidence of microplastics in human placenta. Environ Int. 2021;146:106274. https://doi.org/10.1016/j.envint.2020.106274.

Rahman A, Sarkar A, Yadav OP, Achari G, Slobodnik J. Potential human health risks due to environmental exposure to nano- and microplastics and knowledge gaps: a scoping review. Sci Total Environ. 2021;757:143872. https://doi.org/10.1016/j.scitotenv.2020.143872.

Rist S, Carney Almroth B, Hartmann NB, Karlsson TM. A critical perspective on early communications concerning human health aspects of microplastics. Sci Total Environ. 2018;626:720–6. https://doi.org/10.1016/j.scitotenv.2018.01.092.

Santibanez-Andrade M, Chirino YI, Gonzalez-Ramirez I, Sanchez-Perez Y, Garcia-Cuellar CM. Deciphering the code between air pollution and disease: the effect of particulate matter on cancer hallmarks. Int J Mol Sci. 2019;21(1):136. https://doi.org/10.3390/ijms21010136.

Schwabl P, Koppel S, Konigshofer P, Bucsics T, Trauner M, Reiberger T, Liebmann B. Detection of various microplastics in human stool: a prospective Case series. Ann Intern Med. 2019;171(7):453–7. https://doi.org/10.7326/M19-0618.

Setti L, Passarini F, De Gennaro G, Barbieri P, Perrone MG, Borelli M, Miani A. SARS-Cov-2RNA found on particulate matter of Bergamo in northern Italy: first evidence. Environ Res. 2020;188:109754. https://doi.org/10.1016/j.envres.2020.109754.

Shang Y, Zhou Q, Wang T, Jiang Y, Zhong Y, Qian G, et al. Airborne nitro-PAHs induce Nrf2/ARE defense system against oxidative stress and promote inflammatory process by activating PI3K/Akt pathway in A549 cells. Toxicol In Vitro. 2017;44:66–73. https://doi.org/10.1016/j.tiv.2017.06.017.

Stanek LW, Brown JS, Stanek J, Gift J, Costa DL. Air pollution toxicology—a brief review of the role of the science in shaping the current understanding of air pollution health risks. Toxicol Sci. 2011;120(Suppl 1):S8–27. https://doi.org/10.1093/toxsci/kfq367.

Tchounwou PB, Yedjou CG, Patlolla AK, Sutton DJ. Heavy metal toxicity and the environment. Exp Suppl. 2012;101:133–64. https://doi.org/10.1007/978-3-7643-8340-4_6.

Turner MC, Krewski D, Pope CA 3rd, Chen Y, Gapstur SM, Thun MJ. Long-term ambient fine particulate matter air pollution and lung cancer in a large cohort of never-smokers. Am J Respir Crit Care Med. 2011;184(12):1374–81. https://doi.org/10.1164/rccm.201106-1011OC.

Turner MC, Andersen ZJ, Baccarelli A, Diver WR, Gapstur SM, Pope CA 3rd, Cohen A. Outdoor air pollution and cancer: An overview of the current evidence and public health recommendations. CA Cancer J Clin. 2020;70:460. https://doi.org/10.3322/caac.21632.

Valavanidis A, Vlachogianni T, Fiotakis K, Loridas S. Pulmonary oxidative stress, inflammation and cancer: respirable particulate matter, fibrous dusts and ozone as major causes of lung carcinogenesis through reactive oxygen species mechanisms. Int J Environ Res Public Health. 2013;10(9):3886–907. https://doi.org/10.3390/ijerph10093886.

Wang Y, Zhang M, Li Z, Yue J, Xu M, Zhang Y, et al. Fine particulate matter induces mitochondrial dysfunction and oxidative stress in human SH-SY5Y cells. Chemosphere. 2019;218:577–88. https://doi.org/10.1016/j.chemosphere.2018.11.149.

Wang Q, Bai J, Ning B, Fan L, Sun T, Fang Y, et al. Effects of bisphenol a and nanoscale and microscale polystyrene plastic exposure on particle uptake and toxicity in human Caco-2 cells. Chemosphere. 2020a;254:126788. https://doi.org/10.1016/j.chemosphere.2020.126788.

Wang X, Hart JE, Liu Q, Wu S, Nan H, Laden F. Association of particulate matter air pollution with leukocyte mitochondrial DNA copy number. Environ Int. 2020b;141:105761. https://doi.org/10.1016/j.envint.2020.105761.

Wang Y, Wu T, Tang M. Ambient particulate matter triggers dysfunction of subcellular structures and endothelial cell apoptosis through disruption of redox equilibrium and calcium homeostasis. J Hazard Mater. 2020c;394:122439. https://doi.org/10.1016/j.jhazmat.2020.122439.

Wang YL, Lee YH, Hsu YH, Chiu IJ, Huang CC, Huang CC, et al. The kidney-related effects of polystyrene microplastics on human kidney proximal tubular epithelial cells HK-2 and male C57BL/6 mice. Environ Health Perspect. 2021;129(5):57003. https://doi.org/10.1289/EHP7612.

Waring RH, Harris RM, Mitchell SC. Plastic contamination of the food chain: a threat to human health? Maturitas. 2018;115:64–8. https://doi.org/10.1016/j.maturitas.2018.06.010.

Wei H, Liang F, Cheng W, Zhou R, Wu X, Feng Y, Wang Y. The mechanisms for lung cancer risk of PM2.5: induction of epithelial-mesenchymal transition and cancer stem cell properties in human non-small cell lung cancer cells. Environ Toxicol. 2017;32(11):2341–51. https://doi.org/10.1002/tox.22437.

Wei H, Yuan W, Yu H, Geng H. Cytotoxicity induced by fine particulate matter (PM2.5) via mitochondria-mediated apoptosis pathway in rat alveolar macrophages. Environ Sci Pollut Res Int. 2021;28(20):25819–29. https://doi.org/10.1007/s11356-021-12431-w.

Wu JZ, Ge DD, Zhou LF, Hou LY, Zhou Y, Li QY. Effects of particulate matter on allergic respiratory diseases. Chronic Dis Transl Med. 2018a;4(2):95–102. https://doi.org/10.1016/j.cdtm.2018.04.001.

Wu W, Jin Y, Carlsten C. Inflammatory health effects of indoor and outdoor particulate matter. J Allergy Clin Immunol. 2018b;141(3):833–44. https://doi.org/10.1016/j.jaci.2017.12.981.

Wu B, Wu X, Liu S, Wang Z, Chen L. Size-dependent effects of polystyrene microplastics on cytotoxicity and efflux pump inhibition in human Caco-2cells. Chemosphere. 2019;221:333–41. https://doi.org/10.1016/j.chemosphere.2019.01.056.

Xiao J, Liang J, Zhang C, Tao Y, Ling G-W, Yang Q-H. Advanced materials for capturing particulate matter: Progress and perspectives. Small Methods. 2018;2(7):1800012. https://doi.org/10.1002/smtd.201800012.

Yan X, Zhang Y, Lu Y, He L, Qu J, Zhou C, et al. The complex toxicity of tetracycline with polystyrene spheres on gastric cancer cells. Int J Environ Res Public Health. 2020;17(8):2808. https://doi.org/10.3390/ijerph17082808.

Yang W, Elankumaran S, Marr LC. Concentrations and size distributions of airborne influenza a viruses measured indoors at a health Centre, a day-care Centre and on aeroplanes. J R Soc Interface. 2011;8(61):1176–84. https://doi.org/10.1098/rsif.2010.0686.

Yang B, Chen D, Zhao H, Xiao C. The effects for PM2.5 exposure on non-small-cell lung cancer induced motility and proliferation. Springerplus. 2016;5(1):2059. https://doi.org/10.1186/s40064-016-3734-8.

Yang X, Feng L, Zhang Y, Hu H, Shi Y, Liang S, et al. Cytotoxicity induced by fine particulate matter (PM2.5) via mitochondria-mediated apoptosis pathway in human cardiomyocytes. Ecotoxicol Environ Saf. 2018;161:198–207. https://doi.org/10.1016/j.ecoenv.2018.05.092.

Yao Y, Pan J, Wang W, Liu Z, Kan H, Qiu Y, et al. Association of particulate matter pollution and case fatality rate of COVID-19 in 49 Chinese cities. Sci Total Environ. 2020;741:140396. https://doi.org/10.1016/j.scitotenv.2020.140396.

Zalasiewicz J, Waters CN, Ivar do Sul JA, Corcoran PL, Barnosky AD, Cearreta A, et al. The geological cycle of plastics and their use as a stratigraphic indicator of the Anthropocene. Anthropocene. 2016;13:4–17. https://doi.org/10.1016/j.ancene.2016.01.002.

Zeb B, Alam K, Sorooshian A, Blaschke T, Ahmad I, Shahid I. On the morphology and composition of particulate matter in an urban environment. Aerosol Air Qual Res. 2018;18(6):1431–47. https://doi.org/10.4209/aaqr.2017.09.0340.

Zettler ER, Mincer TJ, Amaral-Zettler LA. Life in the "plastisphere": microbial communities on plastic marine debris. Environ Sci Technol. 2013;47(13):7137–46. https://doi.org/10.1021/es401288x.

Zhang R, Liu C, Zhou G, Sun J, Liu N, Hsu P-C, et al. Morphology and property investigation of primary particulate matter particles from different sources. Nano Res. 2018;11(6):3182–92. https://doi.org/10.1007/s12274-017-1724-y.

Zhang Y, Lu J, Wu J, Wang J, Luo Y. Potential risks of microplastics combined with superbugs: enrichment of antibiotic resistant bacteria on the surface of microplastics in mariculture system. Ecotoxicol Environ Saf. 2020;187:109852. https://doi.org/10.1016/j.ecoenv.2019.109852.

7 Recent Trends of Air Pollution and Its Implication on Human Health: A Spatio-Temporal Study of Two Coastal Cities of Goa

Ritwik Nigam, Alvarinho Luis, and Mahender Kotha

7.1 Introduction

In the name of development, anthropogenic activities such as setting up of industries to boost the production of finished products and constructing structures for housing increasing population have led to a critical situation in the urban and semi-urban regions, especially in developing countries (Sharma and Sharma and Patil 2016b). Environmental pollution and degradation are internationally recognized public health issue with numerous aspects linked to social, economic, and legislative concerns. Urban air quality degradation is the most critical issue amid the environmental degradation affecting human health (Tandon et al. 2008) which ultimately influences the quality of life (Dadhich et al. 2018). Air pollution is considered as the main risk to human health irrespective of gender and age. Air pollution is ranked among the top five global risk factors of mortality (HEI 2019) through degraded and inefficient lung functioning, cardiovascular disease and neuropsychological issues, and, most critically, death (Gupta 1999; WHO 2018). According to the United Nations Environment Programme, about 1.1 billion people experience inhalation problems due to contaminated air worldwide. The report of the medical science investigation revealed that the air particles concentration is linked to a range of impacts on human health, especially cardio and respiratory system.

Anthropogenic air pollution is one of the significant contributors to the public health hazards worldwide, which accounts for about 9% of the deaths per year (Louati et al. 2018; Son and Louati 2016). The global disease assessment revealed that 0.99 million and 0.78 million premature deaths occurred due to particulate matter (PM) 2.5 in 1990 and 2016, respectively (GBD MAPS Working Group 2018). In Asia, the alarming air pollution levels were recorded in 2015, which led to 35% deaths (Sharma and Patil 2016a). India exhibits a grimmer situation in terms of pollution Exceedance Factor (EF), which is the ratio of yearly average concentration of a pollutant to its standard deviation. It is estimated that 0.62 million deaths occurred due to high air pollution and deteriorating AQ in India (Lim et al. 2012). During 1998, 27% of 640 districts of India were found to be above the annual standard value of 40 μg/m^{3}of air pollution, which deteriorated in 2010, when air pollution in 45%

R. Nigam · M. Kotha (✉)
School of Earth, Ocean and Atmospheric Sciences (SEOAS), Goa University, Taleigao Plateau, Goa, India
e-mail: mkotha@unigoa.ac.in

A. Luis
Earth System Science Organization-National Centre of Polar and Ocean Research, Ministry of Earth Sciences, Govt. of India, Headland Sada, Vasco-da-Gama, Goa, India

P. K. Padhy et al. (eds.), *Air Quality and Human Health*,
https://doi.org/10.1007/978-981-97-1363-9_7

of the Indian districts exceeded the annual standard value (Guttikunda et al. 2019).

Air pollution comprises of various gaseous and dust particles such as Carbon dioxide (CO_2), Sulphur dioxide (SO_2), Oxides of Nitrogen (NO_x), Carbon Monoxide (CO), Ozone, Methane, PM, etc. Their major sources are power stations, the chemical and fertilizer industries, metallurgical and other industrial plants, construction activities, open-cast mining, fossil fuel-based vehicles, and municipal incineration. These possess different intensities of adverse effects on human health. SO_2, which is formed as a result of the combustion of fuels like petrol, diesel, and coal that contain sulphur and also produced by the thermal power stations (Darryl 2017), causes significant health complications such as substantial reduction in lungs efficiency, asthma, inflammation, and irritation in the respiratory tract and eyes, cardiovascular disease, etc. PM is another ubiquitous air pollutant dominating the urban and semi-urban regions, which occurs as a cocktail of solids and liquid particles of various sizes floating in the air. The primary source of PM is dust particles, diesel soot, while the gaseous form of sulphate and nitrate transforms physio-chemically and become the secondary source of the PM (Limaye and Salvi 2010).

The PM has different effects on human health depending on the size (Kim et al. 2015). The $PM_{2.5}$ can remain in the atmosphere for a longer time (days to weeks); as such, they can be transported over longer distances. These have a very high level of penetrability in the lungs, which can eventually cause premature deaths. The coarser PM_{10} quickly settle down due to gravity or are washed out by precipitation. It is well recognized that SO_2 is very harmful for the human body, which can lead to a host of health side effects such as lung dysfunction, inflammation of the respiratory tract and mucous membranes, which can lead to asthma attacks and irritation of the eyes.

Researchers have found an association between mortality and SO_2 (Hedley et al. 2002). NO_x and SO_2 pollutants are primary factors that can induce mortality (Dales et al. 2004). NO_x is a combination of NO_2 and NO pollutants, whose main sources are vehicle exhausts. The PM, which is a combination of chemical compositions, solids, and various sizes of liquid particles, is the result of direct emissions like diesel soot, trucks, cars, stone crushing, and burning waste in open. Sources of the PM secondary particles are gases with a physicochemical transformation, like sulphate and nitrate (Limaye and Salvi 2010). Basically, PM has been classified into three categories depending on the particle size: PM 2.5 (diameter < 2.5 μm), PM 2.5–10 (coarse particles), and PM 10 (diameter < 10 μm) (Kim et al. 2015). Natural sources of PM are volcanoes, oceans spray, and ground (Limaye and Salvi 2010). According to WHO, PM particles can make their way into the respiratory tract with a high penetrability, thereby affecting lung functioning. Compared to other pollutants, PM is the cause of many major diseases, and is more dangerous to human health and life.

Goa is the smallest Indian state located on the west India coast and forms a narrow strip of low land sandwiched between the Arabian Sea to the west and the Western Ghats to the east. It is generally regarded as a tropical heaven on earth due to its pristine sandy beaches, laid-back lifestyle and pollution-free environment. However, during the recent times, the state has also come under the menace of air pollution. Majority of the area of the state was under a good amount of vegetation; however, since the past four decades, haphazard urbanization and rapid economic growth due to extensive tourism activities, open-cast mining, and construction activities have degraded the environment. The compounded annual growth rate of Green House Gases (GHG) emission between 1980 and 2005 for Goa was found to be 5.59% for CO_2. It provides clear evidence of subsequent growing energy consumption and future demand in Goa (Jamir and De 2013). This can be attributed to rapid growth in tourism, transport, and mining activities. A quick increment in vehicular traffic is becoming a significant source of air pollution in Goa, as fuel consumption exhibited a growth of 8.5%

(Sharma and Patil 2016b) with 70,000 new vehicles being added on the Goan roads annually (Economic Times 2013). Contribution of gasoline consumption to CO_2 emissions shows the highest growth of 8.5% depicting tremendous increase in vehicular activity. Consequently, air quality which was not a priority issue in the region in the past is gradually becoming an important concern. While gaseous pollutants are within the limits, many regions in the state have shown violations of standards of PM_{10} concentrations. There are multiple sources of PM_{10} pollution in Goa—transportation, ports, iron-ore mining, residential sector energy use, industries, road dust re-suspension etc. Vasco Da Gama, known as the transport capital of Goa, is located on the headland facing the Arabian Sea and is experiencing serious threats of air pollution mainly due to the proximity to the sole seaport of Goa (Marmugao Trust Port) where extensive import of coal and export of iron-ore takes place. On the other hand, the capital city of Panjim, once a town with "Susegad" lifestyle, is fast becoming just like several other polluted Indian cities.

Monitoring air quality plays a vital role in controlling the levels of pollution (Camastra et al. 2019) at different locations in order to mitigate the effect of air pollution on human health. The current work presents the air pollutants trend in the major coastal cities (Panjim and Vasco-Da-Gama) of Goa state over the period of 2017–2021. Panjim is the Goa's capital which has about 90,000 vehicles entering every day. The air pollution in Vasco-da-Gama city is due to burning of biomass material for water heating and cooking, while Marmugao Port Trust contributes less than 10% of the total PM_{10}/$PM_{2.5}$ pollutions in the city (Herald 2019). SO_2, NO_x and PM_{10} were the three pollutants demonstrated in the analysis using the EF method. The data were collected by the Central Pollution Control Board, and the Ministry of Environment, Forest, and Climate Change, India. The goals are to infer how pollution is trending during monsoon and non-monsoon months in these cities.

7.2 Study Area

Goa is situated along the Arabian Sea in the Konkan region of the west coast of India (Fig. 7.1). The spatial coordinates of the state lie between 14°53′54"N to 15°40′00"N latitudes and 73°40′33"E to 74°20′13"E longitudes. It covers a total land area of 3702 km^2. Goa's topography is highly variable. The coastal tract is dotted with world famous sandy beaches and rocky cliffs and headlands of 105-km in length, while the Western Ghats form its eastern boundary. The plateau region sandwiched in between is rich in mineral (iron-ore) deposits.

The state receives heavy to very heavy rainfall of about 2900 mm during the southwest monsoon period. The population was estimated at 1.59 million in 2020. The total vehicles registered in Goa increased from 0.58 million in 2007–1.25 million in 2020 at the rate of 0.135 million per year (https://www.statista.com/statistics/1116894/number-of-vehicles-in-goa-india/).

Fig. 7.1 Study area showing the locations of Panjim and Vasco-da-Gama cities. (Modified redrawn version as JPG file by the authors of the blank outline map (in vector Graphic -.svg format)). (Source: Nigam et al. 2022)

Goa is a tiny state, which has around 62% of the urbanized population. However, the coastal tract is under pressure of high migration and rapid unsustainable development. In this regard, two cities Panjim and Vasco Da Gama have been chosen by also considering the availability of continuous pollution data from 2017 to 2021 provided by the Goa State Pollution Control Board. Panjim, the state administrative capital and a heritage city sprawling on the banks of Mandovi estuary and Vasco Da Gama located in South Goa district on a headland, is generally considered as the transport capital of Goa as it hosts the railway terminus, the sole airport and the sea-port of the state.

7.3 Data and Methods

In the present study, four-year monthly air pollutant data (SO_2, $PM_{2.5}$ and PM_{10}) for the two different stations located at Altinho, Panjim and Marmugao Port Trust, Vasco Da Gama, were downloaded from the Goa State Pollution Control Board (GSPCB) portal (http://goaspcb.gov.in/). The GSPCB provides pollutant data only from the manual monitoring network, due to which only the, and pollutant data were available.

The present study has applied the method given by the Central Pollution Control Board (CPCB), the EF of the air pollutants. EF is the "proportion of the yearly average concentration of a pollutant and its particular standard" (Haque and Singh 2017). In other words, it is applied to show the intensity and the status of air quality at a particular place in comparison to annual standards of the pollutants given by the CPCB.

$$EF = \frac{\text{Yearly average concentration of the pollutant}}{\text{Yearly standard for that particular pollutant}} \tag{7.1}$$

Based on the above Eq. (7.1), the air pollution levels can be categorized into four classes as follows: EF < 0.5: Low pollution; EF between 0.51–0.99: Moderate pollution (M); EF between 1–1.5: High pollution (H); and EF > 1.5: Critical pollution (C).

Yearly concentrations of $PM_{10,}$ $PM_{2.5,}$ SO_2 and NO_2 were downloaded from the GSPCB website for the year 2017, 2018, 2019 and 2020. Later, the datasets were divided on the basis of monsoon (June–September) and non-monsoon seasons (October–May) to assess the pollution in two different seasons which further provided significant differences in the amount of pollutants. In addition, yearly EF was also calculated to determine the difference through the years. Finally, the line of best fit and linear regression analysis were also conducted to determine current and future trends of pollutant concentrations.

7.4 Results

Panjim city has observed high (H) to critical (C) levels (1.88, 1.38, 1.55 and 1.06) of PM_{10} concentrations during all the non-monsoon periods between 2017 and 2020 consequently this trend led to similar E.F values annually as well. The monsoon period has also been found to be concerning with high level (1.16, 1.01, 1.13) of PM_{10} during these 4 years, however it was under moderate level during the 2020 period (0.85). $PM_{2.5}$ has been found to be in more or less uniform pattern under moderate (M) E.F level (0.94 & 0.99) during 2017 and 2018. However, it was found to have increased to (1.17) during 2019 which further reduced to moderate level (0.7) in 2020. Sulphur dioxide was found to be well under control as its E.F value never increased above 0.15 (during monsoon 2017). SO_2 E.F level has decreased significantly since the monsoon of 2019. Similar to SO_2, nitrogen dioxide has also been found to be under low level of the E.F. only the exception was during the 2017 non-monsoon period when it entered under moderate level (0.52) of E.F. The annual trend of NO_2 has observed uniform reduction in the E.F values 0.48 (2017) to 0.26 (2020). The general trend of all the four air pollutants in Panjim city has been found to be significantly and uniformly reducing during these 4 years (Figs. 7.2 and 7.3, Tables 7.1 and 7.2).

Vasco Da Gama has also observed predominantly critical (C) levels (1.65, 1.51, 1.85 and

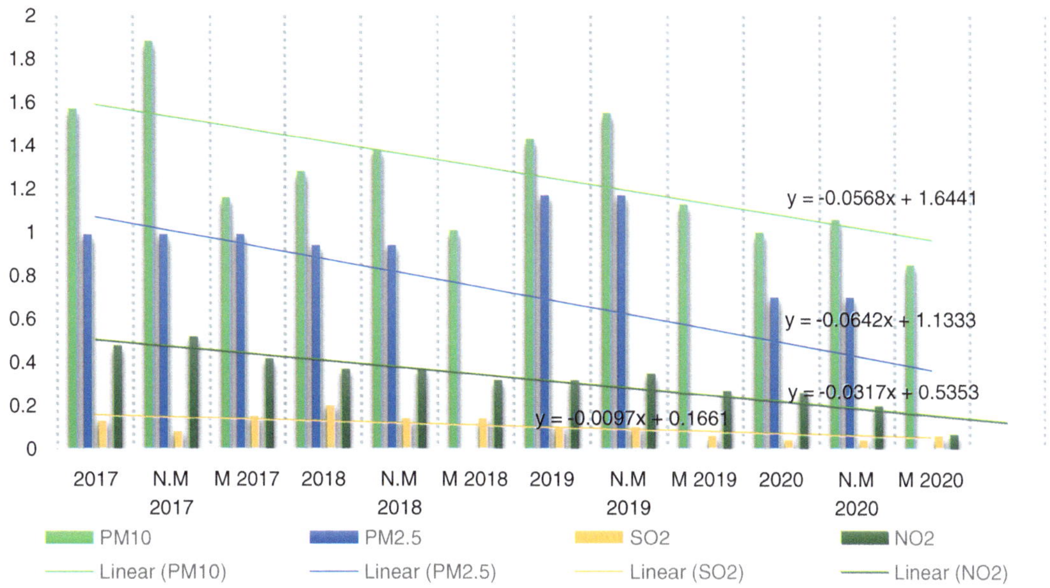

Fig. 7.2 Bar graph showing Exceedance Factor of each of the period (N.M- Non-Monsoon & M- Monsoon) in Panjim city along with the line of best fit and the linear regression equation

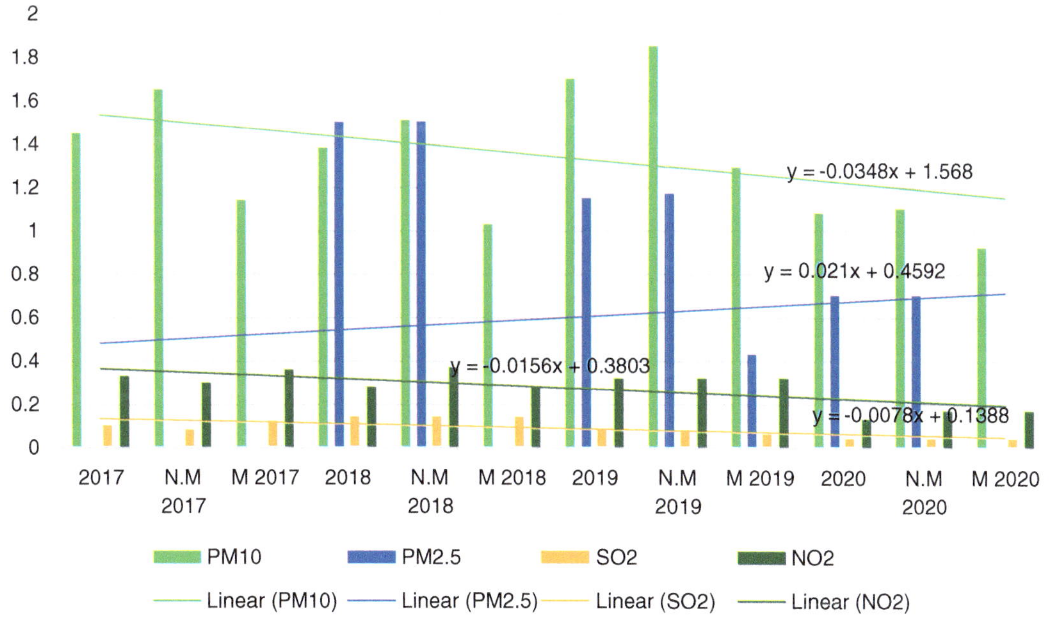

Fig. 7.3 Bar graph showing Exceedance Factor of each of the period (N.M- Non-Monsoon & M- Monsoon) in Vasco Da Gama city along with the line of best fit and the linear regression equation

1.10) of PM_{10} concentrations during all the non-monsoon periods between 2017 and 2020 consequently this periodic trend influenced the annual E.F values significantly due to which annual E.F values were observed as (1.45, 1.38, 170 and 1.10) during the similar period. The monsoon period was also found to be under high level of E.F (1.14, 1.03, 1.29 and 0.92) of PM_{10} during these 4 years, however it was also observed that monsoon period values were significantly lower

Table 7.1 Exceedance Factor value and level of each of the period (N.M-Non Monsoon & M-Monsoon and Annual) in Panjim city (NA refers to non-availability of pollutant data)

	PM_{10}		$PM_{2.5}$		SO_2		NO_2	
Period	E.F	E.F level	E.F	E.F level	E.F	E.F level	E.F	E.F level
2017	1.57	C	0.99	M	0.13	L	0.48	L
N.M 2017	1.88	C	0.99	M	0.08	L	0.52	M
M 2017	1.16	H	0.99	M	0.15	L	0.42	L
2018	1.28	H	0.94	M	0.2	L	0.37	L
N.M 2018	1.38	H	0.94	M	0.14	L	0.37	L
M 2018	1.01	H	NA	NA	0.14	L	0.32	L
2019	1.43	H	1.17	H	0.1	L	0.32	L
N.M 2019	1.55	C	1.17	H	0.1	L	0.35	L
M 2019	1.13	H	NA	NA	0.06	L	0.27	L
2020	1	H	0.7	M	0.04	L	0.26	L
N.M 2020	1.06	H	0.7	M	0.04	L	0.2	L
M 2020	0.85	M	NA	NA	0.06	L	0.07	L

Table 7.2 Exceedance Factor value and level of each of the period (N.M- Non Monsoon & M- Monsoon and Annual) in Vasco Da Gama city (NA refers to non-availability of pollutant data)

	PM_{10}		$PM_{2.5}$		SO_2		NO_2	
Period	E.F	E.F level	E.F	E.F level	E.F	E.F level	E.F	E.F level
2017	1.45	H	NA	NA	0.1	L	0.33	L
N.M 2017	1.65	C	NA	NA	0.08	L	0.3	L
M 2017	1.14	H	NA	NA	0.12	L	0.36	L
2018	1.38	H	1.5	C	0.14	L	0.28	L
N.M 2018	1.51	C	1.5	C	0.14	L	0.37	L
M 2018	1.03	H	NA	NA	0.14	L	0.28	L
2019	1.7	C	1.15	H	0.08	L	0.32	L
N.M 2019	1.85	C	1.17	H	0.08	L	0.32	L
M 2019	1.29	H	0.43	L	0.06	L	0.32	L
2020	1.08	H	0.7	M	0.04	L	0.13	L
N.M 2020	1.1	H	0.7	M	0.04	L	0.17	L
M 2020	0.92	M	NA	NA	0.04	L	0.17	L

in comparison to non-monsoon period. $PM_{2.5}$ pollutant data was not available during 2017 and monsoon periods of 2018 and 2020. However, the critical level of 1.50 during 2018, high E.F of 1.17 during 2019 and low E.F. of 0.7 during 2020 suggest a continuous decline in the $PM_{2.5}$ concentrations. Sulphur dioxide was found to be well under control as its E.F value never increased above 0.14 (during all three periods of 2018). SO_2 E.F levels have been decreasing significantly since 2018. Similar to SO_2, nitrogen dioxide has also been found to be under low level of the E.F with the maximum E.F level hovered around 0.32–0.37 which decreased abruptly during 2020. The general trend of $PM_{2.5}$, SO_2 and NO_2 pollutants in Vasco Da Gama has been found to be significantly and steadily reducing during these 4 years. PM_{10} showed cyclic pattern with highs during odd years and low values during even years.

7.5 Discussion

Air pollution is one of the greatest menaces of modern civilization. Anthropogenic activities around the world are leaving no place un-spoilt with pollution because all kind of human activities from mining to manufacturing to services leave their significant footprints behind in the form of unwanted increment in gaseous or tiny particulate concentrations in the atmosphere. Highly urbanized areas without any doubt are generally susceptible to high amount of air pollution mainly due to industries and vehicular movement; however, these pollutants have grappled many of the un-spoilt small towns mainly across the Indian subcontinent. From the Himalayas to the Western Ghats to coastal areas, a major portion of these previously pristine regions are under the grip of air pollution now. The significant reasons behind this phenomenon can be attributed to unsustainable development in the once remote areas, haphazard construction in small towns and migration of population in search of better place to live. PM or particulate matter is microscopic solid or liquid matter suspended in Earth's atmosphere. PM impacts climate as well as precipitation that adversely affects human health.

The critical to high exceedance level of PM_{10} revealed from the present study strongly supports the report given by the Goa State Pollution Control Board (GSPCB) in 2018, that the major cities in the state such as; Panjim and Vasco Da Gama have been contributing to far higher amount of PM_{10} whose annual permissible concentration limit is 40 $\mu g/m^3$. During the initial period of 2017–2020 time-frame as well, the air quality degradation was found to be alarming in Panjim which verifies the fact further. The major contributors $PM_{2.5}$ and PM_{10} have far more significant contribution in producing air pollution than the other two pollutants (SO_2 and NO_2) in the city. The significant availability of the particulate matters as the pollutants can be attributed to dust particles along with aerosols. Aerosols are ubiquitously present in Panjim city due to its proximity to the Mandovi estuary and the Arabian Sea which maintains the humid climate in the city causing high amount of humidity thus, abundance of aerosols in the atmosphere. On the other hand, significant increment in the construction activities vehicular movement (Navhind Times 2018) and the unsustainable city expansion (Greater Panjim now covers previously green fields of Taleigao, Santa Cruz, Old Goa, Bambolim, etc.) has led to the generation of high amount of pollution due to dust. Therefore, with the abundance of liquid particles (aerosols) and solid particles (dust) in the city, PM_{10} is the most significant pollutant. In addition to the current study, various anecdotes also suggest the similar phenomenon about Panjim. According to the Navhind Times report, the air pollution is clearly visible to the naked eye over Panjim city when passing through the Atal bridge. Exposure to particulate matters can cause several types of respiratory diseases such as coughing, Asthma, Bronchitis, Silicosis, etc. as well as vascular diseases such as high blood pressure, heart attack to premature deaths (WHO 2013). However, there is a sigh of relief that the level of particulate matters is gradually decreasing in the city since 2017 along with the other dangerous pollutants Sulphur dioxide and Nitrogen dioxide. The overall trend in pollutant reduction could be attributed to the effective work of Dept. of Transport by keeping strict vigilance on polluting vehicles and to the Dept. of Agriculture, for the open drives against open burning of leaves and municipal wastes in the city areas (Times of India 2020).

Although overall air pollution is gradually decreasing in the city however, the other form of pollution in the city is on the rapid rise. At the Mandovi estuary mouth, the untreated waste from the resorts and the sediments from the upstream mining activities have caused a significant surge in concentrations of faecal coliform and other pathogen bacteria, which has made the coastal and estuarine waters highly adverse for activities such as bathing and fishing (Mohandass et al. 2010) subsequently impacting human health and nearby environment.

On the other hand, Vasco Da Gama has predominantly been found to be on the top of the list of most polluted towns in Goa. GSPCB in 2018 stated the reason behind critical pollution status in the city which includes the high amount of infrastructure construction, as well as the explosion in the number of vehicles plying on the roads of this congested city devoid of space (Navhind Times 2018). In addition, one of the largest concentrations of the slums across the state is in Vasco, who are devoid of modern home appliances and sustainable living standards. It is strongly believed and experienced since the turn of this century that the coal handling operations at the Marmugao Port located in Vasco city is the major source of pollution pertaining to Particulate Matters ($PM10_{10}$ and $PM_{2.5}$) because the coal used to be transported on the trucks passing through the sole major road of the town and the coal filled railway wagons mostly stationed in the town waiting for their turn. This phenomenon is clearly indicated in the present study as the increasing trend of $PM_{2.5}$ and "High" to "Critical" level of PM_{10} between the 2017 and 2020 period. Moreover, the GSPCB in 2017 had issued show cause notice to the MPT (Marmugao Port Trust) in relation to unsustainable handling of coal at the port site. However, the Mining has been stalled in Goa ever since the Honorable Supreme Court cancelled 88 mining leases order in the state on February 7, 2018, hence this has directly impacted the reduction in particulate matter concentrations because of the ban on mining, far less number of trucks are plying leading to considerable reduction in pollution. However, the increment in pollutants during 2019 compared to 2018 has been mainly due to resumption in construction activities (flyovers) in Varunapuri, Vasco and Baina (Navhind Times 2021) which was again halted in 2020 due to COVID-19 lockdown consequently led to reduction again hence supporting the study results.

The grave concern of coal dust is about the serious negative impact on the health of city dwellers. The dust possesses high risk of getting accumulated in the respiratory tracts of the sufferers and in some cases, children are getting badly affected by this deadly coal dust (Incredible Goa 2019). Coal dust usually comprises of small particulate matters that settle into the respiratory tracts through the nose into the lungs. It may also cause allergies as well as it could make the victims more susceptible to different respiratory diseases and infections. In fact, Vasco Da Gama MLA, Carlos Almeida has already suspected that the high number of morbidity in the town during COVID-19 peak time was mainly due to the exposure of locals to coal particles which made them more vulnerable to respiratory diseases than other people in the state (Times of India 2021c). This factor has already been explained by Dr. Arora at Health-way Hospital, Old Goa stated that "the long-term exposure to coal dust could cause chronic obstructive pulmonary disease (COPD), and that workers exposed to it over a long period even risk catching pneumoconiosis—a disease caused by dust accumulation in the lung sacs. Children are at a greater risk as they tend to be more vulnerable to the pollution" (Incredible Goa 2019).

The other indirect but very significant health implication of coal dust is through the leakage and spillage from the barges during transportation into the Vasco bay. This bay is the fishing ground of villagers from Kharewado, Baina and Desteiro. This poses a great danger of ingestion of coal dust and other pollutants by the local population through edible fishes. Adding to the crisis, these villagers are threatened to be displaced from their village due to expansion of the port area, which will further deteriorate the pollution levels in Vasco Da Gama. In addition to critical level of air pollution in the city, Sea-foam has been found at the city beaches. The scientists from the National Institute of Oceanography strongly advised to treat this phenomenon seriously which could have hazardous health implications for tourists as well as local population living in the proximity. They further revealed that sea-foam formation and garbage dumping near the beaches have led to excessive eutrophication in the seawater where higher inflow of fresh water in the proximity has also been found. This could aggravate the growth of pathogen bacteria in the region, which are known to have antibiotic resistance, a serious concern for human health and environment (Times of India 2021b).

In December 2021, some of the allegations by the residents of the city got proved to be right because the Indian Institute of Technology, Bombay (who was assessing the source apportionment of the air pollution in Vasco Da Gama) gave its pollution source apportionment findings that, along with the coal handling activities, vehicles re-suspended dust due to poor road conditions, marine aerosol and biomass and solid waste burning have been found to be the major source of pollution the city. Moreover, vehicular re-suspended dust has been found to be the larger source of particulate matter pollution than the coal handling (The Times of India 2021a) which proves that ban on mining led to reduction in heavy vehicles plying on the city road hence, subsequent reduction has been found in the particulate matter since 2018 (ban on mining). In addition, the reason of increment in pollution during 2019 is already described above.

Although the ban on mining by the Honorable Supreme Court is still in place, which is led to significant subsequent reduction in pollutants since 2018, however, the above concerns could become reality again, if the ban gets lifted in the future.

7.6 Recommendation and Conclusion

The results obtained in the present study strongly suggest that Vasco Da Gama is rapidly reaching at its tipping point where further, it might become unbearable for the local population to survive. It is prerequisite for the state govt. and urban local body to come forward and decide the limits of the sea port to import the coal and transport it through Vasco Da Gama city as well as significant improvement in roads condition is the need of the hour. In addition, slum areas which have sprouted rapidly in the vicinity of the town need to be checked strictly for the use of biomass burning along with decongestion of several areas is the need of the hour. On the other hand, Panjim city is fast turning into a metro city where all kinds of economic activities are rapidly sprouting. This needs to be checked to maintain the status-quo of "Susagad" lifestyle in the town, which is ideal for overall good quality of life. Particulate matters in both the towns need to be further reduced by road maintenance, plantation along the roads, sustainable construction activities and strictly checking vehicle emissions. Otherwise, the other two pollutants SO_2 and NO_2 would also cross their permissible if particulate matters could not be managed, as SO_2 and NO_2 are also emitted by vehicular and industrial emissions. Goa is an exotic tourist destination where the expansion of tourism industry is inevitable and would never stop. However, the development in this sector along with mining and cargo services will undoubtedly bring further degradation of natural resources, environment and atmosphere in this fragile region. Currently, the state of Goa is in moderate stage of air pollution (as compared to other polluted Indian cities of New Delhi, Mumbai, Kanpur, Vapi, Ahmedabad, Guwahati, etc.). The trends of pollution concentration have been running low since the last year but that is due to COVID-19 induced lockdown, which even lowered the pollutant concentrations in some of the highest polluted cities of Ankleshwar and Vapi (Nigam et al. 2021). Perhaps, it is regarded as the temporary relief from air pollutants however, this might not be the case in 2021 and further down the road. Thus, cutting in pollution emission is the only way forward to return the lost glory of Panjim and Vasco Da Gama who had high quality of life just some decades back.

Acknowledgements We thank the Goa State Pollution Control Board for providing air pollution data and keeping vigil on the negligence of authorities towards pollution in the state. We also acknowledge the research facilities provided by SEOAS, Goa University and the Ministry of Earth Sciences, Government of India.

References

Camastra F, Ciaramella A, Son LH, Riccio A, Staiano A (2019) Fuzzy similarity-based hierarchical clustering for atmospheric pollutants prediction. Springer International Publishing, Cham https://doi.org/10.1007/978-3-030-12544-8_10.

Dadhich AP, Goyal R, Dadhich PN. Assessment of Spatio-temporal variations in air quality of Jaipur City, Rajasthan, India. Egypt J Remote Sens Space Sci. 2018;21(2018):173–81. https://doi.org/10.1016/j.ejrs.2017.04.002.

Dales R, Burnett RT, Smith-Doiron M, Stieb DM, Brook JR. Air pollution and sudden infant death syndrome. Pediatrics. 2004;113:e628–31. https://doi.org/10.1542/peds.113.6.e628.

Darryl DM. 2017. https://indiaclimatedialogue.net/2017/11/24/india-biggest-emitter-so2/.

Economic Times. Goa tourism: drug trade & infrastructure woes ail the sector; in dire need of a boost. Business News, Industry, Services, Travel, Goa tourism. 2013. https://economictimes.indiatimes.com/industry/services/travel/goa-tourism-drug-trade-

infrastructure-woes-ail-the-sector-in-dire-need-of-a-boost/articleshow/27740123.cms?from=mdr. Accessed 27 Oct 2021.
GBD MAPS Working Group. Burden of disease attributable to major air pollution sources in India. Special report 21. Boston: Health Effects Institute; 2018.
Gupta AB. Vehicular air pollution and asthma. Asthma Sanjeevani. 1999;5(2):3–5.
Guttikunda SK, Nishadh KA, Jawahar P. Air pollution knowledge assessments (APnA) for 20 Indian cities. Urban Clim. 2019;27:124–41. https://doi.org/10.1016/j.uclim.2018.11.005.
Haque S, Singh RB. Air pollution and human health in Kolkata, India: a case study. Climate. 2017;5(4):77. https://doi.org/10.3390/cli5040077.
Health Effects Institute (HEI). State of Global Air Report 2019 India-Specific Findings. 2019.
Hedley AJ, Wong C-M, Thach TQ, Ma S, Lam T-H, Anderson HR. Cardiorespiratory and all-cause mortality after restrictions on sulphur content of fuel in Hong Kong: an intervention study. Lancet. 2002;360:1646–52. https://doi.org/10.1016/s0140-6736(02)11612-6.
Herald. Pollution in Vasco 'Many Times Better' than Other Towns: IIT-B. Herald, Goa. 2019. https://www.heraldgoa.in/Goa/Pollution-in-Vasco-%E2%80%98many-times-better%E2%80%99-than-other-towns-IITB/144451. Accessed 3 Nov 2021.
Incredible Goa. The coal dust is the new Hazard for the residents of Vasco Da Gama in Goa. Incredible Goa, awareness, Focus, Society. 2019. https://www.incrediblegoa.org/focus/the-coal-dust-is-a-new-hazard-for-the-residents-of-vasco-da-gama-in-goa/. Accessed 1 Nov 2021.
Jamir T, De US. Trend in GHG emissions from north-east and west coast regions of India. Environ Res Eng Manage. 2013;5:37–47.
Kim K-H, Kabir E, Kabir S. A review on the human health impact of airborne particulate matter. Environ Int. 2015;74:136–43. https://doi.org/10.1016/j.envint.2014.10.005.
Lim SS, Vos T, Flaxman AD, Danaei G, Shibuya K, Adair-Rohani H, AlMazroa MA, Amann M, Anderson HR, Andrews KG, et al. A comparative risk assessment of burden of disease and injury attributable to 67 risk factors and risk factor clusters in 21 regions, 1990–2010: a systematic analysis for the global burden of disease study 2010. Lancet. 2012;380:2224–60.
Limaye S, Salvi S. Ambient air pollution and the lungs: what do clinicians need to know? Breathe. 2010;6:234–44.
Louati A, Son LH, Chabchoub H. Smart routing for municipal solid waste collection: a heuristic approach. J Ambient Intell Humaniz Comput. 2018;10:1865–84. https://doi.org/10.1007/s12652-018-0778-3.
Luis AJ, Nigam R, Pandya K, & Kotha M. Environmental assessment of air pollution scenario (2014–2020) in the tourist state of Goa, India, Under Review. Air Quality, Atmosphere and Health. 2023.
Mohandass C, Kumar SJ, Ramaiah N, Vethamony P. Dispersion and retrievability of water quality indicators during tidal cycles in coastal Salaya, Gulf of Kachchh (West coast of India). Environ Monit Assess. 2010;169(1–4):639–45. https://drs.nio.org/drs/bitstream/handle/2264/3710/Environ_Monit_Assess_169_639a.pdf?sequence=1. Accessed 30 Oct 2021.
Navhind Times. Air quality worsening in Goa's cities and towns. The Navhind Times, Goa News. 2018. https://www.navhindtimes.in/2018/06/06/goanews/air-quality-worsening-in-goas-cities-and-towns/. Accessed 31 Oct 2021.
Navhind Times. MPT, Vasco, Panaji witnessed 'poor air quality' during April 2019—march 2020. The Navhind Times, Goa News. 2021. https://www.navhindtimes.in/2021/03/23/goanews/mpt-vasco-panaji-witnessed-poor-air-quality-during-april-2019-march-2020/. Accessed 20 Nov 2022.
Nigam R, Pandya K, Luis AJ, et al. Positive effects of COVID-19 lockdown on air quality of industrial cities (Ankleshwar and Vapi) of Western India. Sci Rep. 2021;11:4285. (2021). https://doi.org/10.1038/s41598-021-83393-9.
Nigam R, Luis AJ, Prasad P, Kuttikar S, Yadav R, Vaz E, Kotha M. Spatio-temporal assessment of COVID-19 lockdown impact on beach litter status and composition in Goa, India. Marine Pollut Bull. 2022;174:113293. https://doi.org/10.1016/j.marpolbul.2021.113293.
Sharma S, Patil KV. Emission scenario and health impacts of air pollutants in Goa. Aerosol Air Qual Res. 2016a;16:2474–87. ISSN: 1680-8584 print/2071-1409 online. https://doi.org/10.4209/aaqr.2015.12.0664.
Sharma S, Patil KV. Emission scenarios and health impacts of air pollutants in Goa. Aerosol Air Qual Res. 2016b;16:2474–87. https://doi.org/10.4209/aaqr.2015.12.0664.
Son, L.H. and Louati, A. (2016) Modeling Municipal Solid Waste Collection: A Generalized Vehicle Routing Model with Multiple Transfer Stations, Gather Sites and Inhomogeneous Vehicles in Time Windows. Waste Management, 52, 34-49. https://doi.org/10.1016/j.wasman.2016.03.041.
Tandon A, Yadav S, Attri AK. City-wide sweeping a source for respirable particulate matter in the atmosphere. Atmos Environ. 2008;42:1064–9.
The Times of India. Vehicular Pollution Bigger Factor Than Dust in Vasco Air. News, City News, Goa News. 2021a. https://timesofindia.indiatimes.com/city/goa/vehicular-pollution-bigger-factor-than-dust-in-vasco-air-finds-iit-b/articleshow/88601079.cms. Accessed 20 Nov 2022.
The Times of India. Sea foam at Vasco beach hazardous to people: National Institute of Oceanography Study. News, City News, Goa News. 2021b. https://timesofindia.indiatimes.com/city/goa/seafoam-at-vasco-beach-hazardous-to-people-nio-study/articleshow/83292397.

cms?utm_source=contentofinterest&utm_medium=text&utm_campaign=cppst. Accessed 5 Nov 2021.

The Times of India. Coal Blamed for Vasco Deaths. News, City News, Goa News. 2021c. https://timesofindia.indiatimes.com/city/goa/coal-blamed-for-vasco deaths/articleshow/82394676.cms?utm_source=contentofinterest&utm_medium=text&utm_campaign=cppst. Accessed 5 Nov 2021.

Times of India. Air quality 'satisfactory', better than last year: Report News, City News, Goa News. 2020. https://timesofindia.indiatimes.com/city/goa/air-quality-satisfactory-better-than-last-year-report/articleshow/78394526.cms. Accessed 20 Nov 2022.

WHO. WHO global ambient air quality database (update 2018). Geneva: World Health Organization; 2018.

World Health Organization (WHO), Europe. Health effects of particulate matters; policy implications for countries in Eastern Europe, Caucasus and Central Asia. Copenhagen: WHO Regional Office for Europe; 2013. https://www.euro.who.int/__data/assets/pdf_file/0006/189051/Health-effects-of-particulate-matter-final-Eng.pdf. Accessed 30 Oct 2021.

Epidemiology of Fine Particulate Air Pollution and Human Health Impacts

8

Kapil Goel and Arunima Sen

8.1 What Are Fine Particle Air Pollutants?

The air we breathe contains a mixture of solid, liquid, and gaseous compounds derived from natural emissions and anthropogenic sources (those generated by human activity). Broadly speaking, the term "air pollutants" refers to compounds that are airborne and occur at levels dangerous to human health, damaging to flora, and destructive to the environment. Although considerably different in their chemical composition, based on some similarities, these pollutants can be grouped into:

1. Gases
2. Organic pollutants
3. Toxic heavy metals
4. Particulate matters

Gaseous pollutants like sulfur dioxide, nitrogen oxides, ozone, volatile organic compounds, and carbon monoxide are produced mainly due to the combustion of fossil fuels and significantly alter the atmosphere's composition. Toxic substances known as persistent organic pollutants (POPs) linger in the environment for a long time, amplifying their effects as they travel up the food chain (biomagnification). The earth's crust naturally contains a variety of toxic heavy metals, such as lead, mercury, and cadmium. They can neither be broken down nor destroyed; instead, they infiltrate human bodies and turn hazardous at increasing doses. Most heavy metals also have a tendency to bioaccumulate and cause harm. Particulate pollutants, often known as particulate matter (PM), are the most harmful contaminant to humans. PM is a heterogeneous mixture of solid particles and liquid droplets suspended in the air. Ambient PM ranges between 0.001 and 100 μm in aerodynamic diameter. The size distributions are important for health impact. While smaller particles with a median aerodynamic diameter of less than 2.5 μm may reach smaller airways and alveoli, coarser particles get deposited in the nose and throat. The three main categories based on size are PM_{10} (diameter up to 10 μm), $PM_{2.5}$ (diameter up to 2.5 μm), and $PM_{0.1}$ (smaller than 0.1 μm) (Fig. 8.1).

Airborne particles smaller than a 30th of the diameter of a human hair that is less than 2.5 μm in aerodynamic diameter ($PM_{2.5}$) are usually called fine particle air pollutants (United States Environmental Protection Agency 2021c). This group of particles also encompasses ultrafine particles having diameters less than 0.1 μm. They enter the smaller airways and have been strongly associated with negative health outcomes. The source of PM may be primary, which is emitted directly into the atmosphere, or secondary

K. Goel (✉) · A. Sen
Department of Community Medicine, School of Public Health, PGIMER, Chandigarh, India

P. K. Padhy et al. (eds.), *Air Quality and Human Health*,
https://doi.org/10.1007/978-981-97-1363-9_8

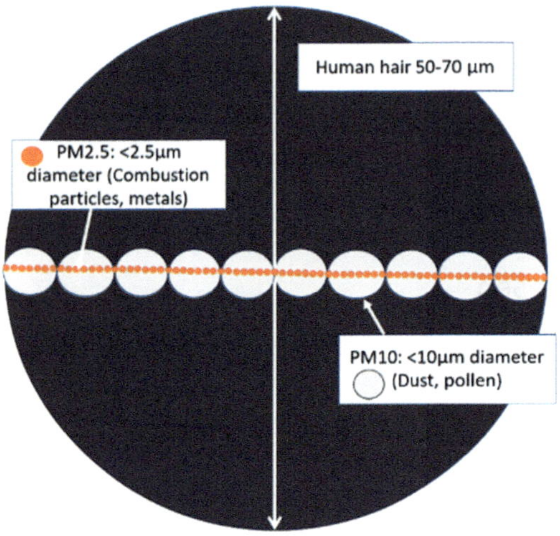

Fig. 8.1 Particulate matter size comparisons

because of the physiochemical reaction of gases. Finer particles are more dangerous to human health, according to data on morbidity and mortality.

8.2 Sources and Exposure

Crucial to estimating the burden of disease is to understand how much pollutants individuals are being exposed to. Ambient air pollution, which occurs outdoors, and indoor (household) air pollution, are the two main categories of air pollution. Both kinds contribute to each other as air moves in and out of buildings.

Vehicles, coal-burning power plants, industries, garbage burning, and other anthropogenic activities all release particulate matter and the chemicals that aid in their secondary creation. Some natural sources also play a role in their creation. Studies on emission sources in India's largest cities have discovered various significant contributing sources, the majority of which are tied to human activity (GBD MAPS Working Group 2018; Sharma and Kumar 2016). In 2015, residential combustion of biomass (24%), coal combustion (around 16% from both industrial and power generation), anthropogenic dust (9%), and open burning of agriculture (5%) were the major sources (GBD MAPS Working Group 2018).

Micrograms of particulate matter (µg) per cubic meter of air (m^3) are used to measure ambient fine particle air pollution ($PM_{2.5}$ concentrations). Data is obtained from reference-grade monitoring stations located all over the world, especially near urban areas, and through satellite and modeling approaches. More than 90% of people worldwide were exposed to annual average $PM_{2.5}$ concentrations above 10 µg/m^3 in 2019, the WHO air quality standard. The highest amounts were recorded in Asia, Africa, and the Middle East (Health Effects Institute 2020a). Exposure to ambient particulate matter pollution is rising globally by more than 1% annually. According to the latest "Global Burden of Disease" study (GBD), exposure to ambient PM increased by 1.46% between 2010 and 2019, while household air pollution from solid fuels decreased by almost 4% (Murray et al. 2020). Global trends show a considerable variation across socio-demographic index (SDI) quintiles. While ambient particulate matter pollution is decreasing in the high SDI quintile, the situation is reversed for the others (low to middle quintiles). However, the disparities in exposure across these regions have remained constant, with South Asia experiencing the highest exposures. The World Health Organization's Global Air Quality Guidelines (AQG), 2021, have updated the recommended annual AQG levels and interim targets for $PM_{2.5}$. The guidelines recommend a series of decreasing concentration targets, starting with interim target 1 at 35 µg/m, interim target 2 at 25 µg/m^3, interim target 3 at 15 µg/m^3, and interim target 4 at 10 µg/m^3. The ultimate recommended level is set at 5 µg/m^3. This stringent standard is designed to substantially lessen the adverse health effects linked to air pollution (World Health Organization 2021) (Fig. 8.2).

When solid fuels such as coal or charcoal, agricultural and animal wastes are burned, a variety of pollutants, including $PM_{2.5}$, are produced, which is known as household air pollution (HAP). HAP exposure is more common in Sub-Saharan Africa and portions of Asia. The two forms of $PM_{2.5}$ exposure have vastly different correlations to socio-demographic change. HAP has a high correlation with SDI and tends to

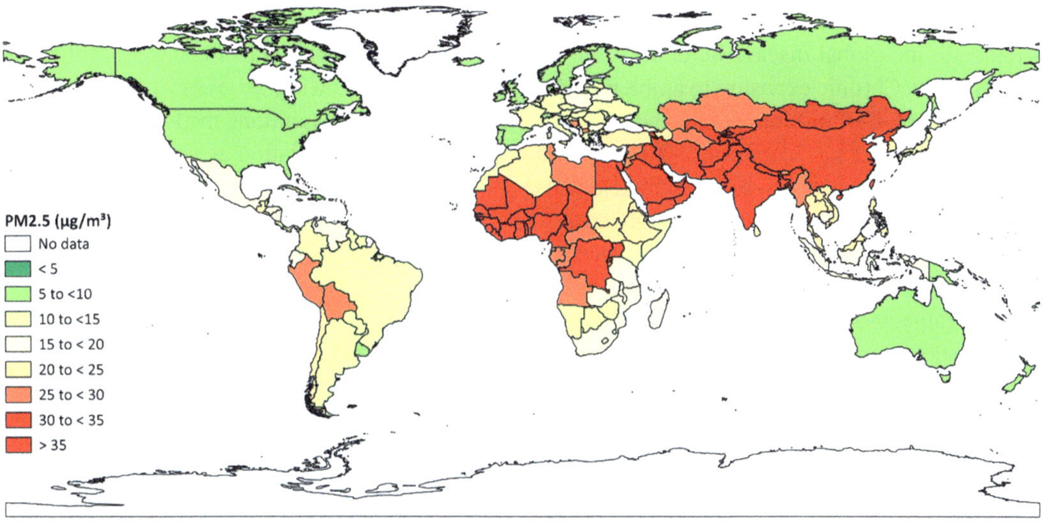

Fig. 8.2 Annual $PM_{2.5}$ concentrations, 2019. Source of data: The Global Health Observatory (World Health Organization 2019)

decline steadily with sociodemographic development. In contrast, as seen in high SDI countries, ambient particulate matter pollution tends to rise with industrialization and subsequently fall with air-quality control (Health Effects Institute 2020a; Pandey et al. 2021). India, Pakistan, and Bangladesh, few of the world's most polluted countries, continue to see increases in ambient $PM_{2.5}$ exposure.

The National Ambient Air Quality Standard (NAAQS) of 40 µg/m^3 for fine PM is exceeded in India., especially in the urban areas of the Indo-Gangetic plain (Ministry of Environment Forest and Climate Change 2019). The country's annual average population-weighted mean concentration of $PM_{2.5}$ was around 92 µg/m^3 in 2019, exceeding the NAAQS by more than two-fold (Pandey et al. 2021). The less developed states in the north and north-eastern parts of the country had a higher burden from HAP. Ambient particulate matter exposure varied significantly between the states, from 15 µg/m^3 in Kerala to 217 µg/m^3 in Delhi. The four states with the highest exposure levels were from the northern region of the country. Campaigns aggressively promoting the move to cleaner fuels helped India reduce its population exposed to HAP from 73% in 2010 to 61% in 2019 (Health Effects Institute 2020a). The eastern and northern states had the highest percentage of people who used solid fuels for cooking, with six states having a proportion above 70%. In addition to the ambient $PM_{2.5}$ present, use of solid fuels like charcoal, coal, and wood contributed an additional 83 µg/m^3 $PM_{2.5}$ in households (Pandey et al. 2021).

8.3 Disease Burden: Global Pattern and Trends

Globally, only high blood pressure, tobacco use, and a poor diet surpassed air pollution as risk factors for premature mortality in 2019, making it the 4th leading risk factor (Health Effects Institute 2020a). A leading cause of early death and disease, it was responsible for approximately three million deaths among women and four million deaths among men. Roughly translating to 12% of all deaths in the year (Health Effects Institute 2020a; Pandey et al. 2021). When viewed in terms of DALYs to reflect the non-fatal disease burden, air pollution continues to be among the top five risks in both genders.

Ambient fine air pollution is the biggest contributor to the global disease burden caused by air pollution. Chronic exposure to ambient $PM_{2.5}$ contributed to 4.14 million deaths and 118 million lost DALYs, accounting for around two-thirds of all air pollution-attributable deaths and more than half DALYs. The highest age-standardized death rates and DALYs belonged to Asian and African countries. While death rates in India and China were 96 and 81 deaths per 100,000 population, respectively, Canada had a death rate of 5.4/100,000, and Norway had 3.8 deaths/100,000 attributable to $PM_{2.5}$ (Health Effects Institute 2020a). Together, India and China were responsible for 58% of deaths attributed to $PM_{2.5}$ in 2019, globally. The number of deaths and DALYs caused by ambient $PM_{2.5}$ has increased during the last decade by 23% and 17%, respectively. However, there has been a decline in age-standardized rates. This decline reflects improvement in treatment outcomes of diseases to which air pollution contributes.

Significant declines in disease burden from air pollution in the last decade stem from reductions in exposure to indoor air pollution. Among the 69 potentially modifiable risk factors, HAP ranked 9th in 2019, a decrease from its ranking of 7 in 2010. Total deaths and age-standardized death rates steadily declined by 24% and 38% respectively. A similar decline was observed in DALYs and DALY rates. Accounting for approximately 4% of deaths and DALYs, worldwide, in 2019, HAP resulted in 2.31 million fatalities and around 92 million DALYs. Years of Life Lost (YLL) contributed to more than 90% of DALYs. The age-standardized death rates attributed to HAP follow a pattern similar to its exposure. China, where significant efforts have been made to replace coal cookstoves, and India, where availability to liquefied petroleum gas (LPG) has been increased nationwide, saw the largest declines in illness burden owing to HAP in South Asia. Currently, India has among the lowest death rates of 60/100,000 population as compared to Nepal (113/100,000), Pakistan (93/100,000), and Bangladesh (79,100,000) attributable to HAP (Health Effects Institute 2020a)

8.4 Disease Burden: India

In India, air pollution was responsible for approximately 18% of total fatalities, that is 1.67 million deaths in 2019. Approximately one million deaths were attributable to ambient PM pollution, and 0.61 million were attributable to HAP. While the crude death rate due to HAP decreased by 64% between 2010 and 2019, those due to ambient $PM_{2.5}$ increased by 115%. Of the total DALYs, 11.5% were due to air pollution, of which 6.7% were from $PM_{2.5}$, and 4.5% were from HAP. Marked disparities in disease burden existed among states. The northern states had the highest $PM_{2.5}$ attributed crude DALY rate, while both the north and north-eastern states had the highest crude DALY rates attributable to HAP. Per capita GDP of states correlated inversely with crude DALY rate attributable to HAP, however for crude DALY rate attributable to $PM_{2.5}$, the correlation is not significant (Pandey et al. 2021) (Figs. 8.3 and 8.4).

In India, COPD accounted for 33% of all DALYs caused by air pollution in 2019. Ischemic heart disease (29%), stroke (16%), and lower respiratory infections (11%) were next in importance. Lung diseases contributed to 39.5%, which included lung cancer (1.3%) in addition to COPD, and lower respiratory infections (Pandey et al. 2021).

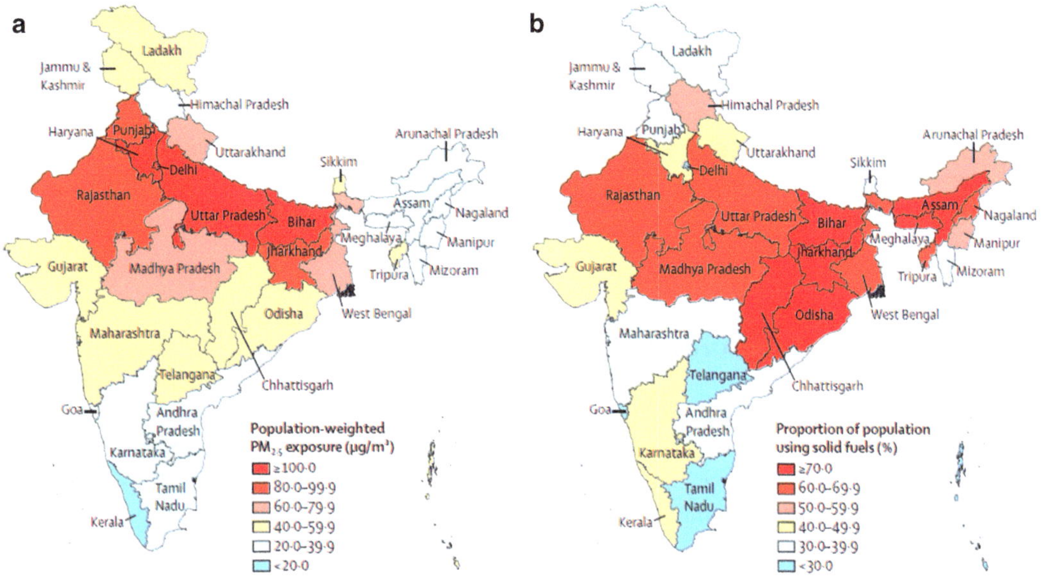

Fig. 8.3 Exposure to $PM_{2.5}$ and solid fuels in India, 2019 Source: Global Burden of Disease Study, 2019 (Pandey et al. 2021) used under a Creative Commons Attribution 4.0 license

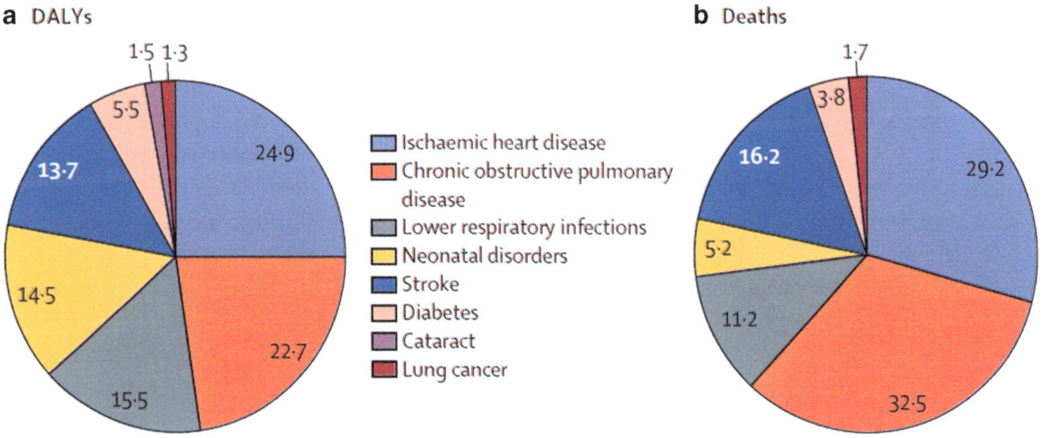

Fig. 8.4 Causes of DALYs (**a**) and deaths (**b**) attributable to air pollution in India, 2019 Source: Global Burden of Disease Study, 2019 (Pandey et al. 2021) used under a Creative Commons Attribution 4.0 license

8.5 Health Effects

Exposure to air pollution, both ambient and household, is associated with a broad range of health problems. Among all, $PM_{2.5}$ is the most significant contributor to the risk to world health, and has the capacity to reach the bloodstream through the alveolar barrier and cause oxidative or inflammatory damage (Manisalidis et al. 2020; Mukherjee and Agrawal 2018). Vulnerable groups comprising of children, pregnant women, the elderly, and those with pre-existing co-morbidities are most affected. The health impacts are mostly determined by factors such as age, lifestyle, medical treatment, climate, and exposure levels. These effects can be grouped into short-term and long-term, direct and indirect, or according to systems affected.

8.6 Short- and Long-term Exposure

Exposure to air pollution for both short and long periods of time, from a few days to years, can have major impacts on one's health, ranging from transient to chronic and even fatal illnesses.

8.6.1 Short-Term Exposure Effects

Short-term episodes of high exposure range from temporary discomforts like irritation of ear, nose, throat, skin, tightness of chest, and breathing difficulties which disappear following a decline in pollution levels to more serious lung and heart problems like asthma, pneumonia, and bronchitis (Manisalidis et al. 2020). Short-term exposure to $PM_{2.5}$ can lead to arrhythmias, heart attacks, and even death in people with heart diseases (Health Effects Institute 2020b).

8.6.2 Long-Term Effects

Long-term exposure to air pollution increases the risk of acquiring chronic heart disease, respiratory conditions, infections of the lungs, lung cancer, diabetes, and other health issues, as well as the possibility of dying prematurely from these conditions. As the scientific evidence regarding the effects of air pollution increase, more diseases get added to the list (Fig. 8.5).

8.7 Effect on Different Systems

8.7.1 Respiratory Disorders

As a result of breathing in air contaminants, the respiratory system is frequently the first to become compromised. The quantity of pollutants inhaled and the area where it is deposited determine the degree of harm. Irritation of the airways from high concentration of pollutants, or continuous exposure at lower concentrations results in nose/throat irritation, voice disturbances, bronchoconstriction, and dyspnea. Air pollution is a major cause of allergic respiratory diseases including allergic rhinitis, and asthma. Allergens become accessible to immune cells because of impaired mucociliary function and damaged mucosal airways. Prenatal exposure influences the development of lung function and predicts adult lung disease. It is associated with wheezing and asthma in childhood. Asthma may lead to emphysema in the long run (El Morabet 2020). $PM_{2.5}$ exposure is associated with increased risks of respiratory morbidity. A higher relative risk indicates people exposed to higher concentrations of $PM_{2.5}$ are at higher risk of developing asthma, COPD, and lung cancer, globally (Sharma et al. 2020). In India, deficit lung function, and higher risks of COPD and lung cancer with an increase in fine particulate air pollution in the country have been reported by various studies (Rajak and Chattopadhyay 2020; Sharma et al. 2020). Every 10 $\mu g/m^3$ increase in fine particle air pollution was found to be associated with an

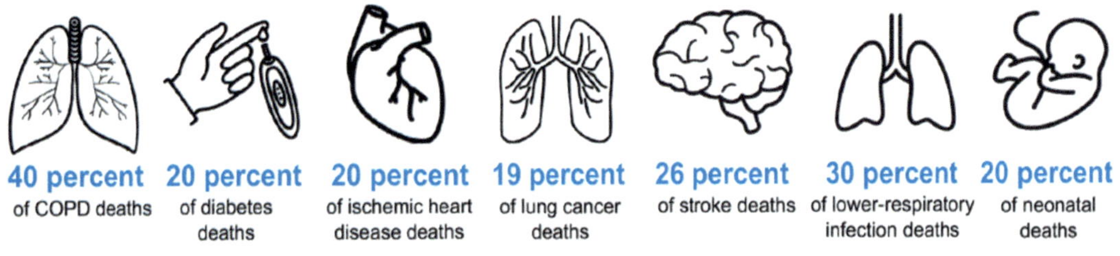

Fig. 8.5 Proportion of deaths (by cause) due to air pollution, globally, 2019. Source: State of Global Air, 2020 (Health Effects Institute 2020b)

increase in the prevalence of asthma after adjusting for other factors, and a 12% greater chance of developing acute lower respiratory infections (ALRI) (Sharma et al. 2020). Although the vast majority of epidemiological evidence comes from data from urban centers, difference in urban–rural population in terms of disease burden exist. Urban residents, and those from industrial towns have a higher prevalence of chronic respiratory symptoms and increased risk for respiratory diseases and lung cancer (Rajak and Chattopadhyay 2020).

Mechanism: Studies on controlled human exposure as well as animal exposure have shown that significant numbers of particles can cause inflammation of the respiratory tract; the severity of the inflammation is influenced by the particle's composition and dose. Inflammation of airways makes them more sensitive to irritants and decreases lung function through vasoconstriction. Repeated exposure promotes chronic inflammation and extracellular matrix reorganization (El Morabet 2020; Manisalidis et al. 2020; United States Environmental Protection Agency 2021b).

8.7.2 Cardiovascular Diseases

Cardiovascular disorders are known to both worsen and emerge as a result of air pollution. The risk of cardiovascular disease mortality, myocardial infarction, and congestive heart failure hospitalization has been closely linked to both short- and long-term exposure to higher concentrations of fine particle pollution. Multi-city epidemiological studies have shown regional heterogeneity in mortality and hospital admissions with respect to $PM_{2.5}$ levels (Liu et al. 2019; United States Environmental Protection Agency 2021a). In the United States, a 10 μg/m^3 increase in fine PM over 2 days was associated with a 2% increase in myocardial infarctions and hospitalizations for heart failure.

Important cardiovascular effects include:

1. Acute coronary syndrome (including myocardial infarction, unstable angina)
2. Stroke
3. Sudden cardiac death
4. Arrhythmia
5. Exacerbation of chronic heart failure

Minor increases in blood pressure have also been associated with acute exposures to PM. Studies have linked air pollution to atherosclerosis, a primary cause of ischemic heart disease (IHD) and stroke (Schraufnagel et al. 2019; United States Environmental Protection Agency 2021a), which may help to explain the wide range of cardiovascular problems that can occur from long-term exposure. The association between fine PM and ischemic mortality appears to depend on the composition and source of $PM_{2.5}$, the most harmful being those produced during coal burning. Relative risk estimates from various studies in India indicated higher risks for IHD in the northern and eastern cities having higher $PM_{2.5}$ levels (Sharma et al. 2020).

8.7.3 Neurological Disorders

Air pollution has been shown to have harmful impacts on both adults and children's central nervous systems. In elderly adults, it is linked to decreased cognitive performance and an increased risk of dementia and stroke. Inhaling PM poses a major risk to the development of the brain and causes neurodegenerative disorders (such as Alzheimer's disease) (El Morabet 2020). Many studies have found association of prenatal/early childhood exposure to fine PM, with delayed psychomotor development and lower intelligence among children (Schraufnagel et al. 2019). While long-term exposure has been linked to a reduced brain capacity, short-term exposure to fine PM increased the incidence of hospitalizations and all-cause mortality in Parkinson's disease (indicator of brain aging). PM's relationship with autism is being currently explored, however there's lack of consistency among studies. Relative risk estimates from studies conducted in India indicate higher risks for stroke in the northern and eastern cities with higher $PM_{2.5}$ levels as compared to other regions. Risk of stroke surpassed risk of all other diseases attributable to fine PM (Sharma et al. 2020).

8.7.4 Others

a. Diabetes and endocrine disorders: Increase in glycosylated hemoglobin levels and diabetes mellitus among both diabetic and non-diabetic individuals have been associated with fine PM pollution. A higher morbidity and mortality of the disease in relation to air pollution has been reported across the world, including India (Schraufnagel et al. 2019; Sharma et al. 2020). Several studies have explored the risk of metabolic syndrome in relation to PM and described increased risks among those exposed (Schraufnagel et al. 2019). Skin and eye: both these external organs are affected by air pollution. Contamination, usually from suspended PM may cause ocular irritation, tearing (acute), or dry eye (chronic), and cataract. Exposure to PM has an effect on skin aging, causing spots, and pigmentation. Pollutant exposure has been associated with skin cancer also (Manisalidis et al. 2020; Schraufnagel et al. 2019).
b. Neonatal disorders: In 2019, deaths of nearly 5 lakh infants in the first 30 days were due to air pollution. A growing body of scientific literature indicates women chronically exposed to particulate air pollution have more risks of preterm-birth, and low-birth-weight (LBW) infants. Globally, of all neonatal deaths attributable to air pollution, 64% is attributed to HAP, and the rest to $PM_{2.5}$ (Health Effects Institute 2020a; Manisalidis et al. 2020) (Table 8.1).

Table 8.1 Diseases and pathobiological processes associated with air pollution (El Morabet 2020; Manisalidis et al. 2020; Rajak and Chattopadhyay 2020; Schraufnagel et al. 2019)

Respiratory system	Cough, phlegm, difficulty breathing, bronchial hyperresponsiveness, exacerbation of respiratory conditions, allergic rhinitis, chronic laryngitis, acute and chronic bronchitis, asthma, COPD, lung cancer, deficient lung function, impeded lung development
Cardiovascular system	Ischemic heart disease, hypertension, congestive heart failure, arrhythmias, atherosclerosis, changes in heart rate, blood pressure, and vascular tone, conduction defect
Nervous system	Stroke, dementia, Parkinson's disease, impaired psychomotor development, cognitive dysfunction, stress, mood disorders, unfavorable emotional symptoms
Reproductive system	Premature birth, low-birthweight, impaired fetal growth, reduced fertility rates, miscarriage, spontaneous abortions, childhood neoplasms, and childhood asthma from intrauterine exposure
Gastrointestinal system	Appendicitis, inflammatory bowel disease, Crohn's disease, gastric cancer, colorectal cancer, hepatic steatosis, hepatocellular carcinoma
Excretory system	Decreased glomerular filtration rate, renal cancer, bladder cancer, prostate hyperplasia, increased mortality in patients undergoing dialysis
Hematology	Leukemia, intravascular coagulation, anemia, sickle cell pain crises
Endocrine and metabolic	Diabetes mellitus, insulin resistance, increased glycosylated hemoglobin, metabolic syndrome, obesity
Others	Eye: conjunctivitis, tearing (acute), dry eye (chronic), blepharitis, cataract Bone and joints: osteoporosis, rheumatic diseases, bone demineralization Sleep: Sleep apnea Overall: shortened life expectancy

References

El Morabet R. Effects of outdoor air pollution on human health. In: Nriagu J, editor. Encyclopedia of environmental health. Amsterdam: Elsevier; 2020. p. 278–86.

GBD MAPS Working Group. Burden of disease attributable to major air pollution sources in India. Special report 21. 2018. https://www.healtheffects.org/publication/gbd-air-pollution-india.

Health Effects Institute. A special report on global exposure to air pollution and its health impacts. State of global air. 2020a. https://www.stateofglobalair.org/.

Health Effects Institute. Health effects of air pollution. State of global air. 2020b. www.stateofglobalair.org.

Liu C, Chen R, Sera F, Vicedo-Cabrera AM, Guo Y, Tong S, Coelho MSZS, Saldiva PHN, Lavigne E, Matus P, Ortega NV, Garcia SO, Pascal M, Stafoggia M, Scortichini M, Hashizume M, Honda Y, Hurtado-Díaz M, Cruz J, Kan H. Ambient particulate air pollution and daily mortality in 652 cities. The New England Journal of Medicine. 2019;381(8):705–15. https://doi.org/10.1056/NEJMOA1817364.

Manisalidis I, Stavropoulou E, Stavropoulos A, Bezirtzoglou E. Environmental and health impacts of air pollution: a review. Frontiers in Public Health. 2020;8:14. https://doi.org/10.3389/FPUBH.2020.00014.

Ministry of Environment Forest and Climate Change. National clean air programme. New Delhi: Government of India; 2019. https://moef.gov.in/wp-content/uploads/2019/05/NCAP_Report.pdf.

Mukherjee A, Agrawal M. A global perspective of fine particulate matter pollution and its health effects. In: de Voogt P, editor. Reviews of environmental contamination and toxicology. Cham: Springer; 2018. p. 6–40.

Murray CJL, Aravkin AY, Zheng P, Abbafati C, Abbas KM, Abbasi-Kangevari M, Abd-Allah F, Abdelalim A, Abdollahi M, Abdollahpour I, Abegaz KH, Abolhassani H, Aboyans V, Abreu LG, Abrigo MRM, Abualhasan A, Abu-Raddad LJ, Abushouk AI, Adabi M, Lim SS. Global burden of 87 risk factors in 204 countries and territories, 1990–2019: a systematic analysis for the Global Burden of Disease Study 2019. The Lancet. 2020;396(10258):1223–49.

Pandey A, Brauer M, Cropper ML, Balakrishnan K, Mathur P, Dey S, Turkgulu B, Kumar GA, Khare M, Beig G, Gupta T, Krishnankutty RP, Causey K, Cohen AJ, Bhargava S, Aggarwal AN, Agrawal A, Awasthi S, Bennitt F, Dandona L. Health and economic impact of air pollution in the states of India: the Global Burden of Disease Study 2019. The Lancet Planetary Health. 2021;5(1):e25–38.

Rajak R, Chattopadhyay A. Short and long term exposure to ambient air pollution and impact on health in India: a systematic review. International Journal of Environmental Health Research. 2020;30(6):593–617.

Schraufnagel DE, Balmes JR, Cowl CT, De Matteis S, Jung S-H, Mortimer K, Perez-Padilla R, Rice MB, Riojas-Rodriguez H, Sood A, Thurston GD, To T, Vanker A, Wuebbles DJ. Air pollution and noncommunicable diseases: a review by the forum of international respiratory societies' environmental committee, part 2: air pollution and organ systems. Chest. 2019;155(2):417.

Sharma S, Kumar A. Air pollutant emissions scenario for India. New Delhi: The Energy and Resources Institute; 2016. https://www.toi.no/getfile.php/1348396-1530775816/Publikasjoner/Airpollutionbook.pdf.

Sharma S, Chandra M, Kota SH. Health effects associated with $PM_{2.5}$: a systematic review. Current Pollution Reports. 2020;6(4):345–67. https://doi.org/10.1007/S40726-020-00155-3.

United States Environmental Protection Agency. Particle pollution and cardiovascular effects. Washington: United States Environmental Protection Agency; 2021a. https://www.epa.gov/pmcourse/particle-pollution-and-cardiovascular-effects.

United States Environmental Protection Agency. Particle pollution and respiratory effects. Washington: United States Environmental Protection Agency; 2021b. https://www.epa.gov/particle-pollution-and-your-patients-health/health-effects-pm-patients-lung-disease.

United States Environmental Protection Agency. Particulate matter (PM) basics. Washington: United States Environmental Protection Agency; 2021c. https://www.epa.gov/pm-pollution/particulate-matter-pm-basics.

World Health Organization. SDG indicator 11.6.2 concentrations of fine particulate matter ($PM_{2.5}$). Geneva: World Health Organization; 2019. https://www.who.int/data/gho/data/indicators/indicator-details/GHO/concentrations-of-fine-particulate-matter-(pm2-5).

World Health Organization. WHO global air quality guidelines. Geneva: World Health Organization; 2021. https://www.who.int/publications/i/item/9789240034228.

Outdoor Air Quality and Assessment of Its Potential Health Risk to Human Being

9

Buddhadev Ghosh, Mallika Chowdhury, Suraj Ghosh, and Pratap Kumar Padhy

9.1 Introduction

Outdoor air pollution is one of the most prominent contributors to global disease burdens such as stroke, heart disease, lung cancer, chronic and acute respiratory diseases including chronic obstructive pulmonary disease (COPD) and asthma (WHO 2021; Turner et al. 2020; Ansari and Ehrampoush 2019). The outdoor air pollution is found in the atmosphere in various forms such as solid (TSP, SPM, PM_{10}, $PM_{2.5}$ and PM_1), gaseous (SO_X, NO_X, VOCs, CO, H_2S, O_3, NH_3, pesticides, etc.), liquid (mist, acid droplet and fog) and radiation (UV-rays). Exposure to outdoor air pollution shows health challenges worldwide because it affects everyone. WHO (2019) reported that about 99% of the world population lived in places where the WHO air quality guidelines levels were exceeded. Worldwide in 2016, exposure to outdoor air pollution in both cities and rural areas was estimated to cause 4.2 million premature deaths, where 91% of premature deaths occurred in low- and middle-income countries (South-East Asia and Western Pacific regions). The numerous epidemiological and toxicological studies reported that outdoor air pollution is significantly associated with various health effects such as allergic rhinitis, sore throat, chronic cough, bronchitis, sinusitis, chest dyspnoea, bronchial asthma, cardiovascular disease, lung cancer, Alzheimer's disease, and others like oxidative stress, change in gene expression, etc. (Dockery 2001; WHO 2003; Chen et al. 2012; Mostofsky et al. 2012; Gangwar et al. 2020). The increased risk burden on global health due to the increasing outdoor air pollution, year after year, may create health epidemics. The ambient air pollutants not only create problems in developed and developing countries but also in underdeveloped countries suffer greatly from it. The governments of countries pay additional costs every year for medical insurance (Grineski et al. 2011), disease treatment (Lim et al. 2012), cardiac rehabilitation (Rich et al. 2012), and providing free of cost medicine and run pollution abatement programs. Most countries have ambient air quality standards to monitor and track counties' air pollution status; India has National Ambient Air Quality Standards (NAAQS).

The objective of this review was to inspect the outdoor air pollution sources, categorization, their possible effects on human health and find out the health risk assessment of pollutants. The health risk assessment is technique or process to estimate the past, present and future risk of exposure to air pollutants. The air pollution-related health risk assessment may be quantitative or qualitative depending on the various factors, i.e., pollutant concentrations, route of

B. Ghosh · M. Chowdhury · S. Ghosh · P. K. Padhy (✉)
Department of Environmental Studies, Visva-Bharati University, Santiniketan, West Bengal, India
e-mail: pkpadhy@visva-bharati.ac.in

P. K. Padhy et al. (eds.), *Air Quality and Human Health*,
https://doi.org/10.1007/978-981-97-1363-9_9

exposure, targeted population, body weight, exposure duration, etc.

9.2 Characterization of Air Pollutants, Sources and Their Possible Health Effects

The pollutant types and their sources play a critical role towards human health. The different kinds of pollutants show various health complications, i.e., respiratory illness, cardiac, haematological and neurological disorders. Table 9.1 shows the pollutants category, their sources and health effects outcomes. The effects of outdoor air pollution on different exposer groups, including adult male and female, children, and pregnant women, are shown in Fig. 9.1. Text mining data are used to express the frequency of potential health outcomes caused by outdoor air pollution effects in Fig. 9.2. This method helps to identify the highest health outcome due to outdoor air pollution exposer.

Table 9.1 Major outdoor air pollutants categories, their sources and health effects

Category	Pollutants	Source	Health effects	References
Gaseous	NO_2	Vehicular emissions, industrial emissions, power plants and the burning of solid biofuels in domestic purposes	Paediatric and adolescent asthma, COPD, stroke, coughing, wheezing or difficulty in breathing and other respiratory disease	Khatri et al. (2021), Al-Hemoud et al. (2021), U.S.EPA (2021a), Chowdhury et al. (2021), Anenberg et al. (2018)
	SO_2	Power plants and industrial activity are the primary sources of SO_2 generation by burning of sulphur-containing fossil fuels. Other sources are volcanos, locomotives, ships, and heavy equipment that high sulphur content burn fuel	Asthma, breathing discomfort, unspecified mental disorders complicating pregnancy, COPD, acute myocardial infarction (AMI), etc.	U.S.EPA (2021b), Kanner et al. (2021), Anenberg et al. (2018), Karambelas et al. (2018)
	CO	Incomplete combustion of fossil fuels, domestic cooking, industrial processing and transportation sectors are the key contributor of CO emissions	Dizziness, confusion, unconsciousness, angina, and high dose cause death, reducing the oxygen transport ability of blood and affecting the cardiovascular system	U.S.EPA (2021c), Peng et al. (2021), Kampa and Castanas (2008), Pénard-Morand and Annesi-Maesano (2004)
	VOCs	VOCs release from both natural (vegetation, biogenic sources) and manmade sources like petroleum, chemical industries and transportation sectors. VOCs emissions from solvents, cleansers, aerosol spray and paints, as well as from the burning of coal, oil and petrol are also the key contributors	VOCs are considered as human carcinogens; they impair lung function, exacerbate asthma, cause neurological disorders, have an adverse influence on reproduction and development, and harm the liver, kidneys, and central nervous system. Short-term effects include conjunctival irritation, pain in the nose and throat, headache, allergic skin reaction, breathing difficulties, decreases in serum cholinesterase levels, nausea, emesis, epistaxis, lethargy, and dizziness	U.S.EPA (2021e), Zheng et al. (2020), Soni et al. (2018), Pénard-Morand and Annesi-Maesano (2004)

Category	Pollutants	Source	Health effects	References
	NH_3	Major sources of NH_3 are agriculture such as animal husbandry (poultry), and fertilizer applications. Industrial operations, vehicle emissions, burning biomass, landfills, and volatilization from soils and seas are other sources of NH_3	The main consequences of NH_3 are irritation of the eyes, nose, and throat, swelling, coughing, wheezing, and constriction of the bronchi, which causes low blood oxygen levels, changed mental state, shortness of breath, acute and chronic respiratory illnesses, and cardiovascular systems	Ma et al. (2021), Younis and Salem (2020), Phillips et al. (2019), Xu et al. (2018), Bray et al. (2018), CDC (2017), Behera et al. (2013)
	H_2S	Natural gas, hot springs, and crude oil are the main H_2S sources. Hydrogen sulphide (H_2S) is naturally produced when bacteria break down organic materials and algae on the ocean's surface. The drilling and refinement of oil and gas, wastewater treatment, coke ovens, tanneries, kraft paper mills and liquid compressed gas are also the sources of H_2S	H_2S inhibits the cytochrome oxidase enzyme system, causing an acid-base imbalance, damage to the neurological system, and death of cardiac and vascular tissues. Skin, eyes, mucous membranes, and respiratory system irritation are additional consequences. H_2S causes quick loss of consciousness, respiratory paralysis, convulsions, and death at high doses	Ulutaş et al. (2021), Bertelsmann et al. (2019), Saeedi et al. (2015), CDC (2014), Doujaiji and Al-Tawfiq (2010), OSHA (2005)
	Tropospheric ozone (O_3)	Oxides of nitrogen (NOx) and volatile organic compounds (VOCs) react chemically with sunlight to generate O_3, a secondary pollutant	Asthma, COPD, oxidative stress, inflammation, and imbalance heart rate and diastolic pressure and myocardial infarction, O_3 induce oxidative damage on cells and the lining fluids of the airways, and immune-inflammatory responses into the lungs	Sahu et al. (2021), U.S.EPA (2021f), Zhang et al. (2019a, b), Liu et al. (2018a, b), Anenberg et al. (2018), Nuvolone et al. (2018)
	PAN	In dry, sunny circumstances, photochemical reactions involving nitrogen oxides (NOx), oxygen (O_3), and volatile organic compounds (VOCs) result in the formation of peroxyacetyl nitrate (PAN), a secondary air pollutant	Lachrymator (watering of the eyes), skin cancer and not much direct effects on human health	Zhang et al. (2015), Movassaghi et al. (2012)

(continued)

Table 9.1 (continued)

Category	Pollutants	Source	Health effects	References
	Pesticides	Agricultural sectors are the main source of the many types of pesticides found in ambient air, such as carbamates, phosphates, and organochlorines	Reproductive problems such as preterm birth, foetal development limitation, neural tube disorders, hypospadias, gastroschisis and anotia, child leukaemia and lymphomas, breast and prostate cancer, autism spectrum disorders, declined IQ level, Parkinson's disease, acute and chronic respiratory effects (child asthma), decrease nerve conduction velocity, heart defects and offspring's increased risk of anencephaly, spina bifida, cleft lip with or without or only	Benka-Coker et al. (2020), Dereumeaux et al. (2020), Raherison et al. (2019), Shaw et al. (2018), Moisan et al. (2015), Carmichael et al. (2014), Yang et al. (2014), Cockburn et al. (2011), Kampa and Castanas (2008)
	PAHs	Burning of fossil fuels, vehicular emissions and incomplete combustion of organic substances are the major anthropogenic sources of PAHs and forest fire, natural combustion vents and volcanos are the main natural source of PAHs	According to USEPA, considered some PAHs are carcinogenic and teratogenic (Benzo (a) Pyrene). PAHs may have genotoxic effects on DNA and interfere with protein function	Ambade et al. (2021), Tarafdar and Sinha (2019), Zheng et al. (2019), Wang et al. (2017), Dat and Chang (2017), Kim et al. (2013), Yan et al. (2019)
Solid	$PM_{(10,\ 2.5\&0.1)}$	The PM comes from both anthropogenic (human) activities such as agricultural activity, industrial processes, burning of wood and fossil fuels (vehicular emissions), construction and demolition activities, and road network dust into the air. Typical examples of natural sources of PM include windblown and suspended terrestrial dust, oceans and seas salt, volcanic eruption, forest fire or wildfires, and volcanic eruption	PM shows eye irritation, respiratory disorder (nose block, sneezing, coughing, hyperacidity, sinusitis, bronchitis, asthma), lung inflammation and injury, reduced lung function capacity, pneumonia, infant death syndrome and cardiovascular diseases. Fine particles may cause alveolar inflammation, reduced heart rate, mild hypoxemia or decline in blood oxygen saturation, increased symptoms of obstructive airway disease (such as COPD, bronchitis, and chest illness). Ultrafine particles can penetrate blood barriers and affect the vascular system as inflammation, oxidative stress, cellular damage, mitochondrial damage, lipid peroxidation, and DNA oxidation	Wang et al. (2021), Wu et al. (2021), Kim et al. (2021), Schraufnagel (2020), CDC (2019), Anenberg et al. (2018), Weichenthal et al. (2016), Yan et al. (2019), Kim and Criner (2013)
	Pb	Ore and metals processing industry, piston-engine machines, waste incinerators, utilities, lead smelters and lead-acid battery manufacturers are the sources of lead	Nervous system, kidney function, immune system, reproductive and developmental systems, effects on haem synthesis and other biochemical processes, lowering IQ, attention deficit hyperactivity disorder (ADHD), learning difficulties, oppositional/conduct disorders, and delinquency	U.S.EPA (2021d), Taylor et al. (2019), Mayans (2019), Pénard-Morand and Annesi-Maesano (2004), Tong et al. (2000)

Category	Pollutants	Source	Health effects	References
Radiation	UV-ray	UV radiation is naturally produced by the sun, and some man-made sources include tanning beds, mercury vapour lighting (used in sports arenas and school gymnasiums), halogen, fluorescent, and certain kinds of lasers	The UV light are classified as UV-C, UV-B and UV-A. In these categories UV-C and B are more dangers. UV radiation causes cataracts (eye damage), immune system suppression, and skin malignancies including melanoma, basal cell carcinoma, and squamous cell carcinoma. On the other hand, UV light generates ROS	U.S.EPA (2022), CDC (2021), Ahmad (2017)
Liquid	Mist (sulfuric acid, nitic acid and other inorganic mist, oil, and metal working fluids mist	Metal pickling, sulfuric acid production, isopropanol production, soap and detergent manufacturing, nitric acid production, ethanol production, copper and zinc refining, phosphate fertilizer production, and lead battery manufacture are the primary sources of sulfuric acid mist. In manufacturing facilities where significant volumes of oil are used to lubricate and cool cutting tools and workpieces, oil mist is produced	Sulfuric acid is a category 1 carcinogen according to the International Agency for Research on Cancer (IARC). Laryngeal cancer risk is raised by strong inorganic acid, which also damages the eyes and causes severe skin burns. Impacts of oil mist on health include skin irritation, laryngeal cancer, asthma, bronchial hyperresponsiveness, congenital heart abnormalities, and impacts on reproductive health in industrial workers	Iwasaki et al. (2020), Siegel et al. (2019), Yang et al. (2019), Semanová and Kučera (2014), McGinnity and Nicol (2014)

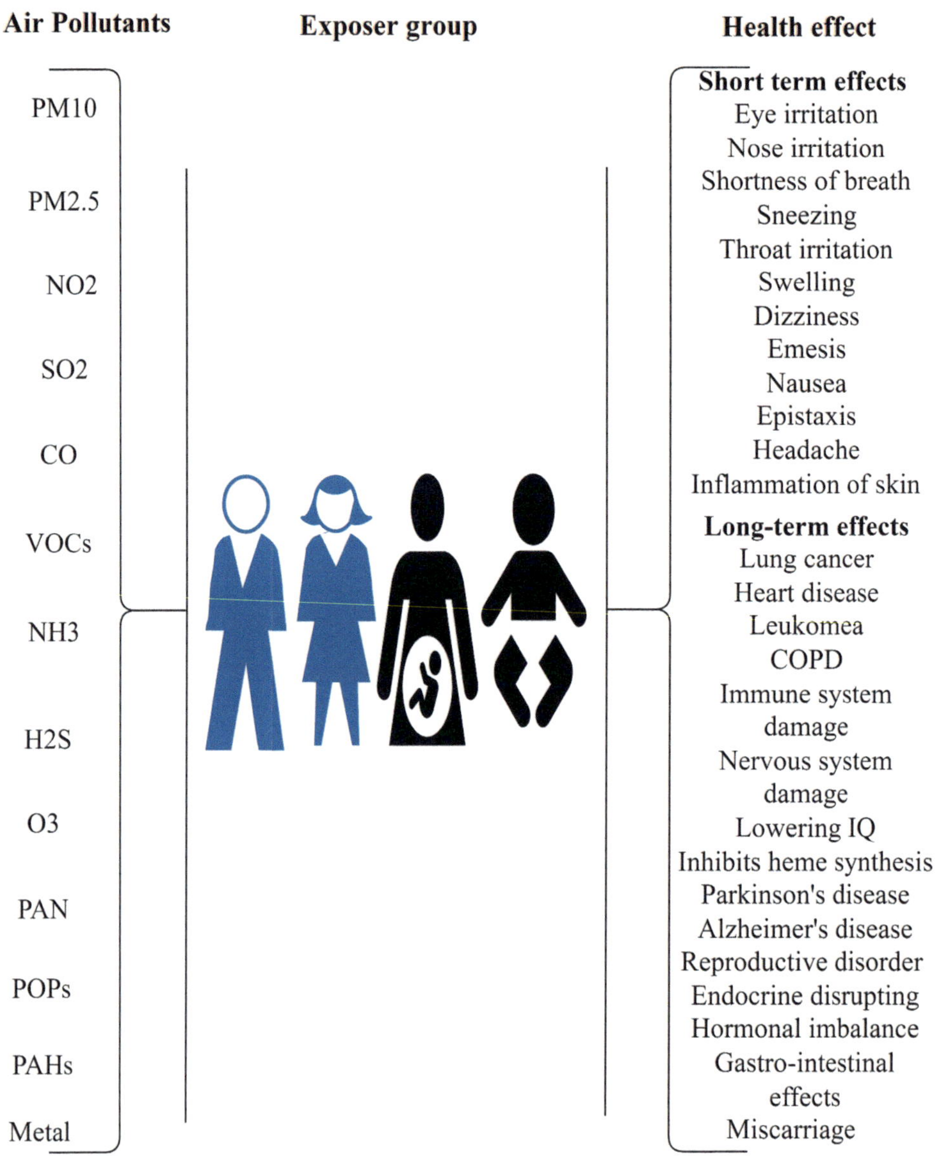

Fig. 9.1 General consequences of short- and long-term health effects due to exposure of outdoor air pollution

Fig. 9.2 Text mining data of possible health outcomes frequency of outdoor air pollution effects

9.3 Health Risk Assessment

Health risk assessments are methods or techniques to estimate the past, present and future risk of exposure to air pollutants in such concentrations. The air pollution-related health risk assessment may be quantitative or qualitative depending on the air pollution load, route of exposure of the targeted population, body weight, exposure duration, age group, reference value, and types of pollutants. There are three primary routes of exposure pathways to pollutants in the human body, i.e., inhalation, ingestion and contact (dermal). The degree of air pollutants exposure is quantified by measuring the amount of the hazardous substance at the point of contact. Risk assessments can provide information about non-carcinogenic and carcinogenic effects on human health. The non-carcinogenic and carcinogenic health effects are assessed by using various formulas or statistical equations (USEPA 2015).

9.3.1 Non-carcinogenic Health Risk Assessment

9.3.1.1 Exposure Risk Assessment

The three main pathways humans are exposed to pollution are ingestion, inhalation, and dermal contact (Zhang et al. 2019a, b; Mundackal and Ngole-Jeme 2022). In order to evaluate the risk of exposure to pollutants, the average daily dosage (ADD) was calculated from the following Eqs. (9.1)–(9.3) by using the US Environmental Protection Agency's health risk handbook guide (U.S.EPA 2015).

$$\mathrm{ADD}(\mathrm{Ing}) = \frac{C \times \mathrm{Ing}R \times \mathrm{EF} \times \mathrm{ED}}{\mathrm{BW} \times \mathrm{AT}} \quad (9.1)$$

$$\text{ADD}(\text{Inh}) = \frac{C \times \text{ET} \times \text{InhR} \times \text{EF} \times \text{ED}}{\text{BW} \times \text{AT} \times \text{PEF}} \tag{9.2}$$

$$\text{ADD}(\text{Dar}) = \frac{C \times \text{Sa} \times \text{SAF} \times \text{Dabs} \times \text{EF} \times \text{ED}}{\text{BW} \times \text{AT}} \tag{9.3}$$

where C is the concentrations of the pollutants (in mg/kg or μg/m^3 or mg/m^3 or μg/L), IngR is the ingestion rate in mg/day for the average daily dosage of ingestion of contaminants, InhR is the inhalation rate in m^3/day (0.79 m^3/day for males and 0.59 m^3/day for females) for an average daily dosage of inhalation of contaminants, Dabs is the dermal absorption factor for an average daily dosage of dermal contact of contaminants, ET is the exposure time in hours/days, PEF is the particle emission factor in m^3/kg, Sa is the surface area of the skin exposed to contaminants in cm^2, SAF is the skin adherence factor in mg/cm^2, ED is the exposure duration in years, EF is the exposure frequency in days/year, AT is the exposer average time in days, and BW is the average body weights of the exposed group in kg.

9.3.1.2 Hazard Quotient or Ratio (HQ or HR)

A hazard quotient or ratio used for non-carcinogenic risk estimation is the ratio of the potential exposure to pollutants like PM-containing metals, BTXS (benzene, toluene, xylenes, and styrene), ambient air pollutants, and other contaminants; the values of HQ (HQ > 1 or <1) show the effectiveness of the substance (Wu et al. 2019; Tohid et al. 2019; Di Vaio et al. 2018). An HQ < 1 indicates no effects, whereas the value of HQ > 1 shows a greater chance of non-carcinogenic effects. The following Eqs. (9.4) and (9.5) help to calculate the HQ of the contaminants.

$$\text{HQ}(\mu g/\text{mg}) = \frac{\text{EC}}{\text{RfC}} \times 1000 \tag{9.4}$$

$$\text{HQ}(\mu g/\text{kg}/\text{day}) = \frac{\text{ADD}}{\text{RfC}} \tag{9.5}$$

$$\text{EC} = \frac{\text{CA} \times \text{ET} \times \text{EF} \times \text{ED}}{\text{AT}} \tag{9.6}$$

where HQ is the hazard quotient, EC is the exposure concentration (μg m^{-3}), RfC = inhalation reference concentration of pollutants (mg/m^3 or ug/m^3; Table 9.2), CA is the concentration of pollut-

Table 9.2 Shows inhalation unit risk and reference concentration of different pollutants (obtained from: U.S.EPA 2017; Tohid et al. 2019; Wu et al. 2019; RAIS 2020; Mundackal and Ngole-Jeme 2022)

Pollutants		IARC category	IUR (m^3/ug)	RfC (mg/m^3)
Metals	Al		–	5.00×10^{-3}
	As	Gr-1	4.30×10^{-3}	1.50×10^{-5}
	Cd	Gr-1	1.80×10^{-3}	1.00×10^{-5}
	Co		9.00×10^{-3}	6.00×10^{-6}
	Cr (VI)	Gr-1	–	1.40×10^{-4}
	Mn		–	5.00×10^{-5}
	Ni	Gr-1	2.60×10^{-4}	5.00×10^{-5}
	V		–	1.00×10^{-5}
Gas	CO		–	9.00×10^{-3}
	CO_2		–	9.00
	NO_2		–	4.00×10^{-2}
	SO_2		–	5.00×10^{-2}
	O_3		–	6.00×10^{-5}
	H_2S		–	2.00×10^{-3}
	TVOC		–	5.60×10^{-4}
VOCs	Benzene	Gr-1	6.00×10^{-6}	3.00×10^{-2}
	Toluene	Gr-3	–	5.00
	Xylenes	Gr-3	–	1.00×10^{-1}
	Styrene	Gr-2B	–	1.00
PAHs	Benzo[a] pyrene	Gr-1	6.00×10^{-4}	2.00×10^{-6}

IARC International Agency for Research on Cancer, *CO, CO_2, H_2S, NO_2, SO_2 and O_3* non-carcinogenic pollutants

ants (μg/m^3 or mg/m^3 or other units), ET is the exposure time (24 h/day), EF is the exposure frequency (assumed to be 350 days/year), ED is the exposure duration (for children 6 years and for adults 24 years), AT is the averaging time (ED × 365 days/year × 24 h/day for non-carcinogens and 70 years × 365 days/year × 24 h/day for carcinogens). These values depend on the targeted specific to calculating exposer risk of air pollutants (Ghosh et al. 2023).

9.3.1.3 Hazard Index (HI)

Hazard Index (HI) is the total of the hazard quotients (HQ) used to determine the health risks associated with various substances and/or exposure routes as well as the HQ used for a single substance.

$$\mathrm{HI} = \sum\nolimits_1^i HQi \qquad (9.7)$$

9.3.2 Carcinogenic Health Risk Assessment

The Carcinogenic Risk (CR) or Lifetime Cancer Risk (LCR) or Incremental Lifetime Cancer Risk (ILCR, likelihood of developing cancer of continuous exposure of contaminants during a 70-year lifetime) can be estimated with the following Eq. (9.8) through the inhalation pathway exposer (Ghosh et al. 2023; Alsafran et al. 2021).

$$\mathrm{ILCR\ or\ CR} = \mathrm{IUR} \times \mathrm{EC} \qquad (9.8)$$

where IUR is inhalation unit risk (m^3/μg; Table 9.2), EC is exposure concentration (μg m^{-3}). The tolerable risk for regulatory purposes ranges from 10^{-6} to 10^{-4} (Ferreira-Baptista and De Miguel 2005).

9.4 Health Risk Outcomes

According to Wu et al. (2019), who examined the health hazards associated with heavy metals in $PM_{2.5}$ in Ningbo, China, they observed that metals such as Mn > As > Ni > Cd > Cr (VI) > Se have no non-carcinogenic risks. For every single metal had HQ values that were lower than safe values (HQ <1) but HI values exceeded the safe levels (HI >1). The risk of cancer from Cr (VI) and As is increased yet within tolerable bounds. Children who live in Hubei Province, a fourth-tier city, maybe in danger of metal exposure, according to Zhang et al.'s 2019 study. They find out the higher concentrations of Pb, Cd and Mn in children's urine samples than in other residents (a nearby city of sampling sites). They also found that inhalation exposure is the main route of exposure pathway of metal(loid)s via aerosol particles than another route of exposure. The non-carcinogenic and carcinogenic health risk of local children of Mn exposure was much higher than the acceptable level. Tohid et al. (2019), studied health risk of ambient air VOCs mainly BTXS in an industrialized city in Iran and they find out the non-carcinogenic and carcinogenic health risk of BTXS. The mean non-carcinogenic risk (HR) for benzene (1.07×10^{-1}), toluene (1.29×10^{-3}), xylene (5.38×10^{-2}) and styrene (8.10×10^{-4}) were in acceptable range but carcinogenic (ILCR) health risk of benzene (7.82×10^{-6}) was higher than the threshold values established by the USEPA and WHO. Another study from Ahvaz, southwest Iran, the inhalation and dermal exposures carcinogenic risks of metals such as Cr and Pb in PM_{10} for children and adults exceeded (>1×10^{-4}) in the study areas (Goudarzi et al. 2018). Garg and Gupta (2019), published a paper on health risk assessment of BTEX emissions in ambient air of Delhi, India which was the same type of work as Tohid et al. (2019). Here they also find out the health risk assessment, the significant potential cancer risks observed by inhalation of benzene which exceeded the acceptable risk. In addition, non-cancer risks (HQ and HI) were less than their acceptable of their studied sites. The non-carcinogenic (HQ) health risk of total metal concentration of Co, Ni, Mn, and Cd exceeded. On the other hand, Cr, Ni, and Cd exceeded the carcinogenic unit risk for the residents of Pune City, India (Jan et al. 2018). The possible health risks of seven heavy metals (Mn, Zn, Cd, Pb, Cu, Ni, and Co) in PM_{10} on children and adults were studied by Ghosh et al. (2018) in semi-urban, urban, and industrial areas of West Bengal, India.

Although they found that the pollutants exposure from ingestion, inhalation, and skin contact were negligible, the pollutants might provide a greater risk to children than adults in the future. Ansari and Ehrampoush used the AirQ$^+$ programme, a WHO-developed tool for assessing the health risks associated with air pollution, in a study they published in a peer-reviewed publication in 2019. The two main inputs used to determine the mortality and morbidity relationships with pollutants are the local baseline incidence (BI) and non-local RR data. They observed that in Tehran, Iran (2017–2018), 6710, 27, 172, 3797, 135 and 1145 people died as a result of natural causes (adults 30+ years), acute lower respiratory infections (ALRI), children 0–5 years old, chronic obstructive pulmonary disease (COPD), ischemic heart disease (IHD), lung cancer (LC) adult 30+ years, and stroke (adults 25+ years). Karambelas et al. (2018) studied on health impacts attributable to $PM_{2.5}$ and O_3 in rural and urban area in northern India and interestingly they observed the majority of premature deaths resulting from $PM_{2.5}$ and O_3 are in rural (383,600) as compared to urban (117,200). An estimated 463,200 persons in urban and rural regions each year died prematurely from $PM_{2.5}$ and 37,800 adults per year passed away from O_3. The other study of ambient BTX health risk assessment in urban and rural settings of a terai region of northern India discovered that the integrated lifetime cancer risk (ILTCR) for benzene was higher than the threshold value 1×10^{-6} where the non-cancer hazard quotient (HQ) for BTX was less than unity (Masih et al. 2018). According to Kumar et al. (2020), in Jamshedpur, India, semi-urban, urban, and rural sites had a carcinogenicity contribution of 60.23%, 57.97%, and 55.45% from atmospheric $PM_{2.5}$-bound polycyclic aromatic hydrocarbons (BaP). According to Boogaard et al. (2019) findings, solid fuel burning indoors and outdoors was the fifth leading cause of death worldwide in 2017 after diet, high blood pressure, tobacco use, and high fasting blood glucose. Age-standardized mortality rates linked to air pollution were highest in East and South Asia and sub-Saharan Africa and lowest in the high-income Asia-Pacific region, North America, and Western Europe. With 2.9 million fatalities (5.2% of all deaths worldwide) attributable to $PM_{2.5}$, which was the eighth-leading risk factor globally, cardiovascular disease and tropospheric ozone each caused an additional 472,000 deaths from chronic respiratory diseases. Kumar and Mishra (2018) conducted a study on human health risks associated with air pollutants along transport corridors in Delhi, India. The study revealed an estimated excess of 281,107 cases of respiratory morbidity, specifically COPD-related hospital admissions and 71,608 cases of cardiovascular morbidity-related hospital admissions across the transport corridors. However, Central Delhi recorded the highest number of cases of respiratory mortality (1948), cardiovascular mortality (7455), total mortality (11,373), respiratory morbidity (COPD hospital admission, 101,421), and cardiovascular morbidity (25,460). On the other hand, North Delhi, West Delhi, and East Delhi districts recorded the least number of excess cases for all health risk parameters.

9.5 Conclusions

One of the biggest threats to human health is outdoor air pollution, which has been linked to a number of illnesses that have high mortality rates and morbidity rates globally. The rapid industrialization, urbanization, transportation, extensive agriculture and heavy fossil fuel-based power consumptions are the key contributor to the increase in outdoor air pollution levels globally. The various complicated diseases such as neurological disorder, damage in cognitive function, blood clotting in the artery, lung function alteration, etc. are the leading cause due to air pollution. The study of air pollution effects and their health risk assessment is a key factor in monitoring the possible health risk outcomes. On the basis of health risk assessment of various pollutants on human health to find out the possible carcinogenic and non-carcinogenic risk. On the other hand, pollutants also categorize on the basis of carcinogenicity and non-carcinogenicity for example Cr (VI), As, Cd and benzene are considered as group 1 carcinogen, and CO_2, SO_2 and NO_2 considered as non-carcinogens. The policy makers and legislators must update all laws and regulations related to air pollution. Still now more research is needed to outdoor air pollution and health outcomes study to assure the how, when, where, and why pollutants effects on human beings. A purposeful effort may be made to raise awareness of the harmful effects of outdoor air pollution through educational institutions, the media, and other organizations.

References

Ahmad SI, editor. Ultraviolet light in human health, diseases and environment, vol. 996. Cham: Springer; 2017.

Al-Hemoud A, Gasana J, Alajeel A, Alhamoud E, Al-Shatti A, Al-Khayat A. Ambient exposure of O_3 and NO_2 and associated health risk in Kuwait. Environ Sci Pollut Res. 2021;28(12):14917–26.

Alsafran M, Usman K, Rizwan M, Ahmed T, Al Jabri H. The carcinogenic and non-carcinogenic health risks of metal (oid) s bioaccumulation in leafy vegetables: a consumption advisory. Front Environ Sci. 2021;380:742269.

Ambade B, Sethi SS, Kumar A, Sankar TK, Kurwadkar S. Health risk assessment, composition, and distribution of polycyclic aromatic hydrocarbons (PAHs) in drinking water of Southern Jharkhand, East India. Arch Environ Contam Toxicol. 2021;80(1):120–33.

Anenberg SC, Henze DK, Tinney V, Kinney PL, Raich W, Fann N, Kuylenstierna JC. Estimates of the global burden of ambient $PM_{2.5}$, ozone, and NO_2 on asthma incidence and emergency room visits. Environ Health Perspect. 2018;126(10):107004.

Ansari M, Ehrampoush MH. Meteorological correlates and AirQ$^+$ health risk assessment of ambient fine particulate matter in Tehran, Iran. Environ Res. 2019;170:141–50.

Behera SN, Sharma M, Aneja VP, Balasubramanian R. Ammonia in the atmosphere: a review on emission sources, atmospheric chemistry and deposition on terrestrial bodies. Environ Sci Pollut Res. 2013;20(11):8092–131.

Benka-Coker W, Hoskovec L, Severson R, Balmes J, Wilson A, Magzamen S. The joint effect of ambient air pollution and agricultural pesticide exposures on lung function among children with asthma. Environ Res. 2020;190:109903.

Bertelsmann A, Knight G, Tiwary A, Calico A. PHA guidance for correlating H2S concentrations in process streams to severity of adverse health outcomes in the event of a leak. J Loss Prev Process Ind. 2019;60:282–7.

Boogaard H, Walker K, Cohen AJ. Air pollution: the emergence of a major global health risk factor. Int Health. 2019;11(6):417–21.

Bray CD, Battye W, Aneja VP, Tong DQ, Lee P, Tang Y. Ammonia emissions from biomass burning in the continental United States. Atmos Environ. 2018;187:50–61.

Carmichael SL, Yang W, Roberts E, Kegley SE, Padula AM, English PB, Shaw GM. Residential agricultural pesticide exposures and risk of selected congenital heart defects among offspring in the San Joaquin Valley of California. Environ Res. 2014;135:133–8.

CDC. 2014. https://www.atsdr.cdc.gov/mhmi/mmg114.pdf. Accessed 16 August 2022.

CDC. 2017. https://www.atsdr.cdc.gov/mhmi/mmg126.pdf. Accessed 16 August 2022.

CDC. 2019. Centers for Disease Control and Prevention. https://www.cdc.gov/air/particulate_matter.html. Accessed 26 August 2022.

CDC. 2021. https://www.cdc.gov/nceh/features/uv-radiation-safety/index.html#:~:text=UV%20exposure%20increases%20the%20risk,cancer%20in%20the%20United%20States. Accessed 26 August 2022.

Chen R, Kan H, Chen B, Huang W, Bai Z, Song G, Pan G. Association of particulate air pollution with daily mortality: the China Air Pollution and Health Effects Study. Am J Epidemiol. 2012;175(11):1173–81.

Chowdhury S, Haines A, Klingmüller K, Kumar V, Pozzer A, Venkataraman C, Lelieveld J. Global and national assessment of the incidence of asthma in children and adolescents from major sources of ambient NO_2. Environ Res Lett. 2021;16(3):035020.

Cockburn M, Mills P, Zhang X, Zadnick J, Goldberg D, Ritz B. Prostate cancer and ambient pesticide exposure in agriculturally intensive areas in California. Am J Epidemiol. 2011;173(11):1280–8.

Dat ND, Chang MB. Review on characteristics of PAHs in atmosphere, anthropogenic sources and control technologies. Sci Total Environ. 2017;609:682–93.

Dereumeaux C, Fillol C, Quénel P, Denys S. Pesticide exposures for residents living close to agricultural lands: a review. Environ Int. 2020;134:105210.

Di Vaio P, Magli E, Caliendo G, Corvino A, Fiorino F, Frecentese F, Perissutti E. Heavy metals size distribution in PM10 and environmental-sanitary risk analysis in Acerra (Italy). Atmos. 2018;9(2):58.

Dockery DW. Epidemiologic evidence of cardiovascular effects of particulate air pollution. Environ Health Perspect. 2001;109(4):483–6.

Doujaiji B, Al-Tawfiq JA. Hydrogen sulfide exposure in an adult male. Ann Saudi Med. 2010;30(1):76–80.

Ferreira-Baptista L, De Miguel E. Geochemistry and risk assessment of street dust in Luanda, Angola: a tropical urban environment. Atmos Environ. 2005;39(25):4501–12.

Gangwar RS, Bevan GH, Palanivel R, Das L, Rajagopalan S. Oxidative stress pathways of air pollution mediated toxicity: recent insights. Redox Biol. 2020;34:101545.

Garg A, Gupta NC. A comprehensive study on spatio-temporal distribution, health risk assessment and ozone formation potential of BTEX emissions in ambient air of Delhi, India. Sci Total Environ. 2019;659: 1090–9.

Ghosh S, Rabha R, Chowdhury M, Padhy PK. Source and chemical species characterization of PM10 and human health risk assessment of semi-urban, urban and industrial areas of West Bengal, India. Chemosphere. 2018;207:626–36.

Ghosh B, De M, Rout TK, Padhy PK. Study on spatio-temporal distribution and health risk assessment of BTEX in urban ambient air of Kolkata and Howrah, West Bengal, India: evaluation of carcinogenic, non-carcinogenic and additional leukaemia cases. Atmos Pollut Res. 2023;14(10):101878.

Goudarzi G, Alavi N, Geravandi S, Idani E, Behrooz HRA, Babaei AA, Mohammadi MJ. Health risk assessment on human exposed to heavy metals in the ambient air PM10 in Ahvaz, southwest Iran. Int J Biometeorol. 2018;62(6):1075–83.

Grineski SE, Staniswalis JG, Bulathsinhala P, Peng Y, Gill TE. Hospital admissions for asthma and acute bronchitis in El Paso, Texas: do age, sex, and insurance status modify the effects of dust and low wind events? Environ Res. 2011;111(8):1148–55.

Iwasaki M, Hirai K, Fukumori K, Higashi H, Inomata Y, Seto T. Characterization of submicron oil mist particles generated by metal machining processes. Aerosol Air Qual Res. 2020;20(6):1469–79.

Jan R, Roy R, Yadav S, Satsangi PG. Chemical fractionation and health risk assessment of particulate matter-bound metals in Pune, India. Environ Geochem Health. 2018;40(1):255–70.

Kampa M, Castanas E. Human health effects of air pollution. Environ Pollut. 2008;151(2):362–7.

Kanner J, Pollack AZ, Ranasinghe S, Stevens DR, Nobles C, Rohn MCH, Sherman S, Mendola P. Chronic exposure to air pollution and risk of mental health disorders complicating pregnancy. Environ Res. 2021;196:110937.

Karambelas A, Holloway T, Kinney PL, Fiore AM, DeFries R, Kiesewetter G, Heyes C. Urban versus rural health impacts attributable to $PM_{2.5}$ and O_3 in northern India. Environ Res Lett. 2018;13(6): 064010.

Khatri SB, Newman C, Hammel JP, Dey T, Van Laere JJ, Ross KA, Norris GA. Associations of air pollution and pediatric asthma in Cleveland, Ohio. Sci World J. 2021;2021:8881390.

Kim V, Criner GJ. Chronic bronchitis and chronic obstructive pulmonary disease. Am J Respir Crit Care Med. 2013;187(3):228–37.

Kim KH, Jahan SA, Kabir E, Brown RJ. A review of airborne polycyclic aromatic hydrocarbons (PAHs) and their human health effects. Environ Int. 2013;60:71–80.

Kim HJ, Herath KHINM, Dinh DTT, Kim HS, Jeon YJ, Kim HJ, Jee Y. Sargassum horneri ethanol extract containing polyphenols attenuates PM-induced oxidative stress via ROS scavenging and transition metal chelation. J Funct Foods. 2021;79:104401.

Kumar A, Mishra RK. Human health risk assessment of major air pollutants at transport corridors of Delhi, India. J Transp Health. 2018;10:132–43.

Kumar A, Ambade B, Sankar TK, Sethi SS, Kurwadkar S. Source identification and health risk assessment of atmospheric PM2. 5-bound polycyclic aromatic hydrocarbons in Jamshedpur, India. Sustain Cities Soc. 2020;52:101801.

Lim SS, Vos T, Flaxman AD, Danaei G, Shibuya K, Adair-Rohani H, Pelizzari PM. A comparative risk assessment of burden of disease and injury attributable to 67 risk factors and risk factor clusters in 21 regions, 1990–2010: a systematic analysis for the Global Burden of Disease Study 2010. Lancet. 2012;380(9859):2224–60.

Liu H, Liu S, Xue B, Lv Z, Meng Z, Yang X, He K. Ground-level ozone pollution and its health impacts in China. Atmos Environ. 2018a;173:223–30.

Liu Y, Yu Y, Liu M, Lu M, Ge R, Li S, Qadeer A. Characterization and source identification of $PM_{2.5}$-bound polycyclic aromatic hydrocarbons (PAHs) in different seasons from Shanghai, China. Sci Total Environ. 2018b;644:725–35.

Ma R, Li K, Guo Y, Zhang B, Zhao X, Linder S, Meng J. Mitigation potential of global ammonia emissions and related health impacts in the trade network. Nat Commun. 2021;12(1):1–13.

Masih A, Lall AS, Taneja A, Singhvi R. Exposure levels and health risk assessment of ambient BTX at urban and rural environments of a terai region of northern India. Environ Pollut. 2018;242:1678–83.

Mayans L. Lead poisoning in children. Am Fam Physician. 2019;100(1):24–30.

McGinnity JJ, Nicol MJ. Sulfuric acid mist: Generation, suppression, health aspects, and analysis. Miner Process Extr Metall Rev. 2014;35(3):149–92.

Moisan F, Spinosi J, Delabre L, Gourlet V, Mazurie JL, Bénatru I, Elbaz A. Association of Parkinson's disease and its subtypes with agricultural pesticide exposures in men: a case–control study in France. Environ Health Perspect. 2015;123(11):1123–9.

Mostofsky E, Schwartz J, Coull BA, Koutrakis P, Wellenius GA, Suh HH, Mittleman MA. Modeling the association between particle constituents of air pollution and health outcomes. Am J Epidemiol. 2012;176(4):317–26.

Movassaghi K, Russo MV, Avino P. The determination and role of peroxyacetil nitrate in photochemical processes in atmosphere. Chem Cent J. 2012;6(2):1–8.

Mundackal A, Ngole-Jeme VM. Evaluation of indoor and outdoor air quality in university academic buildings and associated health risk. Int J Environ Health Res. 2022;32(5):1076–94.

Nuvolone D, Petri D, Voller F. The effects of ozone on human health. Environ Sci Pollut Res. 2018;25(9):8074–88.

OSHA. 2005. https://www.osha.gov/sites/default/files/publications/hydrogen_sulfide_fact.pdf. Accessed 27 August 2022.

Pénard-Morand C, Annesi-Maesano I. Air pollution: from sources of emissions to health effects. Breathe. 2004;1(2):108–19.
Peng W, Liu X, Taghizadeh-Hesary F. Carbon monoxide emission and air quality analysis based on an improved double-weight fuzzy comprehensive evaluation. Front Public Health. 2021;9:790383.
Phillips FA, Naylor T, Forehead H, Griffith DW, Kirkwood J, Paton-Walsh C. Vehicle ammonia emissions measured in an urban environment in Sydney, Australia, using open path Fourier transform infra-red spectroscopy. Atmos. 2019;10(4):208.
Raherison C, Baldi I, Pouquet M, Berteaud E, Moesch C, Bouvier G, Canal-Raffin M. Pesticides exposure by air in vineyard rural area and respiratory health in children: a pilot study. Environ Res. 2019;169:189–95.
RAIS. 2020. https://rais.ornl.gov/tutorials/toxvals.html#2.4%20Derivation%20of%20Inhalation%20RfDs%20and%20Slope%20Factors. Accessed 29 August 2022.
Rich DQ, Zareba W, Beckett W, Hopke PK, Oakes D, Frampton MW, Utell MJ. Are ambient ultrafine, accumulation mode, and fine particles associated with adverse cardiac responses in patients undergoing cardiac rehabilitation? Environ Health Perspect. 2012;120(8):1162–9.
Saeedi A, Najibi A, Mohammadi-Bardbori A. Effects of long-term exposure to hydrogen sulfide on human red blood cells. Int J Occup Environ Med. 2015;6(1):20–5. https://doi.org/10.15171/ijoem.2015.482.
Sahu SK, Liu S, Liu S, Ding D, Xing J. Ozone pollution in China: background and transboundary contributions to ozone concentration & related health effects across the country. Sci Total Environ. 2021;761:144131.
Schraufnagel DE. The health effects of ultrafine particles. Exp Mol Med. 2020;52(3):311–7.
Semanová P, Kučera M. Health effects from occupational exposure to metalworking fluid mist. In: Key engineering materials, vol. 581. Bäch: Trans Tech Publications Ltd.; 2014. p. 112–8.
Shaw GM, Yang W, Roberts EM, Kegley SE, Stevenson DK, Carmichael SL, English PB. Residential agricultural pesticide exposures and risks of spontaneous preterm birth. Epidemiology. 2018;29(1):8.
Siegel M, Rocheleau CM, Johnson CY, Waters MA, Lawson CC, Riehle-Colarusso T, National Birth Defects Prevention Study. Maternal occupational oil mist exposure and birth defects, national birth defects prevention study, 1997–2011. Int J Environ Res Public Health. 2019;16(9):1560.
Soni V, Singh P, Shree V, Goel V. Effects of VOCs on human health. In: Air pollution and control. Singapore: Springer; 2018. p. 119–42.
Tarafdar A, Sinha A. Health risk assessment and source study of PAHs from roadside soil dust of a heavy mining area in India. Arch Environ Occup Health. 2019;74(5):252–62.
Taylor MP, Isley CF, Glover J. Prevalence of childhood lead poisoning and respiratory disease associated with lead smelter emissions. Environ Int. 2019;127:340–52.
Tohid L, Sabeti Z, Sarbakhsh P, Benis KZ, Shakerkhatibi M, Rasoulzadeh Y, Darvishali S. Spatiotemporal variation, ozone formation potential and health risk assessment of ambient air VOCs in an industrialized city in Iran. Atmos Pollut Res. 2019;10(2):556–63.
Tong S, Schirnding YEV, Prapamontol T. Environmental lead exposure: a public health problem of global dimensions. Bull World Health Organ. 2000;78(9):1068–77.
Turner MC, Andersen ZJ, Baccarelli A, Diver WR, Gapstur SM, Pope CA III, Cohen A. Outdoor air pollution and cancer: an overview of the current evidence and public health recommendations. CA Cancer J Clin. 2020;70(6):460–79.
U.S.EPA. 2015. https://www.epa.gov/sites/default/files/2015-09/documents/rags_a.pdf. Accessed 6 August 2022.
U.S.EPA. 2017. https://iris.epa.gov/static/pdfs/0136tr.pdf. Accessed 6 August 2022.
U.S.EPA. 2021a. https://www.epa.gov/no2-pollution/basic-information-about-no2#Effects. Accessed 6 August 2022.
U.S.EPA. 2021b. https://www.epa.gov/so2-pollution/sulfur-dioxide-basics#effects. Accessed 6 August 2022.
U.S.EPA. 2021c. https://www.epa.gov/co-pollution/basic-information-about-carbon-monoxide-co-outdoor-air-pollution#Effects. Accessed 6 August 2022.
U.S.EPA. 2021d. https://www.epa.gov/lead-air-pollution/basic-information-about-lead-air-pollution#health. Accessed 6 August 2022.
U.S.EPA. 2021e. https://www.epa.gov/indoor-air-quality-iaq/volatile-organic-compounds-impact-indoor-air-quality. Accessed 6 August 2022.
U.S.EPA. 2021f. https://www.epa.gov/ground-level-ozone-pollution/ground-level-ozone-basics#effects. Accessed 6 August 2022.
U.S.EPA. 2022. https://www.epa.gov/sunsafety/health-effects-uv-radiation. Accessed 6 August 2022.
Ulutaş K, Kaskun S, Demir S, Dinçer F, Pekey H. Assessment of H2S and BTEX concentrations in ambient air using passive sampling method and the health risks. Environ Monit Assess. 2021;193(7):1–10.
Wang J, Xu H, Guinot B, Li L, Ho SSH, Liu S, Cao J. Concentrations, sources and health effects of parent, oxygenated-and nitrated-polycyclic aromatic hydrocarbons (PAHs) in middle-school air in Xi'an, China. Atmos Res. 2017;192:1–10.
Wang X, Xu Z, Su H, Ho HC, Song Y, Zheng H, Cheng J. Ambient particulate matter (PM1, PM2.5, PM10) and childhood pneumonia: the smaller particle, the greater short-term impact? Sci Total Environ. 2021;772:145509.
Weichenthal S, Lavigne E, Evans G, Pollitt K, Burnett RT. Ambient PM2.5 and risk of emergency room visits for myocardial infarction: impact of regional PM2.5 oxidative potential: a case-crossover study. Environ Health. 2016;15(1):1–9.

WHO. Health aspects of air pollution with particulate matter, ozone and nitrogen dioxide: report on a WHO working group, Bonn, Germany 13–15 January 2003 (No. EUR/03/5042688). Copenhagen: WHO Regional Office for Europe; 2003.

WHO. 2019. https://www.who.int/news-room/fact-sheets/detail/ambient-(outdoor)-air-quality-and-health. Accessed 26 September 2022.

WHO. 2021. https://www.euro.who.int/en/health-topics/environment-and-health/air-quality/activities/airq-software-tool-for-health-risk-assessment-of-air-pollution. Accessed 26 September 2022.

Wu Y, Lu B, Zhu X, Wang A, Yang M, Gu S, Chen L. Seasonal variations, source apportionment, and health risk assessment of heavy metals in PM2. 5 in Ningbo, China. Aerosol Air Qual Res. 2019;19(9):2083–92.

Wu C, Yan Y, Chen X, Gong J, Guo Y, Zhao Y, Xiang H. Short-term exposure to ambient air pollution and type 2 diabetes mortality: a population-based time series study. Environ Pollut. 2021;289:117886.

Xu RT, Pan SF, Chen J, Chen GS, Yang J, Dangal SRS, Tian HQ. Half-century ammonia emissions from agricultural systems in Southern Asia: Magnitude, spatio-temporal patterns, and implications for human health. GeoHealth. 2018;2(1):40–53.

Yan D, Wu S, Zhou S, Tong G, Li F, Wang Y, Li B. Characteristics, sources and health risk assessment of airborne particulate PAHs in Chinese cities: a review. Environ Pollut. 2019;248:804–14.

Yang W, Carmichael SL, Roberts EM, Kegley SE, Padula AM, English PB, Shaw GM. Residential agricultural pesticide exposures and risk of neural tube defects and orofacial clefts among offspring in the San Joaquin Valley of California. Am J Epidemiol. 2014;179(6):740–8.

Yang JH, Koedrith P, Kang DS, Kee NK, Jung JH, Lee CM, Seo YR. A putative adverse outcome pathway relevant to carcinogenicity induced by sulfuric acid in strong inorganic acid mists. J Cancer Prevent. 2019;24(3):139.

Younis F, Salem E. Respiratory health disorders associated with occupational exposure to bioaerosols among workers in poultry breeding farms. Environ Sci Pollut Res. 2020;27(16):19869–76.

Zhang G, Mu Y, Zhou L, Zhang C, Zhang Y, Liu J, Yao B. Summertime distributions of peroxyacetyl nitrate (PAN) and peroxypropionyl nitrate (PPN) in Beijing: understanding the sources and major sink of PAN. Atmos Environ. 2015;103:289–96.

Zhang H, Mao Z, Huang K, Wang X, Cheng L, Zeng L, Jing T. Multiple exposure pathways and health risk assessment of heavy metal (loid)s for children living in fourth-tier cities in Hubei Province. Environ Int. 2019a;129:517–24.

Zhang JJ, Wei Y, Fang Z. Ozone pollution: a major health hazard worldwide. Front Immunol. 2019b;10:2518.

Zheng L, Ou J, Liu M, Chen Y, Tang Q, Hu Y. Seasonal and spatial variations of PM10-bounded PAHs in a coal mining city, China: distributions, sources, and health risks. Ecotoxicol Environ Saf. 2019;169:470–8.

Zheng H, Kong S, Yan Y, Chen N, Yao L, Liu X, Qi S. Compositions, sources and health risks of ambient volatile organic compounds (VOCs) at a petrochemical industrial park along the Yangtze River. Sci Total Environ. 2020;703:135505.

Fine Particulate Matter-Related Disability-Adjusted Life Years in Indian Union Territories

10

N. Manojkumar, B. Srimuruganandam, and K. Krishnamurthi

10.1 Introduction

Air pollution in India is becoming a major problem due to the rapid urbanization and increasing population density. The significant sources of air pollution in all Indian cities are developmental activities, industrial emissions, biomass burning, and tailpipe emissions (Srimuruganandam and Shiva Nagendra 2012; Gao et al. 2016; Jaiprakash and Habib 2017). Other sources include fugitive road dust emissions, wind-carried dust particles, and sea salt particles especially in coastal areas (Borgie et al. 2016; Sharma et al. 2016). All cities will not have common sources and thus air pollution varies in spatial and temporal scales (Wang et al. 2016; Leena et al. 2017). This shows the need for air pollution concentration assessment in each city. The Central Pollution Control Board (CPCB) of India has laid down the National Ambient Air Quality Standards (NAAQS) in the year 2009. Various air pollutants are considered viz., ammonia, carbon monoxide, lead, nitrogen dioxide, ozone, particulate matter (PM), and sulphur dioxide. Among this PM is identified to cause detrimental human health effects of air pollution (Pope et al. 2004; Maleki et al. 2016; Lo et al. 2017; Li et al. 2017).

Unlike other gaseous pollutants, PM doesn't have a definite chemical composition and physical structure (Hughes et al. 1999; Xia and Gao 2010; Alharbi et al. 2015). PM is composed of soot, elemental carbon, ions, elements and volatile compounds (Smolik et al. 2003; Srimuruganandam and Shiva Nagendra 2011). Because of this complex nature, PM has a wide range of effects on human health. So far, the reported evidence of PM health effects are asthma (Von Klot et al. 2002), chronic obstructive pulmonary disease (Ling and van Eeden 2009), lung cancer (Santibáñez-Andrade et al. 2017) and stroke (Kettunen et al. 2007). Globally, PM is studied in two aerodynamic sizes of 2.5 and 10 μm which are commonly referred to as $PM_{2.5}$ and PM_{10}, respectively. $PM_{2.5}$ being smaller than PM_{10} can penetrate deeper into the lungs and may reach even the alveoli. Thus, $PM_{2.5}$ has more potential to cause health effects than PM_{10} (Manojkumar et al. 2019).

Several literatures have highlighted the relationship of $PM_{2.5}$ with premature mortality (Maji et al. 2018; Al-Hemoud et al. 2018). Hence, it is important to estimate the total years of life lost (YLL) among the given population. Further, the assessment of expected life remaining (ELR) and years of life that can be saved when air pollution

N. Manojkumar · B. Srimuruganandam (✉)
Centre for Clean Environment, Vellore Institute of Technology, Vellore, Tamil Nadu, India
e-mail: bsrimuruganandam@vit.ac.in

K. Krishnamurthi
Environmental Health Cell, CSIR-National Environmental Engineering Research Institute, Nagpur, India
e-mail: k_krishnamurthi@neeri.res.in

P. K. Padhy et al. (eds.), *Air Quality and Human Health*,
https://doi.org/10.1007/978-981-97-1363-9_10

is reduced to stipulated standards (YLS) are crucial for directive decisions. Earlier studies focused on major cities, however there is no study related to all union territories (Maji et al. 2016; Manojkumar and Srimuruganandam 2021). Thus, the present study aims to estimate the $PM_{2.5}$-related YLL, ELR and YLS in the union territories of India using AirQ+ software.

10.2 Methodology

10.2.1 Study Area and Data Collection

India has seven union territories viz., Andaman and Nicobar Islands, Chandigarh, Dadra and Nagar Haveli, Daman and Diu, National Capital Region—Delhi, Lakshadweep, and Puducherry. The CPCB of India is periodically monitoring various air pollutants and further check for compliance with stipulated NAAQS. Also, the monitored data is made available free in the CPCB website (CPCB 2019). While assessing the data availability, only the year 2016 has $PM_{2.5}$ manual monitoring data for four union territories viz., Chandigarh, Dadra and Nagar Haveli, Daman and Diu, and National Capital Region—Delhi. Hence for the present study $PM_{2.5}$ data of 2016 is selected. Further, the population and death data are essential for quantifying YLL, ELR and YLS. The required population data and age-specific death statistics are downloaded from the Office of the Registrar General and Census Commissioner, India website (ORGCC 2019).

10.2.2 AirQ+ Software

AirQ+ software (version 1.2) is developed by the World Health Organization, Europe and the first version is released on 19 May 2016 at Germany. This software can be used for assessing health risk in any region of the world. AirQ+ software assesses the human health risk caused by several air pollutants viz., $PM_{2.5}$, PM_{10}, NO_2, SO_2, and O_3. The software can quantify the hospital admissions and mortality caused by several health endpoints viz., respiratory diseases, cardiovascular diseases, and stroke. Further disability-adjusted life years can be estimated in terms of YLL. Apart from YLL, ELR and YLS can also be calculated. For all these assessments the required input parameters are annual mean concentration, population data, age-specific death data, relative risk, and cut-off value. Annual mean concentration of Chandigarh, Dadra and Nagar Haveli, Daman and Diu, and Delhi are 123.40 $\mu g\ m^{-3}$, 37 $\mu g\ m^{-3}$, 34 $\mu g\ m^{-3}$, and 118.42 $\mu g\ m^{-3}$, respectively. The population and death statistics of the study regions are given in Table 10.1. The relative risk values are set to default values as given in AirQ+ software. These default values in AirQ+ software are based on several cohort studies. The cut-off value of 10 $\mu g\ m^{-3}$ is considered as per the World Health Organization (WHO) recommendations. The protocol followed in AirQ+ software is shown in Fig. 10.1.

Table 10.1 Population and death statistics used in the study

	Chandigarh				Dadra and Nagar Haveli				Daman and Diu				Delhi			
	Population		Deaths		Population		Deaths		Population		Deaths		Population		Deaths	
Age	M	F	M	F	M	F	M	F	M	F	M	F	M	F	M	F
0–4	43,347	38,507	695	417	18,902	17,659	79	53	10,020	9138	18	21	734,336	646,864	6241	4014
5–14	101,395	83,263	417	130	37,444	33,808	12	16	19,128	16,699	4	9	1,724,933	1,459,186	1617	1293
15–24	128,988	96,674	784	354	47,574	29,831	75	43	46,164	18,734	53	20	1,876,529	1,554,906	3806	2845
25–34	110,319	90,721	854	435	41,453	29,073	95	53	37,358	19,854	56	21	1,635,651	1,467,374	6278	3227
35–44	82,834	69,759	1273	489	25,416	18,042	125	45	20,639	12,512	90	30	1,289,313	1,127,128	9435	3796
45–54	59,001	48,293	1559	685	12,886	10,406	164	59	9295	6933	136	34	871,449	727,233	12,700	5842
55–64	33,275	27,451	1565	851	6320	6462	207	94	4553	4811	122	39	507,288	473,623	16,466	8712
65–69	8250	7589	1412	727	1755	1961	102	64	1182	1571	63	48	139,398	133,054	7988	5071
≥70	13,053	12,385	1549	1278	1810	2535	264	240	1714	2586	205	277	199,254	203,720	21,744	18,027

Note: *M* male, *F* female

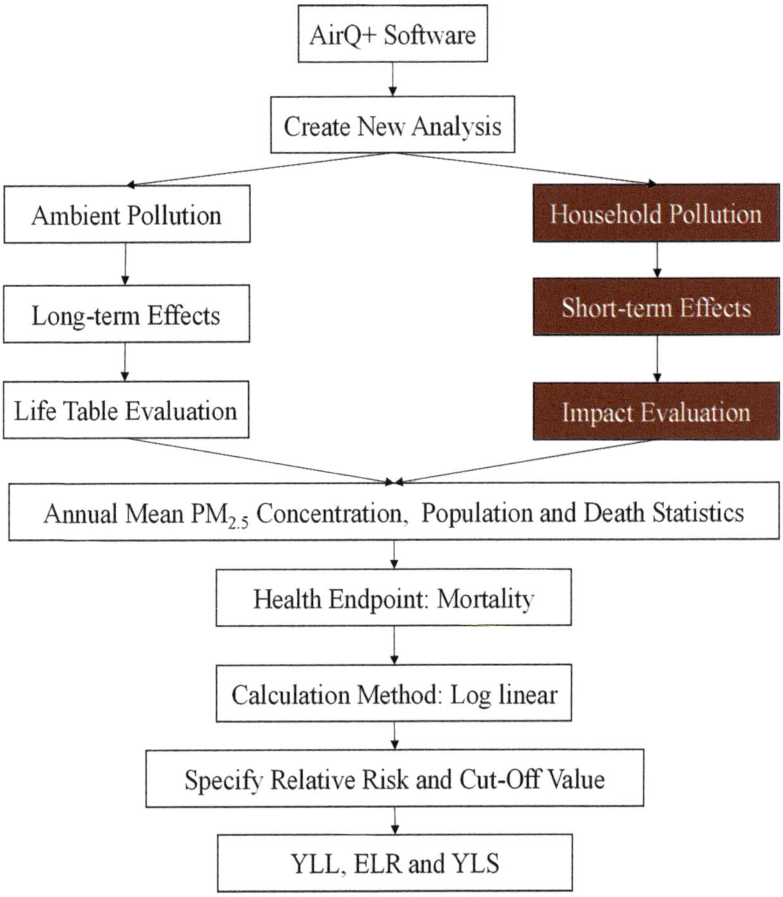

Fig. 10.1 Protocol followed in AirQ+ software. Note: The boxes highlighted in brown colour are the options available in AirQ+ but not used for the present study

10.3 Results and Discussion

10.3.1 Years of Life Lost (YLL)

The AirQ+ software estimates YLL with a 95% confidence interval for 0–100 years (all age groups) and 0–64 years (except elderly group). YLL results of the male and female population in selected union territories are summarized in Table 10.2. Delhi has the highest YLL (all ages) of 20,667 (14,942–24,965) years and 12,655 (9149–15,286) years in male and female population, respectively. In case of Chandigarh, the YLL (all ages) is 2499 years for male and 1327 for the female population. Daman and Diu are found to have the lowest YLL (all ages) in both male and female population. YLL of non-elderly population (0–64 age) in selected union territories range from 32 to 13,545 years in male and 12 to 7122 years in female. The elderly group (65–100 age) contribution in total YLL is found to be 29% (37%), 33% (46%), 36% (65%), and 34% (44%) in male (female) population of Chandigarh, Dadra and Nagar Haveli, Daman and Diu, and Delhi, respectively.

Similar study from Nanjing, China is estimated with a YLL of 714 for male and 522 for female (Lu et al. 2015). In case of Wuxi, China, the mean YLL for male and female is 385 and 542, respectively (Zhu et al. 2017). In Beijing, YLL varied between 299 and 1504 years (Guo et al. 2013). YLL in Tehran, Iran ranges from 67,970 (CI: 45,072–88,695) to 102,072 (CI: 67,934–132,754) (Faridi et al. 2018). Another study (Röösli et al. 2005) in Switzerland recorded YLL between 40,600 (CI: 24,500–68,000) and 46,200 (CI: 21,800–59,100). Daily YLL estimate in male and female population of Tianjin, China is 64.9 ± 43.7 years and 55.9 ± 41.4 years, respec-

Table 10.2 YLL in male and female population

Union territory	Gender	Age	Central	Lower	Upper
Chandigarh	Male	All ages	2499	1815	3008
		Age 0–64	1767	1283	2127
	Female	All ages	1327	963	1597
		Age 0–64	831	603	1000
Dadra and Nagar Haveli	Male	All ages	84	56	109
		Age 0–64	57	38	73
	Female	All ages	50	34	65
		Age 0–64	27	18	35
Daman and Diu	Male	All ages	50	34	65
		Age 0–64	32	22	42
	Female	All ages	34	22	43
		Age 0–64	12	8	15
Delhi	Male	All ages	20,667	14,942	24,965
		Age 0–64	13,545	9793	16,362
	Female	All ages	12,655	9149	15,286
		Age 0–64	7122	5149	8603

Note: The central, lower and upper values correspond to respective confidence interval limits of central (1.062), lower (1.04) and upper (1.083) relative risk values

tively. Also, a total YLL of 351,644 and 89,487 years are related to the elderly (≥65 years) and non-elderly (<65 years) population (Zeng et al. 2017). In Tallinn, Estonia, the total YLL is estimated as 3859 (CI: 1023–6636) with 988 YLL per 1 lakh population (Orru et al. 2009).

10.3.2 Estimated Life Remaining (ELR)

ELR is quantified separately for male and female population and shown in Figs. 10.2 and 10.3, respectively. Within the male and female population, the ELR is quantified with different age groups viz., infant (≤1 year), children (2–19 years), adult (20–64 years), and elderly (≥65 years). This age group classification is proposed by the WHO. The ELR in male group is observed to be high in Daman and Diu whereas the low in Chandigarh. ELR in the female infant group ranges from 58 years (Chandigarh) to 72 years (Daman and Diu). Also, mean ELR of all the four union territories is found to be 67 years in female infant group. In both male and female population, the ELR is high in less polluted union territories (Dadra and Nagar Haveli, Daman and Diu) for all age groups. While considering all study areas, the ELR in male and female child range from 37 to 64 years and 44 to 70 years, respectively. In case of adult age group, the ELR is 47 years (53 years) for 20 years old male (female) in Daman and Diu. In the same place, the ELR is gradually reduced to 11 years (13 years) for 64 years old people in the male (female) population. Further, it is also observed that elderly age group has recorded the lowest ELR in all age groups of the male and female population.

10.3.3 Years of Life Saved (YLS)

AirQ+ also estimates the years of life saved in each age group when air pollutant levels are reduced to stipulated $PM_{2.5}$ standards. As per the WHO recommendations, the annual mean $PM_{2.5}$ standard of 10 μg m^{-3} is adopted for YLS calculation. The average years of life saved in each age group of male and female population is given in Table 10.3. The average YLS of considered union territories varies from 1 to 13 years in the male population and 1 to 12 years in the female population. It is observed that the average YLS is high in the infant age group. Both adult and child age groups are recorded with a

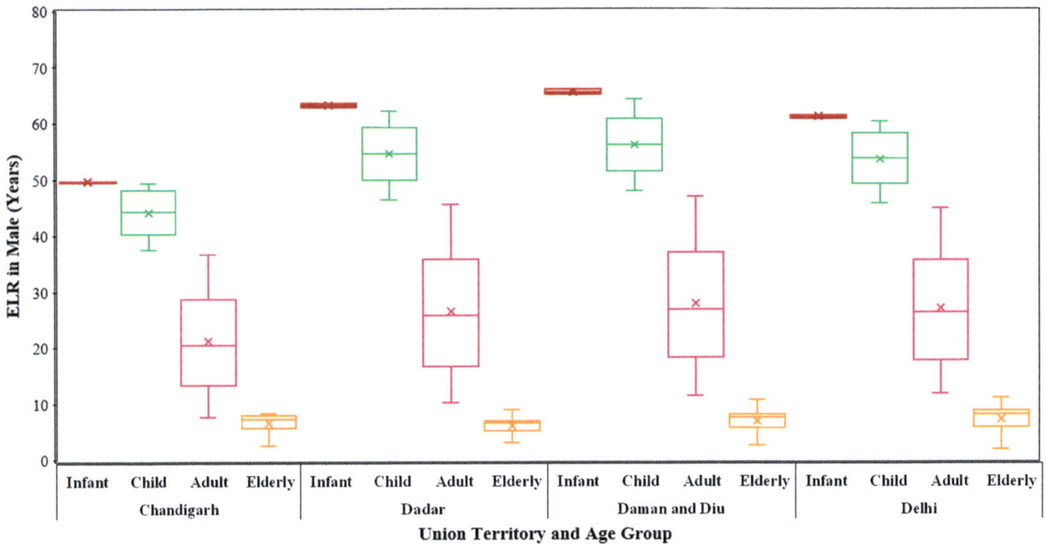

Fig. 10.2 ELR in different age groups of male population

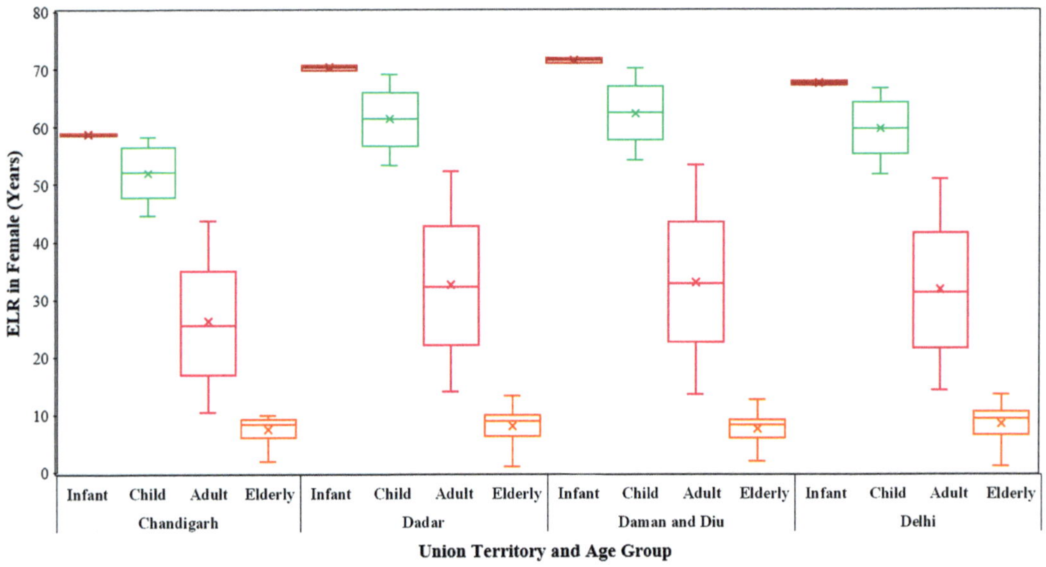

Fig. 10.3 ELR in different age groups of female population

maximum YLL of 11 years (9) years in male (female) population. Further, the male and female elderly age group are found to have the lowest YLS average in all union territories. It is understood that places with low $PM_{2.5}$ pollution (Dadra and Nagar Haveli, Daman and Diu) are having minimum YLS. Also, it is important to note that high levels of $PM_{2.5}$ led to have maximum YLS in Chandigarh and Delhi.

An earlier study (Michael et al. 2015) used manually monitored and satellite PM data, and estimated the average YLS of the Indian population to be 3.2 years. Another recent study from India (Kalpana et al. 2018) showed the association of air pollution reduction to standard levels with an increase of 0.9 years among Indian population. Another study (Pope et al. 2009) showed that a decrease of 10 μg m^{-3} will have increased

Table 10.3 YLS in different age groups of male and female population

	Age Group	Union Territory											
		Chandigarh			Dadar			Daman and Diu			Delhi		
Gender		Central	Lower	Upper	Central	Lower	Upper	Central	Lower	Upper	Central	Lower	Upper
Male	Infant	13	8	17	2	1	3	2	1	3	11	7	14
	Child	11	7	15	2	1	3	2	1	2	10	6	13
	Adult	9	5	12	2	1	2	2	1	2	8	5	11
	Elder	4	2	5	1	1	1	1	0	1	3	2	4
Female	Infant	12	8	16	2	1	3	2	1	2	10	7	13
	Child	11	7	14	2	1	3	2	1	2	9	6	12
	Adult	9	6	12	2	1	2	1	1	2	8	5	11
	Elder	4	2	5	1	1	1	1	0	1	3	2	4

Note: The central, lower and upper values correspond to respective confidence interval limits of central (1.062), lower (1.04) and upper (1.083) relative risk values life expectancy of 0.77 years. A similar study (Correia et al. 2014) in the United States between the years 2000 and 2007 estimated the YLS of 0.35 years. Another study (Ebenstein et al. 2017) found a total gain of 3.7×10^3 million years for China population when PM is reduced to standard levels.

10.4 Conclusion

PM and its effect on human health are well linked by numerous epidemiological studies. India is one among the Asian countries and is experiencing severe PM pollution. However, the relation of PM with YLL and ELR is not assessed in India. Thus, the knowledge on YLL due to premature mortality is crucial. Also, ELR and YLS of different gender and age will give useful information for policymakers. Also, the union territories of India are rarely considered for particulate matter studies. Hence this study aims to quantify the YLL, ELR, and YLS in four Indian union territories using AirQ+ software. The required input data viz., $PM_{2.5}$ concentration and age-specific population statistics are acquired from CPCB and Office of the Registrar General and Census Commissioner, India, websites respectively. Results showed that YLL and YLS are very high in Delhi and Chandigarh. The elderly group has contributed 29–36% (37–65%) of total YLL in male (female) population. It is also observed that places with low $PM_{2.5}$ levels are having high ELR. Thus, it is evident that if the present $PM_{2.5}$ levels are reduced to WHO standards (10 μg m^{-3}), then the ELR of both male and female will be increased.

References

Alharbi B, Shareef MM, Husain T. Study of chemical characteristics of particulate matter concentrations in Riyadh, Saudi Arabia. Atmos Pollut Res. 2015;6(1):88–98.

Al-Hemoud A, Gasana J, Al-Dabbous A, Al-Shatti A, Al-Khayat A. Disability adjusted life years (DALYs) in terms of years of life lost (YLL) due to premature adult mortalities and postneonatal infant mortalities attributed to $PM_{2.5}$ and PM_{10} exposures in Kuwait. Int J Environ Res Public Health. 2018;15(11):1–15.

Borgie M, Ledoux F, Dagher Z, Verdin A, Cazier F, Courcot L, Courcot D. Chemical characteristics of $PM_{2.5-0.3}$ and $PM_{0.3}$ and consequence of a dust storm episode at an urban site in Lebanon. Atmos Res. 2016;180:274–86.

Correia AW, Iii CAP, Dockery DW, Avenue H, Building H, Wang Y. The effect of air pollution control on life expectancy in the United States: an analysis of 545 US counties for the period 2000 to 2007. Epidemiology. 2014;24(1):23–31.

CPCB. Central Pollution Control Board, Ministry of Environment Forest and Climate Change, Government of India website. 2019. https://app.cpcbccr.com/ccr/#/caaqm-dashboard-all/caaqm-landing. Accessed 20 Jan 2019.

Ebenstein A, Fan M, Greenstone M, He G, Zhou M. New evidence on the impact of sustained exposure to air pollution on life expectancy from China's Huai. Proc Natl Acad Sci U S A. 2017;114(39):10384–9.

Faridi S, Shamsipour M, Krzyzanowski M, Künzli N, Amini H, Azimi F, Naddafi K. Long-term trends and health impact of $PM_{2.5}$ and O_3 in Tehran, Iran, 2006–2015. Environ Int. 2018;114:37–49.

Gao Y, Lee SC, Huang Y, Chow JC, Watson JG. Chemical characterization and source apportionment of size-resolved particles in Hong Kong sub-urban area. Atmos Res. 2016;170:112–22.

Guo Y, Li S, Tian Z, Pan X, Zhang J, Williams G. The burden of air pollution on years of life lost in Beijing, China, 2004-08: retrospective regression analysis of daily deaths. BMJ. 2013;347:1–10.

Hughes LS, Allen JO, Kleeman MJ, Johnson RJ, Cass GR, Gross DS, Prather KA. Size and composition distribution of atmospheric particles in Southern California. Environ Sci Technol. 1999;33(20):3506–15.

Jaiprakash, Habib G. Chemical and optical properties of $PM_{2.5}$ from on-road operation of light duty vehicles in Delhi city. Sci Total Environ. 2017;586:900–16.

Kalpana B, Dey S, Gupta T, Dhaliwal RS, Brauer M, Bhansali. The impact of air pollution on deaths, disease burden, and life expectancy across the states of India: the Global Burden of Disease Study 2017. Lancet Planet Health. 2018;3(1):e26–39.

Kettunen J, Lanki T, Tiittanen P, Aalto PP, Koskentalo T, Kulmala M, Pekkanen J. Associations of fine and ultrafine particulate air pollution with stroke mortality in an area of low air pollution levels. Stroke. 2007;38(3):918–22.

Leena PP, Vijayakumar K, Anilkumar V, Pandithurai G. Analysing temporal variability of particulate matter and possible contributing factors over Mahabaleshwar, a high-altitude station in Western Ghats, India. J Atmos Sol Terr Phys. 2017;164:105–15.

Li G, Xue M, Zeng Q, Cai Y, Pan X, Meng Q. Association between fine ambient particulate matter and daily total mortality: an analysis from 160 communities of China. Sci Total Environ. 2017;599–600(38):108–13.

Ling SH, van Eeden SF. Particulate matter air pollution exposure: role in the development and exacerbation of chronic obstructive pulmonary disease. Int J Chron Obstruct Pulmon Dis. 2009;4:233–43.

Lo WC, Shie RH, Chan CC, Lin HH. Burden of disease attributable to ambient fine particulate matter exposure in Taiwan. J Formos Med Assoc. 2017;116(1):32–40.

Lu F, Zhou L, Xu Y, Zheng T, Guo Y, Wellenius GA, Zheng X. Short-term effects of air pollution on daily mortality and years of life lost in Nanjing, China. Sci Total Environ. 2015;536(5):123–9.

Maji KJ, Dikshit AK, Deshpande A, Speldewinde PC. Human health risk assessment due to air pollution in ten urban cities in Maharashtra, India. Cogent Environ Sci. 2016;12:1–16.

Maji KJ, Arora M, Dikshit AK. Premature mortality attributable to $PM_{2.5}$ exposure and future policy roadmap for 'airpocalypse' affected Asian megacities. Process Saf Environ Prot. 2018;118:371–83.

Maleki H, Sorooshian A, Goudarzi G, Nikfal A, Baneshi MM. Temporal profile of PM_{10} and associated health effects in one of the most polluted cities of the world (Ahvaz, Iran) between 2009 and 2014. Aeolian Res. 2016;22:135–40.

Manojkumar N, Srimuruganandam B. Health effects of particulate matter in major Indian cities. Int J Environ Health Res. 2021;31:258–70.

Manojkumar N, Srimuruganandam B, Shiva Nagendra SM. Application of multiple-path particle dosimetry model for quantifying age specified deposition of particulate matter in human airway. Ecotoxicol Environ Saf. 2019;168:241–8.

Michael G, Janhavi N, Rohini P, Nicholas R, Anant S, Anish S. Lower pollution, longer lives. Econ Polit Wkly. 2015;8:40–6.

ORGCC. Office of the Registrar General & Census Commissioner, Ministry of Home Affairs, Government of India. 2019. http://www.censusindia.gov.in/. Accessed 25 Jan 2019.

Orru H, Teinemaa E, Lai T, Tamm T, Kaasik M, Kimmel V, Forsberg B. Health impact assessment of particulate pollution in Tallinn using fine spatial resolution and modeling techniques. Environ Health. 2009;8(1):1–9.

Pope CA, Burnett RT, Thurston GD, Thun MJ, Calle EE, Krewski D, Godleski JJ. Cardiovascular mortality and long-term exposure to particulate air pollution: epidemiological evidence of general pathophysiological pathways of disease. Circulation. 2004;109(1):71–7.

Pope A, Majid E, Douglas D. Fine particulate air pollution and US county life expectancy. N Engl J Med. 2009;360(4):376–86.

Röösli M, Künzli N, Braun-Fahrländer C, Egger M. Years of life lost attributable to air pollution in Switzerland: dynamic exposure-response model. Int J Epidemiol. 2005;34(5):1029–35.

Santibáñez-Andrade M, Quezada-Maldonado EM, Osornio-Vargas Á, Sánchez-Pérez Y, García-Cuellar CM. Air pollution and genomic instability: the role of particulate matter in lung carcinogenesis. Environ Pollut. 2017;229:412–22.

Sharma SK, Sharma A, Saxena M, Choudhary N, Masiwal R, Mandal TK, Sharma C. Chemical characterization and source apportionment of aerosol at an urban area of Central Delhi, India. Atmos Pollut Res. 2016;7(1):110–21.

Smolik J, Zdimal V, Schwarz J, Lazaridis M, Havr V. Size resolved mass concentration and elemental composition of atmospheric aerosols over the Eastern Mediterranean area. Atmos Chem Phys. 2003;3:2207–16.

Srimuruganandam B, Shiva Nagendra SM. Chemical characterization of PM_{10} and $PM_{2.5}$ mass concentrations emitted by heterogeneous traffic. Sci Total Environ. 2011;409(17):3144–57.

Srimuruganandam B, Shiva Nagendra SM. Source characterization of PM_{10} and $PM_{2.5}$ mass using a chemical mass balance model at urban roadside. Sci Total Environ. 2012;433:8–19.

Von Klot S, Wölke G, Tuch T, Heinrich J, Dockery DW, Schwartz J, Peters A. Increased asthma medication use in association with ambient fine and ultrafine particles. Eur Respir J. 2002;20(3):691–702.

Wang K, Tian H, Hua S, Zhu C, Gao J, Xue Y, Hao J. A comprehensive emission inventory of multiple air pollutants from iron and steel industry in China: temporal trends and spatial variation characteristics. Sci Total Environ. 2016;559(19):7–14.

Xia L, Gao Y. Chemical composition and size distributions of coastal aerosols observed on the US East Coast. Mar Chem. 2010;119:77–90.

Zeng Q, Wu Z, Jiang G, Wu X, Li P, Ni Y, Pan X. The association between ambient inhalable particulate matter and the disease burden of respiratory disease: an ecological study based on ten-year time series data in Tianjin, China. Environ Res. 2017;157:71–7.

Zhu J, Zhang X, Zhang X, Dong M, Wu J, Dong Y, Zhou W. The burden of ambient air pollution on years of life lost in Wuxi, China, 2012–2015: a time-series study using a distributed lag non-linear model. Environ Pollut. 2017;224:689–97.

Application of Biomarkers in Assessing Human Health Impacts of Air Pollution

11

Soumya Niyogi and Markus Hecker

11.1 Introduction

Air pollution causes a huge number of fatalities every year around the world. It has been suggested that air pollution contributes to approximately 8.8 million extra deaths worldwide every year, and about 40–80% of these deaths have been linked to various cardiovascular diseases (CVD) (Burnett et al. 2018). Recent epidemiological studies have provided extensive evidence linking air pollution with a wide variety of adverse health conditions and diseases pertaining to respiratory and cardiovascular systems, especially among vulnerable demographic groups such as the children, the elderly, and those with preexisting compromised cardiopulmonary conditions such as asthma (Kim et al. 2018, 2020; Münzel et al. 2020; Shen et al. 2017). These adverse effects predominantly include reduced functionality of the lung, chronic respiratory and cardiovascular inflammation, chronic obstructive pulmonary disease (COPD), stroke and heart attacks, and lung cancer. All these effects have been associated with exposure to at least one major air pollutant that is routinely found in the ambient air in large cities around the world. These air pollutants include suspended particulate matter (PM), ozone (O_3), nitrogen oxides (NO_x), sulphur dioxide (SO_2), carbon monoxide (CO), volatile organic carbon compounds including polyaromatic hydrocarbons (PAHs), and trace metals such as lead (Pb) (Cohen et al. 2017). Although each of these air pollutants can induce serious health implications for humans, exposure to fine (*diameter* $\leq 2.5\ \mu m$) and ultrafine (*diameter* $\leq 0.1\ \mu m$) particles are believed to be the leading cause of cardiorespiratory illnesses in the developing world. However, the current understanding of the health effects of PM is limited, and most of the epidemiological evidence of linking adverse health effects to PM exposure is not conclusive. Moreover, the cellular and molecular mechanisms by which PM, exclusively and in mixture with other air pollutants, elicit toxic effects in humans are yet to be fully understood. Nevertheless, the use of conventional biochemical and novel molecular biomarker assessments can provide important insights into the cellular perturbations that lead to the onset of adverse health effects during chronic exposure to air pollutants. Furthermore, the advent of next generation, systems biology molecular techniques (e.g., omics technologies) provides exciting new avenues to study the effects of air pollution on human health. These new tools can provide new knowl-

S. Niyogi (✉)
Department of Biology and Toxicology Centre, University of Saskatchewan, Saskatoon, SK, Canada
e-mail: som.niyogi@usask.ca

M. Hecker
School of Environment and Sustainability and Toxicology Centre, University of Saskatchewan, Saskatoon, SK, Canada
e-mail: markus.hecker@usask.ca

P. K. Padhy et al. (eds.), *Air Quality and Human Health*,
https://doi.org/10.1007/978-981-97-1363-9_11

edge on the interactions of air pollutants with genes, proteins and other biomolecules that are currently unknown, and thus, develop a more holistic and in-depth mechanistic understanding of the toxicity of these pollutants. Ultimately, this advanced understanding can then be used to develop toxicity models that can predict potential future adverse health outcomes early during exposure, enabling preventive interactions and implementation of effective mitigation strategies.

11.2 Biomarkers and Their Application in Epidemiological Investigations

Knowledge of individual exposures and effects is essential in successful epidemiologic investigations to avoid mischaracterization of exposures or effects (Barr et al. 2020). However, intra- and inter-individual variation in predictors of exposures (e.g., behaviours, domestic and/or occupational activity patterns), genetic susceptibility, and differences in the toxicokinetics of pollutants make quantitative assessments of exposure and/or effects often very challenging (Cohen Hubal et al. 2019; Lioy and Smith 2013; Vincent 2012). Biomarkers are widely recognized as critical tools for understanding environmental exposures, susceptibility, and early physiological responses leading to disease outcomes (Barr et al. 2005; Needham et al. 2005). Biomarkers can be particularly useful for characterizing exposure to environmental pollutants and their modes of toxic action in organisms. Biomarkers are typically defined as the cellular, biochemical, and molecular alterations that are measurable in biological samples such as human tissues, cells, or fluids (Mayeux 2004). The World Health Organization (WHO) uses a broader definition of biomarkers and defines biomarkers as any measurement reflecting an interaction between a biological system and a potential hazard, which may be of a chemical, physical, or biological nature, that corresponds with quantifiable molecular, biochemical, and/or physiological responses (WHO 2001). In practice, biomarkers include tools and approaches that measure time-integrated biological responses which can aid in understanding the cause and/or prediction of progression or regression of eventual apical effects in organisms that can result from exposure to toxicants.

Biomarkers can be classified into three different categories: biomarkers of exposure, biomarkers of effects, and biomarkers of susceptibility. A biomarker of exposure relates exposure to a pollutant to the level of that pollutant (or its metabolites, or the product of its interaction with target biomolecules) in body tissue or fluid. For example, a biomarker of exposure may include concentrations of trace metals in the blood, which can indicate an internal or absorbed dose. It may also include PAH-metabolites in the blood or other tissues, which can be a measure of effective internal dose. Moreover, it may include the measurement of metal- or PAH-DNA adducts and specific enzymes or proteins that directly interact with metals (e.g., metallothionein) or PAHs (e.g., CYP P450 Phase I enzymes). On the other hand, a biomarker of effect pertains to an early health or clinical effect associated with possible health impairment or onset of chronic illness or disease (Califf 2018). Biomarkers of effects typically include measurements of the impaired structural and/or functional integrity of proteins and enzymes (e.g., protein carbonyls, decreased enzyme activity), membrane lipids (e.g., lipid peroxidation) and DNA (e.g., DNA base oxidation), and altered expression of genes or cellular responses (e.g., systemic inflammation, alterations in DNA methylation) that can be linked with risk of disease (e.g., neurological disorders, cancer).

In modern biomonitoring and epidemiological investigations, both biomarkers of exposure and biomarkers of effect are employed together, which is critical to establish the possible linkage between exposure to pollutants and associated adverse outcomes in impacted populations. Moreover, such investigations also require employing a wide array of biomarkers of effects that can complement and corroborate each other to acquire a robust and reliable understanding of potential adverse health outcomes. Finally, biomarkers of susceptibility are indicators of increased sensitivity to pollutant exposure and/or effects among individuals, which are determined by their genetic disposition. These biomarkers include the evaluation of alterations in metabolism or toxicokinetics resulting in increased sensitivity of target biomolecules and/or internal dose (Califf 2018). Biomarkers of susceptibility are seldom employed in biomonitoring and epidemiological investigations as they require sophisticated molecular techniques, and thus, are not always cost-effective and feasible. In this chapter, we will discuss various types of biomarkers of effects and their suitability and limitations in assessing the adverse health implications of air pollutants in humans.

11.3 Physiological and Biochemical Markers for Assessing the Impacts of Air Pollutants

11.3.1 Biomarkers of Oxidative Stress

Oxidative stress occurs due to the accumulation of reactive oxygen species (ROS) in biological systems. ROS can be produced endogenously through various natural biochemical processes, but their endogenous production increases during exposure to various natural and synthetic contaminants as their metabolism in the body produce ROS. In fact, air pollutants such as particulate matter (PM), especially fine ($PM_{2.5}$, $PM < 2.5\,\mu m$) and ultrafine ($PM_{0.1}$, $PM < 0.1\,\mu m$) particles, ozone, nitrogen oxides, and transition metals, are potent oxidants and can generate ROS (MØller and Loft 2010; Valavanidis et al. 2008). Oxidative stress can be defined as a physiological condition where prooxidant forces exceed the antioxidant defence systems leading to an imbalance between them ultimately leading to significant tissue damage. Free radicals contain one or more unpaired electrons which render them extremely reactive. Because of their high reactivity, they can extract electrons from the biomolecules that they interact with including lipids, proteins, and DNA to gain stability, and thereby can cause cellular damage by oxidizing critical biomolecules. Thus, oxidative damage to membrane lipids (lipid peroxidation), proteins (protein carbonyls) and deoxyribonucleic acids (DNA oxidation) are commonly used as biomarkers of pollutant-induced oxidative stress in biological systems. Malondialdehyde can be excessively produced following ROS-mediated lipid peroxidation and it is commonly used to measure the level of oxidative damage to the cell membranes. Similarly, protein carbonyls are measured to evaluate oxidative damage to cellular proteins as many of the oxidative modifications of proteins involve some type of carbonylation. Thus, irreversible ROS-mediated lipid and protein damage results in elevated malonaldehyde and protein carbonyl levels, which have been linked with various bronchopulmonary and cardiovascular diseases such as COPD and chronic heart disease (reviewed by Delfino et al. 2011). Moreover, a molecule that loses electrons to ROS can also turn into a free radical itself and can trigger a chain reaction cascade leading to the disruption of cellular structure and functions (Phaniendra et al. 2015).

Oxidative stress and inflammation have been implicated as one of the primary causes of increased cardiovascular and bronchopulmonary diseases and mortality in urban populations impacted by air pollution. Moreover, oxidative stress is known to be involved in the pathogenesis of various other human diseases, including neurobehavioural diseases such as Alzheimer's disease and Parkinson's disease, and malignancies in critical organs such as the lung (Hahad et al. 2020; Rao et al. 2018; Fiorito et al. 2018). Oxidative stress can activate redox-sensitive pathways that often lead to inflammation and cell death. However, the susceptibility of the target organ to oxidative damage also depends upon its ability to upregulate protective ROS scavenging defence mechanisms. Cellular ROS scavenging systems involve both enzymatic antioxidants such as catalase and glutathione peroxidase (scavenger of hydrogen peroxide and lipid hydroperoxides), superoxide dismutase (scavenger of superoxide anion), and glutathione reductase (maintains the cellular thiol balance) as well as non-enzymatic antioxidants such as thiols (e.g., glutathione and thioredoxin), vitamin E and C, melatonin, and carotenoids. Specifically, reduced glutathione (GSH) is the most commonly used index of cellular thiol redox status. It is a major intracellular scavenger of free radicals and neutralizes peroxides in conjunction with glutathione peroxidase and glutathione reductase, thus maintaining the redox balance of cells. Furthermore, GSH detoxifies highly reactive electrophile compounds either directly or via conjugation reactions catalysed by the enzyme glutathione-s-transferase.

Both enzymatic and non-enzymatic antioxidants in the lung are the first line of defence against ROS (Kelly and Tetley 1997). The composition and quantity of antioxidants in respiratory epithelium and the blood represent an important determinant of individual responsiveness to air pollutants. To defend against oxidative damage induced by air pollutants, cells upregulate the synthesis of antioxidants via the activation of transcriptional factor Nrf2 (Kang et al. 2005). Previous studies have demonstrated the protective role of Nrf2 against oxidative injury to human lung cells induced by PAHs and Metals adsorbed to PMs (Pardo et al. 2019, 2020; Canella et al. 2018). Furthermore, the knockdown of Nrf2 was found to greatly increase the cytotoxicity and cell death associated with reduced glutathione levels and reduced expression of enzymatic antioxidant genes (Rubio et al. 2011). Although the interrelationship between the antioxidant levels in the respiratory epithelium and blood levels is not well understood, the susceptibility of the lung to pollutant-induced oxidative damage is determined largely by its capacity to upregulate protective ROS scavenging systems via Nrf2 activation. On the other hand, airborne PMs have been suggested to suppress the upregulation of antioxidant moieties involved in oxidative defence response likely via the inactivation of Nrf2 (Radan et al. 2019). For example, exposure to PM_{10} has been found to enhance ROS production and suppress the activity of enzymatic antioxidants like superoxide dismutase and glutathione reductase and decrease glutathione in a human lung epithelial cell line (Chirino et al. 2010). Thus, elevated or reduced activity and/or quantity of cellular enzymatic and non-enzymatic antioxidants can be used as biomarkers of cellular oxidative stress. Interestingly though, oxidative stress has also a useful role in physiologic adaptation and in the regulation of intracellular signal transduction (Schieber and Chandel 2014). Oxidative compounds play a critical role in inflammation and tissue repair processes, and therefore, they represent some of the critical cellular defence mechanisms against pathogens and malignant cells as well as tissue repairing and remodelling (Lugrin et al. 2014). However, hyperactivity of the cellular inflammation response, which is often induced by exposure to various airborne contaminants, can also lead to the overproduction of ROS and induce oxidative damage to the lungs and other target organs (Miller 2020).

Although each air pollutant can induce its own specific toxicity in the respiratory and cardiovascular systems, PMs, ozone, and nitrogen oxides all cause oxidative tissue damage, either through direct effects on cellular lipids and proteins or indirectly through the activation of intracellular oxidant pathways (Møller and Loft 2010; Valavanidis Favia et al. 2008; Jeng 2010). Several studies have attributed the oxidative potential of fine and ultrafine particles to organic carbon compounds, such as quinones and PAHs. Airborne PMs originating from incomplete combustion processes often contain major reactive and toxic compounds such as substituted (e.g., methyl naphthalene) and unsubstituted (e.g., benzo-a-pyrene, phenanthrene) PAHs, and their nitrated derivatives (e.g., 1-nitropyrene, 3-nitrofluoranthene, dinitro pyrene) (Lin et al. 2015a, b). PAHs can undergo metabolic activation resulting in the generation of diol epoxides, catalysed by the Phase I enzyme cytochrome P450 (CYP450), which can form adducts with DNA, and induce free radical production catalysed by CYP450 peroxidases. Furthermore, the metabolic activation of PAHs in the lung and liver can also lead to the generation of highly reactive PAH-quinones and ROS through the activation of the enzyme CYP4501A1 (Ma and Ma 2002). Human exposure to PAHs and volatile organics like benzene via vehicular emission has been reported to induce a greater loss of DNA integrity and a higher frequency of oxidized lesions in lymphocyte DNA relative to the non-exposed subjects (Bagryantseva et al. 2010). Airborne fine and ultrafine particles also include diesel exhaust particles, which have been reported to cause lipid peroxidation and oxidative DNA damage via the induction of intracellular ROS production (Andersson et al. 2009). Diesel exhaust particles typically consist of a carbon core or organic droplets with adsorbed organic compounds, such as PAHs, quinones, and redox-active transition metals.

Transition metals such as iron, lead, mercury, cadmium, silver, nickel, vanadium, chromium, manganese, and copper are found in airborne $PM_{2.5}$ and $PM_{0.1}$ (ultrafine particles). These soluble elements get adsorbed on the surface of suspended PMs and then leach out when inhaled deep into the lung and eventually get absorbed by the alveolar epithelium. The redox-active transition elements (e.g., iron, copper, chromium) can catalyse Fenton's reaction (e.g., $Fe^{2+} + H_2O_2 + H^+ \rightarrow Fe^{3+} + OH^{\bullet} + H_2O$) and can produce highly reactive hydroxyl radicals and thus can cause oxidative damage to the alveolar epithelial cells as well as other types of target cells including blood cells (Vidrio et al. 2008; Sørensen et al. 2005). In contrast, redox-inactive transition metals (e.g., cadmium, lead, mercury) can also induce oxidative injury to cells essentially by depleting the pool of cellular antioxidative capacity because of their high affinity of binding to thiols which occupy the active sites of enzymatic antioxidants and thereby rendering them inactive (Gray et al. 2014). An elevated level of oxidized lipids (malonaldehyde) in human urine has been found to be associated with transition metals associated with $PM_{2.5}$ and PM_{10}, thereby indicating the role of PM in inducing oxidative stress. The same study also suggested that PM-bound PAHs are also responsible for inducing oxidative stress in humans, at least in part (Bae et al. 2010). Furthermore, PM-bound transition metals such as vanadium and chromium (VI) have also been reported to cause oxidative DNA damage in human lymphocytes (Sørensen et al. 2005).

In addition to PMs, photochemical oxidants (ozone and peroxyacetyl nitrate), secondary air pollutants formed by the action of sunlight in the atmosphere, also contain reactive hydrocarbons and nitrogen oxides which are potent oxidants. Thus, exposure to elevated levels of these air pollutants can also lead to oxidative stress. For example, ozone induces cellular oxidative stress through the formation of hydrogen peroxides and ozonides, which triggers the activation of Nrf2, and genes involved in the proinflammatory response (Mumby et al. 2019). Elevated exposure to ozone has been found to induce apoptosis, oxidative DNA damage, and cytotoxicity in human alveolar epithelial type I-like cells (Kosmider et al. 2010). Moreover, ozone is also an activator of transcription factors such as protein-1 fos and c-jun, and these 2 transcription factors combine to form activator protein-1 (AP-1), which has been strongly linked with the proliferation and transformation of tumour cells (Yan et al. 2016).

11.3.2 Biomarkers of Inflammation and Carcinogenicity

Oxidative stress can trigger redox-sensitive pathways that trigger different cellular processes including inflammation and carcinogenesis. Inflammation is a critical protective mechanism which removes the pathogens and damaged cells and produces ROS able to induce cell death. In the early phase of inflammation, oxidant stress does not directly cause cell damage and can activate the transcription of genes involved in cellular defence mechanisms including antioxidant genes. This preconditioning effect of ROS increases the resistance against future inflammatory oxidant stress and facilitates tissue repair processes. However, exposure to air pollutants such as $PM_{2.5}$ and $PM_{0.1}$ amplifies the inflammatory process and consequently enhances ROS production leading to tissue injury (Kelly and Fussell 2015). Alveolar macrophages are a crucial component of the pulmonary defence mechanism and are involved in removing and processing of inhaled PMs.

Alveolar macrophages and lung epithelial cells absorb PMs through phagocytosis, which in turn triggers the production and release of proinflammatory proteins (e.g., cytokines and chemokines) that contribute to lung immune response leading to oxidative stress and systemic inflammation. These effects are mediated by the redox-sensitive mitogen-activated protein kinase (MAPK) and nuclear factor kappa B (NF-κB) that are responsible for the expression of proinflammatory proteins (Kelly and Fussell 2015). These proinflammatory mediators released from the lung stimulate the migration of polymorphonuclear neutrophils and monocytes into the circulation, further contributing to the oxidative stress and inflammation response. Myeloperoxidases released by activated neutrophils, monocytes, and macrophages can induce ROS production leading to lung damage and arterial tissue (Fiorito et al. 2018), and its activity can be used as a marker of chronic inflammation.

Human immune cells such as macrophages and T helper cells synthesize and secrete a type of small molecular weight (≈6–70 kDa) proteins known as cytokines (e.g., interleukins) which play a critical role in mediating inflammation response. Cytokines are classified based on their functions, which include both proinflammatory (e.g., IL-1β, IL-6, TNF-α, IL-8) and anti-inflammatory (e.g., IL-10, TGF-b) responses in addition to other types of immune functions such as anti-infectious (IFNa, IFNb, IFNg, TNFa), activating B cells (IL-3, IL-4, IL-5, IL-6, IL-21), activating T cells (IL-2, IL-4, IL-10, IL-13, IL-15), mast cell activating (IL-3, IL-4, IL-13, IL-5) and hematopoietic (GMCSF, IL-3, G-CSF, M-CSF, IL-5, IL-7) (Kindt et al. 2007). Several in vivo and in vitro studies documented elevated levels of circulating proinflammatory cytokines such as IL-1β, IL-6, IL-8 and TNF-α following exposure to PMs, ozone, nitrogen dioxide and volatile organic carbon compounds (reviewed by Saraswathy et al. 2018). Thus, elevated levels of these cytokines in the plasma can be used as markers of air pollution-induced chronic inflammation, which is a precursor of various diseases including COPD, cardiovascular and neurological diseases, and carcinogenesis.

Although inflammation is a beneficial response, unregulated chronic inflammation can induce malignant cell transformation in the surrounding tissue leading to carcinogenicity. The inflammatory response shares various molecular targets and signalling pathways with the carcinogenic process. For example, chronic inflammation can increase the risk of cancer by supplying bioactive molecules such as cytokines and chemokines from cells infiltrating the tumour microenvironment, thereby promoting cell proliferation and inhibition of apoptosis. Tumour necrosis factor (TNF-α) is an inflammatory cytokine that has been implicated in carcinogenesis, due to its participation in chronic inflammatory diseases (Popa et al. 2007), and thus, was suggested to be a marker of carcinogenicity. This cytokine exerts its function by binding to its receptors (TNF-αR-1 and TNF-αR-2) leading to the activation of cellular signalling cascades (e.g., NF-κB and MAPK pathways) that facilitate cell survival, differentiation and proliferation and transcription of anti-apoptotic factors (Szlosarek et al. 2006). Sustained elevated levels of TNF-α have also been suggested to promote carcinogenicity by increasing ROS and RNS (reactive nitrogen species) generation, which can induce DNA damage, leading to mutations and tumorigenesis (Balkwill 2006; Woo et al. 2000). In addition, interleukin-6 (IL-6), another proinflammatory cytokine, is also known to have protumorigenic effect and has thus been proposed as a malignancy predictor, with sensitivity and specificity of about 60–70% and 58–90%, respectively, in human studies. Elevated serum IL-6 levels have been consistently reported in humans with systemic cancers as compared to healthy individuals with benign diseases (see Heikkilä et al. 2008 for review). IL-6 plays a key role in promoting cell proliferation and inhibition of apoptosis, by binding to its receptor (IL-6Rα) and coreceptor gp130 (glycoprotein 130), which activate the Janus kinase (JAK)-signal transducer and activator of transcription (STAT) pathway. The JAK/STAT pathway has been implicated in the cancer initiation and progression pathway (Hodge et al. 2005). IL-6 has also been suggested to induce tumorigenesis by hypermethylation of tumour suppressor genes (Gasche et al. 2011), which is a frequent epigenetic event in various human cancers. Like TNF-α, IL-6 has been suggested to promote carcinogenesis by the conversion of noncancer cells into tumour stem cells (Kim et al. 2013).

The association between exposure to air pollutants and the risk of lung cancer has been evaluated by multiple studies, which have established a strong linkage between PM exposure ($PM_{2.5}$ and PM_{10}) and lung cancer (Shahadin et al. 2018; Raaschou-Nielsen et al. 2016; Chen et al. 2015). In the pathogenesis of carcinogenicity in the lung, the key mechanisms are believed to be damage to DNA in the form of base oxidation induced by pollutant-driven ROS formation and bulky adducts formation from biotransformation of PAHs, in addition to chronic inflammation, leading to chromosomal damage and mutations (Rossner et al. 2013). DNA adducts can occur due to the interaction of carcinogenic air pollutants such as PAHs or their metabolites which covalently bind with sites in DNA. Even though adducts can be removed by repair proteins, some can remain unrepaired and contribute to tumorigenesis by causing nucleotide substitutions, deletions, and chromosome rearrangements during replication (Hecht 2008). Several studies reported positive correlations between exposure to air pollution or chemicals in polluted air and the formation of DNA adducts in exposed human beings with a wide range of occupations such as bus and taxi drivers, traffic enforcement personnel, students including school children, and residents of industrial areas (see Demetriou and Vineis 2015 for review). Furthermore, oxidized DNA bases can potentially damage the integrity of the genome leading to carcinogenic mutations. For example, 8-oxo-7,8-dihydroguanine (8-oxoGua), one of the crucial lesions, leads to GC to TA transversion unless it is repaired before DNA replication. The markers of guanine oxidation (8-oxoGua) and/or its tautomeric counterpart 8-oxo-7,8-dihydro-2′-deoxyguanosine (8-oxodG) have been suggested to be potential biomarkers of carcinogenesis (Valavanidis et al. 2013). Increased excretion of 8-oxoGua has been reported in studies where individuals were exposed to high levels of air pollution (Buonaurio et al. 2022).

DNA methylation is a cellular process where a methyl group is added to the nucleotides. An increase or decrease of the methylated sites in DNA can alter the expression of genes and thereby affect the genome stability. Alterations in DNA methylation and the associated upregulation or downregulation of gene expression have been implicated in carcinogenesis (Valavanidis et al. 2013). Several studies have examined the effects of air pollutants on methylation patterns and reported that both short-term and long-term exposure to air pollutants lead to global hypomethylation (see Demetriou and Vineis 2015 for review). Most of these studies focused on long interspersed element-1 (LINE-1) and Alu elements methylation as proxies to whole genome methylation. LINE-1 and Alu elements are repetitive and mobile sequences in the genome, and their methylation correlates with the rate of overall DNA methylation in the cell. LINE-1 methylation was consistently found to be altered by exposure to air pollutants (Barchitta et al. 2018; Ding et al. 2016; Tarantini et al. 2009). Moreover, global methylation in healthy individuals was found to be decreased following exposure to ambient air pollutants (De Prins et al. 2013). Despite the small number of available studies, the replication of findings suggests exposure to air pollution influences global methylation levels, and these epigenetic changes can contribute to carcinogenesis at least as much as genetic changes. More importantly, alterations in DNA methylations, like inflammatory proteins, can be detected in human blood samples much earlier than the diagnosis of carcinogenesis, and thus, can be used as potential preclinical biomarkers.

11.4 Application of New Approaches (Omics) for Identifying Novel Biomarkers

The above-discussed traditional biomarker approaches have been extensively applied to investigate the potential impacts of air pollution on human health. However, one of the key concerns with some of these approaches is that at the time responses are observed significant progression of disease often has already occurred, which then can be difficult to cure or ameliorate (Rai et al. 2018). Furthermore, traditional research in toxicology that forms the basis for developing biomarkers and how they are related to health outcomes relies on extensive live animal testing, which violates current commitments by several jurisdictions such as the European Union and the USA to move towards an animal-free testing regime within the next decade (European Commission 2010). Most importantly, traditional biomarker studies are based on hypothesis testing that requires prior knowledge of the specific toxicological hazards associated with exposure, limiting the diagnostic power of these targeted approaches for the assessment of complex environmental exposure scenarios such as air pollution. To address these shortcomings, stakeholders from academia, industry, governments, and NGOs have requested a paradigm shift in the toxicological sciences to one that is routed in mechanistic toxicity data (Basu et al. 2019; NRC 2007) (Fig. 11.1).

The advent of high-content and high-throughput 'omics' technologies, such as sequence-by-synthesis whole genome (DNAseq), transcriptome (RNAseq) and epigenome analyses, as well as ultra-high resolution chromatography mass spectroscopy (URMS) shotgun proteomics or metabolomics, and advanced bioinformatics has enabled unprecedented, holistic analysis of biological systems in response to the exposure with chemical contaminants and other environmental stressors (Brockmeier et al. 2017; Portugal et al. 2022). These technologies have transformed the way of scientific inquiry from one that relies on the a priori formulation of specific hypotheses to be tested to one that enables the unbiased generation of new knowledge that guides the generation of a more comprehensive understanding of the molecular processes that drive health outcomes (Brockmeier et al. 2017; NRC 2007). Accordingly, the past two decades have witnessed a significant increase in toxicogenomics studies in support of environmental and human health risk assessments (reviewed in Portugal et al. 2022). In fact, since the early 2000s there have been efforts to collate the increasing toxicogenomics information generated by scientists into public databases such as the Comparative Toxicogenomics Database (CTD), the most recently curated version of which now contains information on over 45 million toxicogenomics relationships for greater than 16,000 chemicals, 50,000 genes, 5000 phenotypes, 7000 diseases and 160,000 exposure events for over 600 species (Davis et al. 2021). These databases have become a critical repository in support of novel biomarker development and research into the specific molecular mechanisms that drive health outcomes of exposure to chemical contaminants.

The majority of toxicogenomic studies to characterize the effects of air pollution have focused on the investigation of individual chemical constituents (e.g., gases, metals, PAHs) or particulate matter (PM) using in vitro or animal models (reviewed in Zavala et al. 2020). Furthermore, a limited number of studies aimed to correlate molecular response patterns with in situ exposure to air pollution in people using a range of 'omics approaches including genom-

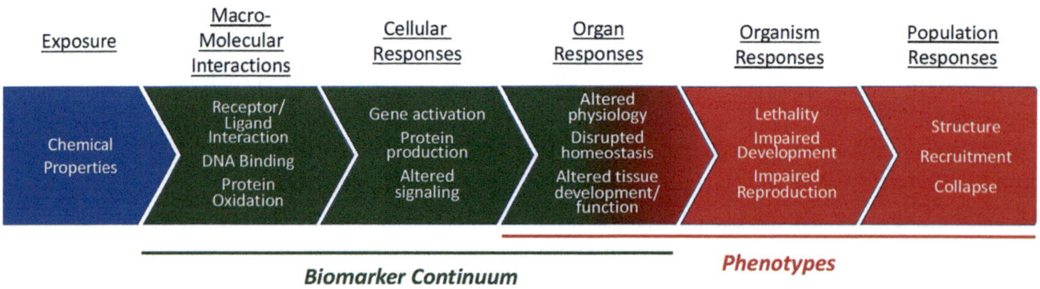

Fig. 11.1 Toxicity pathway framework describing the progression of the cascade of biological key events along organizational levels in response to exposure with a chemical contaminant

ics, transcriptomics, epigenomics, proteomics and metabolomics (reviewed in Holloway et al. 2012; Wadhwa et al. 2019). Most of these studies have focused on either genetic (e.g., single nucleotide polymorphisms [SNPs]), epigenetic (e.g., changes in DNA methylation), transcriptomics (gene expression changes) or metabolomics (changes in metabolite abundances) analyses using blood or other tissue samples (e.g. endobronchial brushings) that can be non-invasively obtained (e.g. van Leeuwen et al. 2006; Vlaanderen et al. 2022). In this section, we briefly summarize emerging 'omics biomarker approaches and their technologies (Fig. 11.2).

11.4.1 Genomics and Epigenetics

Several genomic approaches are used to identify the genetic susceptibility of humans to diseases due to environmental exposure to chemical contaminants. Among these, somatic genetic (e.g., SNPs) and epigenetic (e.g., DNA methylation) changes to the genome are among the most common endpoints investigated.

SNPs can be caused by somatic mutations as a consequence of exposure to environmental stressors such as chemical contaminants during the life cycle of a cell (reviewed in Holloway et al. 2012). The susceptibility to somatic SNPs can be affected by several factors including inherited genetics associated with certain race/ethnic groups or different locations. As such, SNPs are the result of environment-gene interactions that can affect the susceptibility of exposed individuals to chemical contaminants. Traditionally, SNPs have been identified using a candidate gene approach that focused on certain genes that were hypothesized to be associated with a disease phenotype. The limitations of these targeted approaches are that they require a priori knowledge about the DNA sites where these mutations are expected to occur. These limitations have recently been overcome by developments in gene array technologies that now enable the unbiased and genome-wide identification of SNPs, which allow for the unbiased identification of specific

Omics Level	***Examples of Endpoints***	***Examples of Technologies***
Genome	➢ Epigenetic changes (e.g. methylation, chromatin, histones) ➢ Genetic changes (mutations)	➢ Bisulfide sequencing; chromatin immunoprecipitation[1] ➢ DNA sequencing; PCR; CRISPR[2]
Transcriptome	➢ Changes in transcription of individual transcripts ➢ Changes in global gene transcription	➢ qPCR; digital PCR ➢ RNA sequencing
Proteome	➢ Changes in abundance of individual proteins ➢ Functional changes (e.g. enzyme activity, receptor binding) ➢ Changes in global proteome abundance	➢ Enzyme-linked or radio immune assays ➢ Substrate-based enzyme activity assays ➢ Mass spectroscopy
Metabolome	➢ Abundance of individual metabolites ➢ Global metabolite abundance	➢ Diverse biochemical assays ➢ Mass spectroscopy; nuclear magnetic resonance
Phenotype	➢ Behavior ➢ Histology ➢ Morphology	➢ Diverse technologies

[1] DeAngelis JT, Farrington WJ 2008. An Overview of Epigenetic Assays. Mol Biotechnol 38(2):179–183.
[2] Hajian R, Balderston S, Tran T *et al.* 2019. Detection of unamplified target genes via CRISPR–Cas9 immobilized on a graphene field-effect transistor. *Nat Biomed Eng* 3:427–437. https://doi.org/10.1038/s41551-019-0371-x

Fig. 11.2 Examples of biomarkers used to characterize exposure and health effects of air pollution genetic sub-types associated with the susceptibility to air pollution (Holloway et al. 2012). For example, a current genome-wide gene-air pollution interaction study of ~300,000 individuals from the UK Biobank using high-throughput Affymetrix Axiom UK BiLEVE and UK Biobank arrays revealed seven new interaction signals that were related to the exposure with $PM_{2.5}$ (Melbourne et al. 2022). Similarly, Gref et al. (2017) identified eight SNPs associated with links between early childhood exposure to air pollution and asthma. These studies are powerful in determining individuals or populations at greatest risk to exposure to certain air pollutants, as has been shown for O_3, particulates and NO_2 (Holloway et al. 2012). Furthermore, identifying the genetic determinants of susceptibility to air pollution is important to support the characterization of the specific mechanisms that may lead to disease outcomes.

In addition to somatic mutations, epigenetic changes to the genome can also contribute to the susceptibility and potential health risks of individuals to air pollution. Other than SNPs, epigenetic changes to the genome are not the consequence of changes/mutations to the DNA sequence but represent chemical modifications of either the DNA or its associated proteins (Wang and Chang 2018). These changes include methylation of specific DNA sequences, modifications to histones, and remodelling of chromatin, among others, that affect the expression of genes (Wang et al. 2004). Epigenetic modifications to the genome result from environment-gene interactions and, in addition to processes such as genetic imprinting and X chromosome silencing, enable rapid adaptation to environmental changes that are inheritable but also reversible. Exposure to environmental stressors including chemical contaminants has been shown to affect DNA methylation patterns. As reviewed by Holloway et al. (2012), exposure to air pollutants originating from coal, wood and cigarette smoke, or PM caused hyper- or hypomethylation of certain genes, in particular those involved with inflammation and immune responses. Similarly, several studies have linked exposure to air pollutants including black carbon, NO_2, NO_x, O_3 and PM across life stages ranging from in utero to adults with methylation of genes involved with antioxidant responses, cortisol synthesis and immune responses (reviewed in Rider and Carlsten 2019). The authors could link these methylation patterns with several health outcomes including reduced growth, oxidative stress, and immune/inflammatory response linked to asthma and other pulmonary diseases. Finally, DNA methylation patterns associated with the exposure to air pollutants including NO_2, O_3, metals, PM, and PAHs have been shown to be associated with cardiovascular and respiratory diseases as well as inflammatory and immune response (Poursafa et al. 2022). There are several targeted and untargeted approaches that are used to identify individual or whole genome DNA methylation sites (reviewed in Rider and Carlsten 2019). In this context, recent developments in high-content sequence-by-synthesis-based technologies are particularly promising as they provide unprecedented opportunities to characterize global DNA methylation patterns that enable identification of novel methylation sites.

11.4.2 Transcriptomics

Changes in the transcription of genes as a response to the exposure with a toxicant are one of the early steps in the molecular cascade that eventually results in the manifestation of a disease phenotype (Fig. 11.1). While effects at this early stage in the response of an organism to a toxic insult do not necessarily result in higher organizational-level outcomes, they provide an initial snapshot of specific molecular perturbations that occur rapidly after exposure, which can serve as an early indication of exposure and potential later health effects before these manifest. Importantly, together with higher organizational level responses (i.e., proteomics, metabolomics, physiological, histological, etc.) alterations in transcriptional responses can be used to identify and establish toxicity pathway models such as Adverse Outcome Pathways (AOPs) that can be used to predict disease phenotypes (Ankley et al. 2010; Figs. 11.1 and 11.2).

As a consequence, transcriptomics studies are increasingly used to inform molecular initiating or early key events in the development of AOPs, which can then be used as early biomarkers to support risk or hazard assessments (Vinken 2019).

While there is a significant body of literature describing the effects of specific air pollutants such as PM, NO_2, PAHs, etc. on gene expression under controlled laboratory studies using in vitro or animal models, only a very few epidemiological or population studies have been conducted that used transcriptomics investigations to assess the effects of air pollution in human populations (Liao et al. 2022; Croft et al. 2021; Roosner Jr. et al. 2015; van Leeuwen et al. 2006). The majority of studies used microarray technologies (e.g., van Leeuwen et al. 2006; Vlaanderen et al. 2022) while more recent works increasingly focus on next-generation RNAseq-based investigations (Liao et al. 2022; Croft et al. 2021; Roosner Jr. et al. 2015). The advantages of the latter technologies are that they can quantitatively capture changes across the entire transcriptome, representing all genes that are expressed at a given time, and thus, enable unbiased characterization of the transcriptional (dys)regulation of biological processes. These studies revealed that certain disease phenotypes were correlated with certain gene expression patterns as well as certain pollutants. While the most commonly observed transcriptional alterations measured in people exposed to air pollution were associated with immune or inflammatory processes, several other pathways including amino acid metabolism, neurodegenerative diseases, genotoxicity, iron homoeostasis and protein folding were dysregulated as well (Vlaanderen et al. 2022; Liao et al. 2022; Croft et al. 2021; Roosner Jr. et al. 2015). Statistical associations were observed between exposure $PM_{2.5}$ and NO_2 and dysregulation of molecular pathways associated with amino acid metabolism (Liao et al. 2022). These authors also reported correlations between exposure to PM_{10} and O_3 and immune response and one carbon metabolism. Other studies found significant changes in immune pathways, iron homeostasis and protein folding that were linked to elevated exposure to black carbon and PM, respectively (Croft et al. 2021). In conclusion, epidemiological or population whole transcriptome studies show promise in identifying specific molecular pathways and processes that are indicative of specific disease phenotypes including asthma and respiratory infection after exposure to certain air pollutants. Importantly, transcriptional responses can be detected during early exposure stages, and thus, can serve as early warning signs of increased risk of disease. However, given the paucity of studies that assessed a very limited number of air pollution scenarios and their association with only a few phenotypes (e.g., asthma), there are significant future research needs to enable robust biomarker discovery.

11.4.3 Proteomics

Proteins are the main building blocks of an organism, and disruption of their function or structure is directly related to the manifestation of disease. In fact, most biomarkers used in human health research are based on biochemical measurements of proteins and/or their activities. However, traditional biomarker studies primarily rely on hypothesis-based targeted proteomics that measure well-described endpoints associated with specific disease outcomes. Only the recent advances in URMS now enable the quantitative characterization of whole proteomes providing unprecedented opportunities for explorative research. This has opened novel opportunities for risk assessments of complex environmental pollution scenarios, including air pollution. However, single or focused panels of proteomics' biomarkers remain the most commonly used approaches in support of epidemiological or population studies to characterize the potential health risks of air pollution. In contrast, only a very small number of studies have utilized non-target or other explorative proteomics methodologies to investigate the potential molecular drivers of toxicity of air pollution in exposed populations. For example, we could identify a single cohort study that assessed the changes in protein abundances in infants exposed to traffic-related pollution in

Stockholm (Sweden) using a 92 systemic inflammation-related protein array (He et al. 2022). This work identified significant associations between certain protein markers and $PM_{2.5}$, and to a lesser extent to PM_{10} and NO_2 that may show promise as novel biomarkers for early-life exposure to air pollution. A different manuscript applied a 1300 protein target hybridization assay (SOMAmers™) in a controlled randomized cross-over study during which candidates were exposed to traffic/diesel exhaust for 4 h (Mookherjee et al. 2022). This study identified significant associations with exposure to $PM_{2.5}$ and changes in abundance of proteins involved with inflammation and cardiovascular disease, and which were proposed as a candidate biosignature for exposure to traffic-related air pollution. While scarce, these examples suggest that explorative high-content proteomics have significant promise in uncovering relevant biomarkers indicative of potential later health effects. Looking forward, the application of URMS-based non-target whole proteome technologies will further advance biomarker discovery to assess the risks of human exposure to air pollution.

11.4.4 Metabolomics

Metabolomics is defined as the study of small biomolecules that are the biochemical intermediates of metabolic processes that constitute the active molecules responsible for proper biological functioning. It represents the last step in the 'omics continuum, and disruption of the metabolome is directly associated with alternations in the physiological state of an organism. Manifestation of changes across the metabolome is the integrated consequence of earlier 'omics responses as well as environmental risk factors (e.g., exposure to pollutants), and thus, can reflect disease state (Rattray et al. 2018). Consequently, global metabolome analysis can provide insights at the systems-biology level of how environmental exposure affects populations at risk. Hence, metabolomics has become an increasingly popular approach to study the potential impacts of air pollution on humans (e.g., Jin et al. 2021).

As for the above-discussed advances in next-generation systems biology technologies to support the study of entire genomes, transcriptomes and proteomes, recent developments in mass spectroscopy (i.e., URMS) now enable the unbiased analysis of global metabolite profiles in an organism in a hypothesis-free manner. There have been several recent studies that have applied such untargeted metabolomics approaches to characterize the potential mechanisms underlying air pollution-associated disease under varying exposure conditions (reviewed in Jin et al. 2021). These studies investigated diverse scenarios ranging from short-term crossover experiments to sub-chronic or long-term cross-sectional and case-control studies. The vast majority of studies characterized the changes in abundances of metabolites in people exposed to traffic-related air pollution (e.g., van Veldhoven et al. 2019; Vlaanderen et al. 2017; Yan et al. 2019) while a few others assessed industrial exposure scenarios (e.g., living in the vicinity of petrochemical plants) (e.g., Chen et al. 2019; Kuo et al. 2012; Yuan et al. 2016). Interestingly, independent of the diverse case scenarios investigated, most of these studies identified common dysregulated pathways implicated by whole metabolome analyses: oxidative stress and inflammation (Jin et al. 2021). While several pollutant groups such as NOx, O_3 or black carbon were identified as potential factors contributing to changes in certain metabolite patterns, the main factor driving the metabolomic responses reported to date appears to be exposure to $PM_{2.5}$. In conclusion, untargeted metabolomics analyses represent an increasingly utilized and powerful approach to support human health risk assessments of air pollution.

11.5 Concluding Remarks

11.5.1 Limitations and Challenges

Significant progress has been made in context with the development and use of both targeted and non-target biomarker approaches to assess the impacts of air pollution on human health. However, there are several remaining limitations and research needs. In particular, the often-complex nature of air pollution due to the large number of diverse chemical constituents present makes it very difficult to pinpoint the specific causes for health risks in exposed populations. This represents an issue when using traditional, focused biomarker panels that may not fully capture the full suite of biological disruptions or disruption potential. As discussed in Sects. 11.3.1 and 11.3.2, oxidative stress, inflammation, and subsequent genotoxic processes are considered the predominant drivers of adverse health outcomes such as pulmonary and cardiovascular diseases and cancer associated with air pollution. However, traditional biomarker panels that were established to assess these effect types do not provide information on their specific mechanisms and causes, which is important to establish effective mitigation strategies. This is of particular concern as there are a myriad of molecular mechanisms triggered by the diverse cocktail of chemicals typical for most air pollution scenarios that converge at general toxic outcomes such as oxidative stress, inflammation, or genotoxicity. Further, many of these biomarkers are indicators of an advanced disease state at the time of analysis that may be difficult to cure.

To address some of these concerns, there have been significant efforts over the past decade to develop new approaches methods that apply systems biology technologies to further our understanding of and assess the risks posed by air pollution, and which are discussed in detail in Sect. 11.4. However, these approaches come with their own challenges. For example, new approach methods are currently limited in context with their utility in support of human health risk assessments because of the lack of harmonized, calibrated, and validated tools that can conclusively and reliably link molecular mechanistic information to apical outcomes (Basu et al. 2019). This is particularly of concern for studies generating information to support regulatory decision processes. Not only is there a lack of validated methods that have been demonstrated to accurately and reliably predict health outcomes of regulatory relevance, but there is also a significant lack of familiarity with and trust in these new approach methods by the regulatory community as well as the public that hampers adoption of such approaches (Vachon et al. 2017). These concerns are further exacerbated by the high complexity of the technologies and data evaluation approaches needed to generate and interpret the "big 'omics data" generated, and which is often not accessible and non-intuitive for non-experts. For example, URMS-based global proteomics and metabolomics analysis require expensive and sophisticated liquid chromatography mass spectroscopy instruments (e.g., ion trap Orbitrap systems) only available in a few specialized laboratories. Similarly, instrumentation required to conduct shotgun genomics and transcriptomics analyses such as Illumina's sequence-by-synthesis sequencers are costly and require highly trained personnel to operate. Most importantly, analysis of the huge amounts of data generated by these omics analyses remains challenging, and there is urgent need for consensus-based, intuitive, and harmonized bioinformatics tools. In addition, most investigations using unbiased 'omics approaches are of a qualitative nature, which further limits their usefulness to risk assessors that require quantitative data to identify toxicity thresholds that drive risks of exposure. Hence, there remain gaps in the predictive capability of 'omics data to assess health effects because they need to be anchored quantitatively to relevant apical outcomes to fully realize their potential in regulatory toxicology. Specific remaining challenges that need to be overcome to unlock the full potential of new approach methods to support risk assessments of air pollution range from technical limitations to lack of understanding, trust, and uptake of these methods by the regulatory community and the public. Technical limitations include, but are not limited to, lack of complete databases required to identify proteins or metabolites associated with mass spectra obtained, non-representative extraction of peptides or metabolites, difficulties in identification of post-translational protein modifications, difficulties in integrating big 'omics data and traditional biomarker and health data, and lack of quantitative risk assessment tools (e.g., Dupree et al. 2020; van Roekel et al. 2019). Social implications include high costs, limited accessibility of technologies and lack of trust by decision-makers and the public (Vachon et al. 2017; Pain et al. 2020; Mondou et al. 2020).

11.5.2 Future Perspectives

Over the past decades significant strides have been made in the development of biomarkers that can effectively and reliably characterize the effects of exposure of people to diverse air pollution scenarios. In particular, recent developments in next-generation 'omics' biotechnologies have revolutionized the way by which we can probe entire biological systems, which allows for the identification and development of novel and relevant biomarkers that can predict future health outcomes with greater certainty, and thus, will enable better decision-making and proactive risk management strategies. While many 'omics technologies are not readily accessible yet, there are several efforts underway to make toxicogenomics tools available in a format that is user-friendly, transparent, consistent, and reliable, so that they can feed directly into regulatory decision-making processes and render chemical management and environmental assessments more effective and ethical (Basu et al. 2019; Mondou et al. 2021). Furthermore, rapid development of artificial intelligence (AI) and machine learning tools may soon overcome current limitations in bioinformatics and enable efficient mining of the "big data" generated by new approach methods. Importantly, these AI approaches may help with the integration of diverse data types (i.e., genomics, transcriptomics, proteomics, metabolomics, biochemical, physiological, histological, apical data) that have been a main challenge for establishing conclusive toxicity pathways models and networks, and which are indispensable for novel biomarker development in support of human health risk assessment of air pollution.

References

Andersson H, Piras E, Demma J, Hellman B, Brittebo E. Low levels of the air pollutant 1-nitropyrene induce DNA damage, increased levels of reactive oxygen species and endoplasmic reticulum stress in human endothelial cells. Toxicology. 2009;262(1):57–64.

Ankley GT, Bennett RS, Erickson RJ, Hoff DJ, Hornung MW, Johnson RD, Mount DR, Nichols JW, Russom CL, Schmieder PK, Serrrano JA. Adverse outcome pathways: a conceptual framework to support ecotoxicology research and risk assessment. Environ Toxicol Chem Int J. 2010;29(3):730–41.

Bae S, Pan XC, Kim SY, Park K, Kim YH, Kim H, Hong YC. Exposures to particulate matter and polycyclic aromatic hydrocarbons and oxidative stress in schoolchildren. Environ Health Perspect. 2010;118(4):579–83.

Bagryantseva Y, Novotna B, Rossner P Jr, Chvatalova I, Milcova A, Svecova V, Lnenickova Z, Solansky I, Sram RJ. Oxidative damage to biological macromolecules in Prague bus drivers and garagemen: impact of air pollution and genetic polymorphisms. Toxicol Lett. 2010;199(1):60–8.

Balkwill F. TNF-α in promotion and progression of cancer. Cancer Metastasis Rev. 2006;25:409–16.

Barchitta M, Maugeri A, Quattrocchi A, Barone G, Mazzoleni P, Catalfo A, De Guidi G, Iemmolo MG, Crimi N, Agodi A. Mediterranean diet and particulate matter exposure are associated with LINE-1 methylation: results from a cross-sectional study in women. Front Genet. 2018;9:514.

Barr DB, Wang RY, Needham LL. Biologic monitoring of exposure to environmental chemicals throughout the life stages: requirements and issues for consideration for the National Children's Study. Environ Health Perspect. 2005;113(8):1083–91.

Barr DB, Puttaswamy N, Jaacks LM, Steenland K, Rajkumar S, Gupton S, Ryan PB, Balakrishnan K, Peel JL, Checkley W, Clasen T. Design and rationale of the biomarker center of the Household Air Pollution Intervention Network (HAPIN) trial. Environ Health Perspect. 2020;128(4):047010.

Basu N, Crump D, Head J, Hickey G, Hogan N, Maguire S, Xia J, Hecker M. EcoToxChip: a next generation toxicogenomics tool for chemical prioritization and environmental management. Environ Toxicol Chem. 2019;38(2):279–88.

Brockmeier EK, Hodges G, Hutchinson TH, Butler E, Hecker M, Tollefsen KE, Garcia-Reyero N, Kille P, Becker D, Chipman K, Colbourne J. The role of omics in the application of adverse outcome pathways for chemical risk assessment. Toxicol Sci. 2017;158(2):252–62.

Buonaurio F, Borra F, Pigini D, Paci E, Spagnoli M, Astolfi ML, Giampaoli O, Sciubba F, Miccheli A, Canepari S, Ancona C. Biomonitoring of exposure to urban pollutants and oxidative stress during the COVID-19 lockdown in Rome residents. Toxics. 2022;10(5):267.

Burnett R, Chen H, Szyszkowicz M, Fann N, Hubbell B, Pope CA III, Apte JS, Brauer M, Cohen A, Weichenthal S, Coggins J. Global estimates of mortality associated with long-term exposure to outdoor fine particulate matter. Proc Natl Acad Sci. 2018;115(38):9592–7.

Califf RM. Biomarker definitions and their applications. Exp Biol Med. 2018;243(3):213–21.

Canella R, Benedusi M, Martini M, Cervellati F, Cavicchio C, Valacchi G. Role of Nrf2 in preventing oxidative stress induced chloride current alteration in human lung cells. J Cell Physiol. 2018;233(8):6018–27.

Chen G, Wan X, Yang G, Zou X. Traffic-related air pollution and lung cancer: a meta-analysis. Thoracic Cancer. 2015;6(3):307–18.

Chen CHS, Kuo TC, Kuo HC, Tseng YJ, Kuo CH, Yuan TH, Chan CC. Metabolomics of children and adolescents exposed to industrial carcinogenic pollutants. Environ Sci Technol. 2019;53(9):5454–65.

Chirino YI, Sánchez-Pérez Y, Osornio-Vargas ÁR, Morales-Bárcenas R, Gutiérrez-Ruíz MC, Segura-García Y, Rosas I, Pedraza-Chaverri J, García-Cuellar CM. PM10 impairs the antioxidant defense system and exacerbates oxidative stress driven cell death. Toxicol Lett. 2010;193(3):209–16.

Cohen AJ, Brauer M, Burnett R, Anderson HR, Frostad J, Estep K, Balakrishnan K, Brunekreef B, Dandona L, Dandona R, Feigin V. Estimates and 25-year trends of the global burden of disease attributable to ambient air pollution: an analysis of data from the Global Burden of Diseases Study 2015. Lancet. 2017;389(10082):1907–18.

Cohen Hubal EA, Wetmore BA, Wambaugh JF, El-Masri H, Sobus JR, Bahadori T. Advancing internal exposure and physiologically based toxicokinetic modeling for 21st-century risk assessments. J Expo Sci Environ Epidemiol. 2019;29(1):11–20.

Croft DP, Burton DS, Nagel DJ, Bhattacharya S, Falsey AR, Georas SN, Hopke PK, Johnston CJ, Kottmann RM, Litonjua AA, Mariani TJ, Rich DQ, Thevenet-Morrison K, Thurston SW, Utell MJ, McCall MN. The effect of air pollution on the transcriptomics of the immune response to respiratory infection. Sci Rep. 2021;11(1):19436.

Davis AP, Grondin CJ, Johnson RJ, Sciaky D, Wiegers J, Wiegers TC, Mattingly CJ. Comparative toxicogenomics database (CTD): update 2021. Nucleic Acids Res. 2021;49(D1):D1138–43.

De Prins S, Koppen G, Jacobs G, Dons E, Van de Mieroop E, Nelen V, Fierens F, Panis LI, De Boever P, Cox B, Nawrot TS. Influence of ambient air pollution on global DNA methylation in healthy adults: a seasonal follow-up. Environ Int. 2013;59:418–24.

Delfino RJ, Staimer N, Vaziri ND. Air pollution and circulating biomarkers of oxidative stress. Air Qual Atmos Health. 2011;4(1):37–52.

Demetriou CA, Vineis P. Carcinogenicity of ambient air pollution: use of biomarkers, lessons learnt and future directions. J Thorac Dis. 2015;7(1):67.

Ding R, Jin Y, Liu X, Zhu Z, Zhang Y, Wang T, Xu Y. Characteristics of DNA methylation changes induced by traffic-related air pollution. Mutat Res Genet Toxicol Environ Mutagen. 2016;796:46–53.

Dupree EJ, Jayathirtha M, Yorkey H, Mihasan M, Petre BA, Darie CC. A critical review of bottom-up proteomics: the good, the bad, and the future of this field. Proteomes. 2020;8(3):14.

European Commission. Directive 2010/63/EU of the European Parliament and of the Council of 22 September 2010 on the protection of animals used for scientific purposes. Off J Eur Union. 2010;50:33–79.

Fiorito G, Vlaanderen J, Polidoro S, Gulliver J, Galassi C, Ranzi A, Krogh V, Grioni S, Agnoli C, Sacerdote C, Panico S. Oxidative stress and inflammation mediate the effect of air pollution on cardio-and cerebrovascular disease: a prospective study in nonsmokers. Environ Mol Mutagen. 2018;59(3):234–46.

Gasche JA, Hoffmann J, Boland CR, Goel A. Interleukin-6 promotes tumorigenesis by altering DNA methylation in oral cancer cells. Int J Cancer. 2011;129(5):1053–63.

Gray DL, Wallace LA, Brinkman MC, Buehler SS, La Londe C. Respiratory and cardiovascular effects of metals in ambient particulate matter: a critical review. Rev Environ Contam Toxicol. 2014:135–203.

Gref A, Merid SK, Gruzieva O, Ballereau S, Becker A, Bellander T, Bergström A, Bossé Y, Bottai M, Chan-Yeung M, Fuertes E. Genome-wide interaction analysis of air pollution exposure and childhood asthma with functional follow-up. Am J Respir Crit Care Med. 2017;195(10):1373–83.

Hahad O, Lelieveld J, Birklein F, Lieb K, Daiber A, Münzel T. Ambient air pollution increases the risk of cerebrovascular and neuropsychiatric disorders through induction of inflammation and oxidative stress. Int J Mol Sci. 2020;21(12):4306.

He S, Klevebro S, Baldanzi G, Pershagen G, Lundberg B, Eneroth K, Hedman AM, Andolf E, Almqvist C, Bottai M, Melén E. Ambient air pollution and inflammation-related proteins during early childhood. Environ Res. 2022;215:114364.

Hecht SS. Carcinogen metabolites as biomarkers. In: Molecular epidemiology of chronic diseases. Wiley; 2008. p. 97–110. 14 p.

Heikkilä K, Ebrahim S, Lawlar D.A. Systematic review of the association between circulating interleukin-6 (IL-6) and cancer. Eur J Cancer. 2008;44:937–45.

Hodge DR, Hurt EM, Farrar WL. The role of IL-6 and STAT3 in inflammation and cancer. Eur J Cancer. 2005;41(16):2502–12.

Holloway JW, Savarimuthu Francis S, Fong KM, Yang IA. Genomics and the respiratory effects of air pollution exposure. Respirology. 2012;17(4):590–600.

Jeng HA. Chemical composition of ambient particulate matter and redox activity. Environ Monit Assess. 2010;169:597–606.

Jin L, Godri Pollitt KJ, Liew Z, Rosen Vollmar AK, Vasiliou V, Johnson CH, Zhang Y. Use of untargeted metabolomics to explore the air pollution-related disease continuum. Curr Environ Health Rep. 2021;8:7–22.

Kang KW, Lee SJ, Kim SG. Molecular mechanism of nrf2 activation by oxidative stress. Antioxid Redox Signal. 2005;7(11–12):1664–73.

Kelly FJ, Fussell JC. Linking ambient particulate matter pollution effects with oxidative biology and immune responses. Ann N Y Acad Sci. 2015;1340(1):84–94.

Kelly FJ, Tetley TD. Nitrogen dioxide depletes uric acid and ascorbic acid but not glutathione from lung lining fluid. Biochem J. 1997;325(1):95–9.

Kim SY, Kang JW, Song X, Kim BK, Yoo YD, Kwon YT, Lee YJ. Role of the IL-6-JAK1-STAT3-Oct-4 pathway in the conversion of non-stem cancer cells into cancer stem-like cells. Cell Signal. 2013;25(4):961–9.

Kim D, Chen Z, Zhou LF, Huang SX. Air pollutants and early origins of respiratory diseases. Chronic Dis Transl Med. 2018;4(2):75–94.

Kim JB, Prunicki M, Haddad F, Dant C, Sampath V, Patel R, Smith E, Akdis C, Balmes J, Snyder MP, Wu JC. Cumulative lifetime burden of cardiovascular disease from early exposure to air pollution. J Am Heart Assoc. 2020;9(6):e014944.

Kindt TJ, Goldsby RA, Osborne BA, Kuby J. Kuby immunology. Macmillan; 2007.

Kosmider B, Loader JE, Murphy RC, Mason RJ. Apoptosis induced by ozone and oxysterols in human alveolar epithelial cells. Free Radic Biol Med. 2010;48(11):1513–24.

Kuo CH, Wang KC, Tian TF, Tsai MH, Chiung YM, Hsiech CM, Tsai SJ, Wang SY, Tsai DM, Huang CC, Tseng YJ. Metabolomic characterization of laborers exposed to welding fumes. Chem Res Toxicol. 2012;25(3):676–86.

Liao J, Gheissari R, Thomas DC, Gilliland FD, Lurmann F, Khandaker TI, Chen Z. Transcriptomic and metabolomic associations with exposures to air pollutants among young adults with childhood asthma history. Environ Pollut. 2022;299:118903.

Lin Y, Ma Y, Qiu X, Li R, Fang Y, Wang J, Zhu Y, Hu D. Sources, transformation, and health implications of PAHs and their nitrated, hydroxylated, and oxygenated derivatives in PM2. 5 in Beijing. J Geophys Res Atmos. 2015a;120(14):7219–28.

Lin Y, Qiu X, Ma Y, Ma J, Zheng M, Shao M. Concentrations and spatial distribution of polycyclic aromatic hydrocarbons (PAHs) and nitrated PAHs (NPAHs) in the atmosphere of North China, and the transformation from PAHs to NPAHs. Environ Pollut. 2015b;196:164–70.

Lioy PJ, Smith KR. A discussion of exposure science in the 21st century: a vision and a strategy. Environ Health Perspect. 2013;121(4):405–9.

Lugrin J, Rosenblatt-Velin N, Parapanov R, Liaudet L. The role of oxidative stress during inflammatory processes. Biol Chem. 2014;395(2):203–30.

Ma JY, Ma JK. The dual effect of the particulate and organic components of diesel exhaust particles on the alteration of pulmonary immune/inflammatory responses and metabolic enzymes. J Environ Sci Health C. 2002;20(2):117–47.

Mayeux R. Biomarkers: potential uses and limitations. NeuroRx. 2004;1:182–8.

Melbourne CA, Erzurumluoglu AM, Shrine N, Chen J, Tobin MD, Hansell AL, Wain LV. Genome-wide gene-air pollution interaction analysis of lung function in 300,000 individuals. Environ Int. 2022;159:107041.

Miller MR. Oxidative stress and the cardiovascular effects of air pollution. Free Radic Biol Med. 2020;151:69–87.

Møller P, Loft S. Oxidative damage to DNA and lipids as biomarkers of exposure to air pollution. Environ Health Perspect. 2010;118(8):1126–36.

Mondou M, Hickey GM, Rahman HT, Maguire S, Pain G, Crump D, Hecker M, Basu N. Factors affecting the perception of New Approach Methodologies (NAMs) in the ecotoxicology community. Integr Environ Assess Manag. 2020;16(2):269–81.

Mondou M, Maguire S, Pain G, Crump D, Hecker M, Basu N, Hickey GM. Envisioning an international validation process for New Approach Methodologies in chemical hazard and risk assessment. Environ Adv. 2021;4:100061.

Mookherjee N, Ryu MH, Hemshekhar M, Orach J, Spicer V, Carlsten C. Defining the effects of traffic-related air pollution on the human plasma proteome using an aptamer proteomic array: a dose-dependent increase in atherosclerosis-related proteins. Environ Res. 2022;209:112803.

Mumby S, Chung KF, Adcock IM. Transcriptional effects of ozone and impact on airway inflammation. Front Immunol. 2019;10:1610.

Münzel T, Miller MR, Sørensen M, Lelieveld J, Daiber A, Rajagopalan S. Reduction of environmental pollutants for prevention of cardiovascular disease: it's time to act. Eur Heart J. 2020;41(41):3989.

National Research Council. Toxicity testing in the 21st century: a vision and a strategy. Washington, DC: National Academies Press; 2007.

Needham LL, Özkaynak H, Whyatt RM, Barr DB, Wang RY, Naeher L, Akland G, Bahadori T, Bradman A, Fortmann R, Liu LS. Exposure assessment in the National Children's Study: introduction. Environ Health Perspect. 2005;113(8):1076–82.

Pain G, Hickey G, Mondou M, Crump D, Hecker M, Basu N, Maguire S. Drivers of and obstacles to the adoption of toxicogenomics for chemical risk assessment: insights from social science perspectives. Environ Health Perspect. 2020;128(10):105002.

Pardo M, Xu F, Shemesh M, Qiu X, Barak Y, Zhu T, Rudich Y. Nrf2 protects against diverse PM2. 5 components-induced mitochondrial oxidative damage in lung cells. Sci Total Environ. 2019;669:303–13.

Pardo M, Qiu X, Zimmermann R, Rudich Y. Particulate matter toxicity is Nrf2 and mitochondria dependent: the roles of metals and polycyclic aromatic hydrocarbons. Chem Res Toxicol. 2020;33(5):1110–20.

Phaniendra A, Jestadi DB, Periyasamy L. Free radicals: properties, sources, targets, and their implication in various diseases. Indian J Clin Biochem. 2015;30:11–26.

Popa C, Netea MG, Van Riel PL, Van Der Meer JW, Stalenhoef AF. The role of TNF-α in chronic inflammatory conditions, intermediary metabolism, and cardiovascular risk. J Lipid Res. 2007;48(4):751–62.

Portugal J, Mansilla S, Piña B. Perspectives on the use of toxicogenomics to assess environmental risk. Front Biosci (Landmark Ed). 2022;27(10):294.

Poursafa P, Kamali Z, Fraszczyk E, Boezen HM, Vaez A, Snieder H. DNA methylation: a potential mediator between air pollution and metabolic syndrome. Clin Epigenetics. 2022;14(1):1–13.

Raaschou-Nielsen O, Beelen R, Wang M, Hoek G, Andersen ZJ, Hoffmann B, Stafoggia M, Samoli E, Weinmayr G, Dimakopoulou K, Nieuwenhuijsen M. Particulate matter air pollution components and risk for lung cancer. Environ Int. 2016;87:66–73.

Radan M, Dianat M, Badavi M, Mard SA, Bayati V, Goudarzi G. In vivo and in vitro evidence for the involvement of Nrf2-antioxidant response element signaling pathway in the inflammation and oxidative stress induced by particulate matter (PM10): the effective role of gallic acid. Free Radic Res. 2019;53(2):210–25.

Rai V, Mukherjee R, Ghosh AK, Routray A, Chakraborty C. "Omics" in oral cancer: new approaches for biomarker discovery. Arch Oral Biol. 2018;87:15–34.

Rao X, Zhong J, Brook RD, Rajagopalan S. Effect of particulate matter air pollution on cardiovascular oxidative stress pathways. Antioxid Redox Signal. 2018;28(9):797–818.

Rattray NJ, Deziel NC, Wallach JD, Khan SA, Vasiliou V, Ioannidis J, Johnson CH. Beyond genomics: understanding exposotypes through metabolomics. Hum Genomics. 2018;12(1):1–14.

Rider CF, Carlsten C. Air pollution and DNA methylation: effects of exposure in humans. Clin Epigenetics. 2019;11(1):1–15.

Roosner P Jr, Tulupova E, Rossnerova A, Libalova H, Honkova K, Gmuender H, Pastorkova A, Svecova V, Topinka J, Sram RJ. Reduced gene expression levels after chronic exposure to high concentrations of air pollutant. Mutat Res. 2015;780:60–70.

Rossner P, Svecova V, Schmuczerova J, Milcova A, Tabashidze N, Topinka J, Pastorkova A, Sram RJ. Analysis of biomarkers in a Czech population exposed to heavy air pollution. Part I: bulky DNA adducts. Mutagenesis. 2013;28(1):89–95.

Rubio V, Zhang J, Valverde M, Rojas E, Shi ZZ. Essential role of Nrf2 in protection against hydroquinone-and benzoquinone-induced cytotoxicity. Toxicol In Vitro. 2011;25(2):521–9.

Saraswathy S, Dixit R, Agarwal P, Krishnan M. Air Pollution and Cytokines. J Med Acad. 2018;1(1):43–9.

Schieber M, Chandel NS. ROS function in redox signaling and oxidative stress. Curr Biol. 2014;24(10):R453–62.

Shahadin MS, Mutalib NSA, Latif MT, Greene CM, Hassan T. Challenges and future direction of molecular research in air pollution-related lung cancers. Lung Cancer. 2018;118:69–75.

Shen Y, Wu Y, Chen G, Van Grinsven HJ, Wang X, Gu B, Lou X. Non-linear increase of respiratory diseases and their costs under severe air pollution. Environ Pollut. 2017;224:631–7.

Sørensen M, Schins RP, Hertel O, Loft S. Transition metals in personal samples of PM2. 5 and oxidative stress in human volunteers. Cancer Epidemiol Biomarkers Prev. 2005;14(5):1340–3.

Szlosarek P, Charles KA, Balkwill FR. Tumour necrosis factor-α as a tumour promoter. Eur J Cancer. 2006;42(6):745–50.

Tarantini L, Bonzini M, Apostoli P, Pegoraro V, Bollati V, Marinelli B, Cantone L, Rizzo G, Hou L, Schwartz J, Bertazzi PA. Effects of particulate matter on genomic DNA methylation content and iNOS promoter methylation. Environ Health Perspect. 2009;117(2):217–22.

Vachon J, Campagna C, Rodriguez MJ, Sirard MA, Levallois P. Barriers to the use of toxicogenomics data in human health risk assessment: a survey of Canadian risk assessors. Regul Toxicol Pharmacol. 2017;85:119–23.

Valavanidis A, Fiotakis K, Vlachogianni T. Airborne particulate matter and human health: toxicological assessment and importance of size and composition of particles for oxidative damage and carcinogenic mechanisms. J Environ Sci Health C. 2008;26(4):339–62.

Valavanidis A, Vlachogianni T, Fiotakis K, Loridas S. Pulmonary oxidative stress, inflammation and cancer: respirable particulate matter, fibrous dusts and ozone as major causes of lung carcinogenesis through reactive oxygen species mechanisms. Int J Environ Res Public Health. 2013;10(9):3886–907.

van Leeuwen DM, van Herwijnen MHM, Pedersen M, Knudsen LE, Kirsch-Volders M, Sram RJ, Staal YCM, Bajak E, van Delft JHM, Kleinjans JCS. Genome-wide differential gene expression in children exposed to air pollution in the Czech Republic. Mutat Res. 2006;600(1–2):12–22.

van Roekel EH, Loftfield E, Kelly RS, Zeleznik OA, Zanetti KA. Metabolomics in epidemiologic research: challenges and opportunities for early-career epidemiologists. Metabolomics. 2019;15:1–7.

van Veldhoven K, Kiss A, Keski-Rahkonen P, Robinot N, Scalbert A, Cullinan P, Chung KF, Collins P, Sinharay R, Barratt BM, Nieuwenhuijsen M. Impact of short-term traffic-related air pollution on the metabolome–results from two metabolome-wide experimental studies. Environ Int. 2019;123:124–31.

Vidrio E, Jung H, Anastasio C. Generation of hydroxyl radicals from dissolved transition metals in surrogate lung fluid solutions. Atmos Environ. 2008;42(18):4369–79.

Vincent JH. Occupational and environmental aerosol exposure assessment: a scientific journey from the past, through the present and into the future. J Environ Monit. 2012;14(2):340–7.

Vinken M. Omics-based input and output in the development and use of adverse outcome pathways. Curr Opin Toxicol. 2019;18:8–12.

Vlaanderen JJ, Janssen NA, Hoek G, Keski-Rahkonen P, Barupal DK, Cassee FR, Gosens I, Strak M, Steenhof M, Lan Q, Brunekreef B. The impact of ambient air pollution on the human blood metabolome. Environ Res. 2017;156:341–8.

Vlaanderen J, Vermeulen R, Whitaker M, Chadeau-Hyam M, Hottenga J, de Geus E, Willemsen G, Penninx BWJH, Jansen R, Boomsama DI. 2022. Impact of long-term exposure to PM2.5 on peripheral bllod gene

expression pathways involved in cell signalling and immune response. Environ Int. 2022;168:107491.
Wadhwa R, Aggarwal T, Malyla V, Kumar N, Gupta G, Chellappan DK, Dureja H, Mehta M, Satija S, Gulati M, Maurya PK, Collet T, Hansbro PM, Dua K. Identification of biomarkers and genetic approaches toward chronic obstructive pulmonary disease. J Cell Physiol. 2019;234:16703–23.
Wang KC, Chang HY. Epigenomics: technologies and applications. Circ Res. 2018;122(9):1191–9.
Wang Y, Fischle W, Cheung W, Jacobs S, Khorasanizadeh S, Allis CD. Beyond the double helix: writing and reading the histone code. In: Reversible protein acetylation: Novartis Foundation Symposium 259, vol. 259. Chichester: Wiley; 2004, April. p. 3–21.
Woo CH, Eom YW, Yoo MH, You HJ, Han HJ, Song WK, Yoo YJ, Chun JS, Kim JH. Tumor necrosis factor-α generates reactive oxygen species via a cytosolic phospholipase A2-linked cascade. J Biol Chem. 2000;275(41):32357–62.
World Health Organization. WHO international programme on chemical safety biomarkers in risk assessment: validity and validations. 2001.
Yan Z, Jin Y, An Z, Liu Y, Samet JM, Wu W. Inflammatory cell signaling following exposures to particulate matter and ozone. Biochim Biophys Acta. 2016;1860(12):2826–34.
Yan Q, Liew Z, Uppal K, Cui X, Ling C, Heck JE, von Ehrenstein OS, Wu J, Walker DI, Jones DP, Ritz B. Maternal serum metabolome and traffic-related air pollution exposure in pregnancy. Environ Int. 2019;130:104872.
Yuan TH, Chung MK, Lin CY, Chen ST, Wu KY, Chan CC. Metabolic profiling of residents in the vicinity of a petrochemical complex. Sci Total Environ. 2016;548:260–9.
Zavala J, Freedman AN, Szilagyi JT, Jaspers I, Wambaugh JF, Higuchi M, Rager JE. New Approach Methods to Evaluate Health Risks of Airpollutants: Critical Design Considerations for In Vitro Exposure Testing. Int J Environ Res Public Health. 2020;17(6):2124.

12 Environmental Emission from Coal-Fired Power Plant and Control Technology

Hridesh Agarwalla, Monalisa Gangopadhyay, R. Ebhin Masto, Siddharth Bari, and Manish Kumar

12.1 Introduction

Globally air pollution has become a major concern to human health. With increasing population, industrialisation and inadequate emission control techniques, air pollution in developing countries is severe (Guo et al. 2018; Cohen et al. 2005). Combustion of fossil fuels like coal is the major contributor to air pollution (WHO 2018). On combustion, coal emits particulate matter ($PM_{2.5}$) gaseous pollutants such as carbon dioxide (CO_2), sulphur dioxide (SO_2), nitrogen oxides (NOx) along with trace elements like mercury, arsenic, chromium, nickel, other heavy metals, poly aromatic hydrocarbons (PAHs) and acid gases (HCl, HF) into the atmosphere. All these pollutants have various impacts on environment and human health. CO_2 is a major contributor to greenhouse gases and leads to global warming whereas PM causes respiratory and cardiovascular morbidity and lung cancer (Hoek et al. 2013; Oberschelp et al. 2019). SO_2 acts as precursor for $PM_{2.5}$ and also causes acid rain. NOx triggers asthma symptoms and is one of the major reasons for smog formation. PAHs are associated with serious diseases such as cancer, malformation and mutation, on the other mercury toxicity and Minamata disease are well known.

Controlling the emissions from power plants has now become a global priority, amidst increasing industrialisation. A recent report utilises projected the impact of $PM_{2.5}$ on mortality, which suggests power plants contribute to ~13% to ambient $PM_{2.5}$ (Cropper et al. 2021). In India, emissions from coal-fired power plants were linked to ~75% of premature cardiopulmonary deaths for persons older than 30 years in the neighbourhood of power plants (Gupta and Spears 2017).

12.2 Indian Scenario

In the last decade, coal-fired thermal plants have been the pillar of growth of the Indian power sector. To meet the energy demand of growing nation, new technologically efficient plants have come up. As of March 2020, the coal-fired power plants (CFPP) constitute ~55% of the installed capacity of the country (370GW). However, at the same time, CFPPs contribute to ~71% of total power generation, which shows the dependency on CFPPs (CEA 2020). In total ~ 647 MT of coal and lignite were consumed for power generation

H. Agarwalla (✉) · M. Gangopadhyay · R. E. Masto
S. Bari · M. Kumar
Environment Emission and CRM Division, CSIR-Central Institute of Mining and Fuel Research, Dhanbad, India

P. K. Padhy et al. (eds.), *Air Quality and Human Health*,
https://doi.org/10.1007/978-981-97-1363-9_12

in 2019–20 and it is expected to grow continuously (Energy statistics 2021). However, various measure was taken to address the environmental issues along with the closure of old plants and the adoption of advanced air pollution control devices.

12.3 Environmental Standards

There are efforts to reduce emissions from power plants. India is committed to the Paris Agreement on climate change, working on long-term policies to reduce greenhouse gas emission (GHG) and keep global warming below 2 °C above pre-industrial levels (Nagendra et al. 2020). This led to the formulation of new threshold limits. Moreover, air pollution control devices were improved or additional pollution controls were planned. As per the MoEFCC's (Ministry of Environment, Forest and Climate Change 2018) notification, now power plants have to comply with the prescribed standards as shown in Table 12.1 (MoEFCC 2018; Bhati 2016). To comply with the new norms, additional air pollution control devices such as flue gas desulphurisation (FGD), catalytic reactors, etc. need to be retrofitted in the power plants.

12.4 Air Emissions

On coal combustion, various gases and particulate matter are generated which were emitted through stack. The concentration of these gases and PM depends on the quality of the coal and Air Pollution Control Device (APCD) installed in the plant. Schematic diagram of various APCDs in power plant is shown in Fig. 12.1. Major pollutants released through stacks are CO_2, SOx, NOx, Hg and PAH.

12.4.1 Carbon Dioxide Emission

Dunn and Flavin (Dunn and Flavin 2002) in their study identified anthropogenic carbon dioxide emission from fossil fuels as the most important contributor to greenhouse gas, which led to climate change. Carbon dioxide constitutes 0.03% of the atmosphere and is important for all living organisms. However, due the industrial activities and deforestation, the balance of CO_2 changes and the average CO_2 concentration has increased from the preindustrial era. As per the Intergovernmental Panel on Climate Change (IPCC), the global average temperature of the earth surface has increased

Table 12.1 Emission standards for power plants

Country		PM (mg/Nm^3)	SOx (mg/Nm^3)	NOx (mg/Nm^3)	Hg (mg/Nm^3)
India					
Old standard		100–350	None	None	None
New standard (wef 2016)	Installed before 2004	100	600 < 500 MW 200 ≥ 500 MW	600	0.03
	Installed between 2004 and 2016	50	600 < 500 MW 200 ≥ 500 MW	300	0.03
	Installed after 2016	30	100	100	0.03
China		30	100	100	0.03
USA		14.5	100	100	0.017

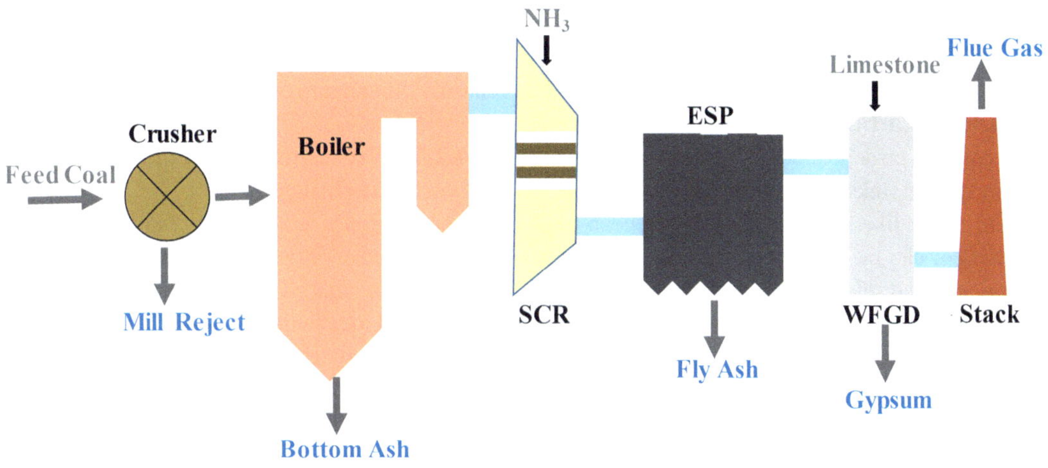

Fig. 12.1 Schematic diagram of power plant with air pollution control devices

by 0.3–0.6 °C from the nineteenth century (Houghton et al. 2001).

Carbon is the major constituent of coal and in thermal power plants coal is combusted in the presence of O_2 to produce CO_2. On complete combustion, 1 kg of carbon produces 3.66 kg of CO_2. In power plants, the carbon present in coal combusted in the presence of oxygen to form carbon dioxide. A small amount of the carbon remains un-burnt and associated with ashes as un-burnt carbon. Depending on the operational conditions, the unburnt carbon in the fly ash and bottom ash varies in the range of 0.5–10% and 2–30% respectively. The CO_2 emission from power plants varies depending on the quality and efficiency of the plant. Total CO_2 emission from power plants in India is shown in Fig. 12.2. The average CO_2 emission per unit of electricity generation from power plants was 0.98 tonne/MWh in 2019–20 in India (CEA 2021).

Government of India recognises the importance of CO_2 mitigation and several efforts have been made to reduce the carbon footprint of power plants. The mitigation strategies include cleaner fuels, increased efficiency of plants, clean coal technologies and carbon sequestration. A combination of low-carbon fuel with efficient system and improved carbon sinks will improve carbon mitigation.

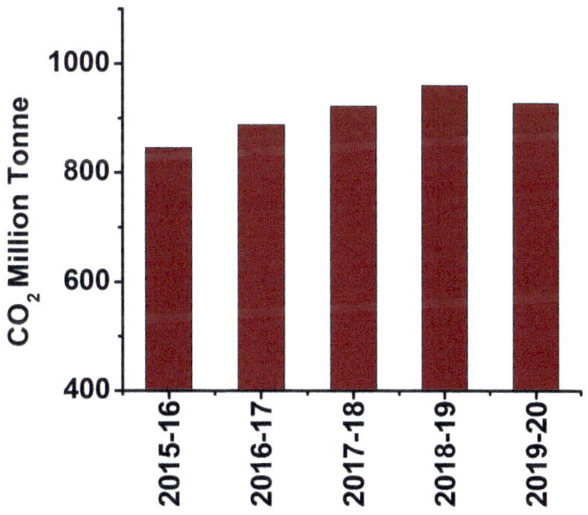

Fig. 12.2 Year-wise CO_2 emission in India from power sector

12.4.1.1 Efficiency Improvement

Carbon dioxide emission from power plants can be reduced by increasing the generation efficiency of the plant. Higher efficiency can be achieved by improved heat rate, reduced auxiliary power consumption and transmission and distribution losses.

12.4.1.2 Clean Coal Technology

For sustainable and eco-friendly use of coal, it is needed to develop technologies which reduce harmful emissions and increase the efficiency of plants known as clean coal technology (CCT). Integrated gasification combined cycle (IGCC) and combined cycle with pulverised fluidised bed combustion (PFBC) are technologies with highest potential to minimise emissions. Reports suggest that CO_2 emission can be reduced from 0.87 kg/Kwh for conventional plant to 0.73 kg/

Kwh for IGCC with hot gas cleaning system (Raghuvanshi et al. 2006).

12.4.1.3 Carbon Dioxide Sinks

CO_2 can also be deposited in the sinks which may be biosphere sinks like oceans, forests, or geosphere sinks like oil reserves, deep ocean and aquifers or can be converted into other materials chemicals and plastics. Naturally, oceans dissolve CO_2 and the amount varies with the best estimate of 7.4 Gt CO_2/year (Raghuvanshi et al. 2006). Afforestation, reforestation and agro forestry are great ways to reduce GHG emissions in a sustainable way. Co-firing of biomass is also an alternative to reduce carbon footprint. Moreover, CO_2 may be sequestered in deep ocean beds of oil and gas fields (IEA 1995).

12.4.2 Sulphur Dioxide Emission

Combustion of fossil fuels is the major contributor to the anthropogenic SO_2 emission. The organic or inorganic sulphur present in coal/fuels gets combusted in the presence of oxygen and is mainly converted to sulphur dioxide. SO_2 has adverse effect on human health and causes many diseases like respiratory irritation, bronchitis, etc. (Zhang et al. 2017). SO_2 is of primary concern for smog and fine particle in air as it forms sulphate aerosols and relates to premature deaths in several countries (Huang et al. 2014, Chakraborty et al. 2016). In India, SO_2 emissions show increasing trend and estimated to be 9.5–12.6 Mt. for the year 2016. It is estimated that 12–16 g SO_2 was emitted on combustion of 1 kg of coal. With the current trend, India may become the largest emitter of sulphur dioxide (Li et al. 2017). Indian coal contains <1% sulphur, however studies show coal-fired power plants contribute to 46–69% of total SO_2 emission (Garg et al. 2002; Lu et al. 2013).

Efforts have been made to reduce the SO_2 emission from power plants and the first emission limits came to light in 2015 by MoEFCC notification. Now the CFPPs are required to limit their SO_2 emission to the prescribed value as shown in Table 12.1. Both pre-combustion and post-combustion techniques are important for SO_2 removal (Sugathana et al. 2018).

12.4.2.1 Pre-Combustion Control Techniques

Pre-combustion techniques mainly revolve around the use of alternate low-sulphur fuels. Otherwise in coal washing, by application of physical and mechanical process sulphur content in coal was reduced. Moreover, during combustion use of sorbents like lime for SO_2 removal is also in application specially for high sulphur-containing CFBC plants combusting lignite.

12.4.2.2 Post-Combustion Control Technology

Flue gas desulphurisation (FGD) is the main technology used for SO_2 removal from flue gas. Two major types of FGD are generally commercialised: (a) Wet FGD, (b) Dry FGD.

Wet FGD

Wet flue gas desulphurisation (WFGD) is the most popular technology for SO_2 removal. In general, WFGD uses a slurry of limestone in the spray chamber, which reacts with SO_2 to form gypsum. In this up to 98% SO_2 removal efficiency was achieved.

Dry FGD

Dry flue gas desulphurisation (DFGD) uses lime slurry in a spray drying chamber and reacts with SO_2 to form $CaSO_4$. At higher temperature of the spray chamber, water evaporates and dry powder of $CaSO_4$ is removed by particulate control system.

12.4.3 Nitrogen Oxide Emission

Nitrogen oxides (NOx) generally consisting of nitric oxide (NO) and nitrogen dioxide (NO_2) are harmful for humans and environment. It is one of the main reasons for photochemical smog, acid rain and ozone accumulation (Omidvarborna et al. 2015). NOx emission is linked to respiratory disease, bronchospasm, and coughing in humans. Also elevated level of NOx in air was reported to reduce crop yield and growth (Hesterberg et al. 2009; Chen et al. 2007). During the combustion of coal in power plant, NO forms at high temperature due to reaction between nitrogen and oxygen (Lu and Streets 2012). In India, power plants are the major source of NOx emission and it is estimated that total NOx emissions were 9.3 Tg $year^{-1}$ in 2010 which increased to 13.9 Tg $year^{-1}$ in 2016 (Itahashi et al. 2019; Janssens-Maenhout et al. 2015).

To bound the NOx emission within the prescribed limit from the Indian power plant, combustion modifications and low NOx burners along with SCR/SNCR are suggested for incorporation.

12.4.3.1 In-Combustion NOx Control

In-combustion techniques include the use of low-NOx burners, over fire air (OFA), low excess air, flue gas recirculation, etc. Low-NOx burners are the most popular technique for NOx control (Vega et al. 2019). In this system, NOx formation is reduced during combustion by controlling a multi-staged mixing of fuel and air with lower flame temperature. In the primary combustion zone ~65% of fuel is injected whereas rest goes downstream of the primary combustion zone. Over fire air advancement of low-NOx burner is to reduce the NOx formation in the primary combustion zone. Use of low excess air could be the simple way to reduce NOx as it does not require any new installation.

12.4.3.2 Post-Combustion Technique

In post-combustion, flue gas can be treated by selective catalytic and non-catalytic reactor (SCR/SNCR) to reduce the NOx formation.

Selective Catalytic Reactor (SCR)

In SCR, the NOx formed in the flue gas was reduced to nitrogen gas by using NH_3 or urea as a reducing agent in the presence of vanadium oxide/titanium oxide catalysts (Bell and Buckingham 2010). The reaction can be done at low temperature and a high NOx removal efficiency can be achieved. The reaction of the reduction of NOx by ammonia and urea is shown in the equations below.

$$4NO + 4NH_3 + O_2 \rightarrow 4N_2 + 6H_2O$$
$$6NO_2 + 8NH_3 \rightarrow 7N_2 + 12H_2O$$
$$NO + 2(NH_2)_2CO + H_2O + O_2 \rightarrow 2N_2 + 6H_2O + 2CO_2$$
$$NO_2 + 4(NH_2)_2CO + 4H_2O \rightarrow 7N_2 + 12H_2O + 4CO_2$$

Selective Non-catalytic Reactor (SNCR)

SNCR also works in a similar fashion to SCR to reduce NOx to N_2. However, the reaction occurs at a high temperature (800 °C) and without the use of a catalyst (Tayyeb et al. 2007).

12.4.4 Particulate Matter (PM)

Particulate matters are of fine particles or small droplets generated due to human activities and goes into the air. Particulate matter is generally classified by the size of the particles like $PM_{2.5}$ for particle <2.5 micron and PM_{10} for particle <10 micron (Zhang 2016). World Health Organisation (WHO) prescribes 10 μg/m^3 as the threshold limit for $PM_{2.5}$ for air quality. In India $PM_{2.5}$ concentrations in different regions were found to be higher than the WHO guideline (Garaga et al. 2018). In India, coal-fired power plants are responsible for ~13% of PM in the atmosphere (Cropper et al. 2021). $PM_{2.5}$ in ambient air is the second largest threat for human health in India and responsible for heart disease, respiratory diseases and lung cancer (EIA 2016). It is estimated that 112,000 deaths were attributed to power plants in India (Cropper et al. 2021). Different techniques were employed to reduce the PM in the flue gas in power plant to meet the emission regulations.

12.4.4.1 Combustion Techniques

Formation of fine particles depends on combustion parameters like temperature and burning time. Both high boiler temperature and burning time increase fragmentation of coal into smaller particles and formation of PM. Moreover, high boiler load tends to reduce $PM_{2.5}$ emissions (Lu and Ren 2014). Further coal type and coal size also have an impact on the PM emission. Combustion optimisation, lower load factor and injection of sorbents into the flame zone are promising ways to reduce PM emission.

12.4.4.2 Post-Combustion Techniques

In post-combustion techniques electrostatic precipitators (ESP), fabric filters and bag filters are used.

Electrostatic Precipitator (ESP)

ESPs are the most widely used PM control technique and in India most of the power plants have installed ESP. An ESP works utilising the electrostatic force of attraction. The flue gas and the particles present in the gas get ionised when passed through high voltage corona. The ionised particles then pass through a series of plates of opposite charge and are deposited on the plate by electrostatic force and removed from gas. The efficiency of ESP can go up to 99% and depends on the coal quality, fly ash resistivity and voltage applied.

Fabric Filter (FF)

Fabric filters are the main alternatives to ESP for large power plants due to their high efficiency, not dependent on the fly ash resistivity and voltage. Fabric filter consists of fabric bags with netted filters to collect dust from the gas in bag house. Fine particles were collected on the filter bags with design efficiency >99.5%.

12.4.5 Mercury Emission

Mercury emission from anthropogenic sources is one of the major concerns for global air pollution. Mercury in air may be present in different forms and may deposit locally (oxidised mercury) or travel long distances (elemental mercury) (Selin 2009). Eventually, when Hg fall back on land or water, it changes into organic form (methyl mercury a potent neurotoxin) due to bacterial activity and enters food chain and bioaccumulates and becomes toxic to human. Mercury's action on human nervous system (neuro-toxicity) and internal organs especially the liver and kidneys are well established (Karagas et al. 2012). In 2013, recognising its effect on human and environment, Minamata Convention on Mercury was adopted which aims to protect human health and environment from anthropogenic mercury emission (Giang et al. 2015).

Stationary combustion of coal is the second largest source of mercury emission globally. In India, the coal combustion is the major source of mercury emission. Coal contains trace amounts of mercury, and during combustion in power plants at high temperature most of the mercury vaporises as elemental Hg (Hg^0) (Moretti and Jones 2012). There it undergoes chemical changes and part of Hg^0 oxides to form oxidised Hg (Hg^{+2}) due to the presence of HCl, O_2 SO_2, etc. In the colder region of the flue gas some of the Hg also get adsorbed on the fly ash particles (Hg^p) and deposited in the ESP (Park et al. 2008).

Indian coal contains trace amount of mercury (10–1000 μg/kg) (Ghosh et al. 1994). In a recent study, Das et al. 2020 reported average mercury concentration in 8 major coalfields in India and reported average mercury concentration of 173 μg/Kg in India which is slightly higher than the previously reported 140 μg/kg (Das, Senapati and Agarwalla 2020, UNEP 2014). Agarwalla et al. 2021 showed mercury distribution in power plants and reported majority of mercury emitted through stack and a fraction of Hg lies with fly ash (Agarwalla et al. 2021, 2023). However, for CFBC plant ~66% of mercury is adsorbed by fly ash and rest are emitted through stack (Agarwalla et al. 2021).

Mercury emission reduction is obligatory for parties ratified in the Minamata convention. The Convention states that each Party shall require the use of best available techniques (BAT) and best environmental practices (BEP).

12.4.5.1 Mercury Emission Control

There are various techniques to control mercury emission from power plant including pre-combustion and post-combustion techniques.

Pre-Combustion Technique

Coal washing and coal blending are the best available pre-combustion technique to reduce the mercury input to the power plant. The study shows significant reduction in mercury concentrations in the washed coal compared to the feed coal mercury content in commercial washery in India. The extent of mercury reduction goes up to 14–40% (Das et al. 2013).

Multi-Pollutant co-Benefit Technique

Air pollution control devices for other pollutants also have co-benefit of reducing mercury emission (Zhao et al. 2017). The ESP can reduce mercury emission by capturing the particulate-bound mercury. Reports show 10–30% of mercury can be removed by ESP only. Moreover, flue gas desulphurisation is also effective in the removal of soluble oxidised mercury from the flue gas stream along with SOx. The efficiency of FGD may be increased by higher oxidised mercury content in the flue gas. Additives like halogens are known to oxidise mercury in the flue gas. De-NOx technology like SCR also has impact on mercury emission, SCR catalyst oxidises elemental Hg which can further be removed in FGD. Overall, a combination of APCDs has a great impact on mercury emission.

Dedicated Mercury Control

Activated carbon injection (ACI) is a dedicated mercury control technology (Zhao et al. 2019). It operates by injecting activated carbon in the flue gas stream which adsorbs mercury and deposited on the PM control devices. ACI has already been demonstrated on full scale. Activated carbons with chemical treatment specially brominated activated carbons have better mercury removal efficiency and lower injection rates.

12.5 Summary

Coal combustion in power plants produces various pollutants like particulate matter, SOx, NOx, CO_2, Hg and others. Use of high ash and low S coals in Indian power plants contributes to higher emission of particulate matter. CO_2, NOx, SOx and Hg emission from power plants depends on coal composition, operational conditions and APCD installed. The impact on environment and human health is documented in various literature.

Measures have been taken by MoEFCC and MoP in the form of stricter norms for emission from power plant. All the power plants are to comply with these norms by 2022. To achieve this, power plants need to increase the efficiency of plant and retrofit required APCD like FGD, SCR/SNCR. CEA has already published a technical guideline for implementation of FGD. Installation of FGD in Indian power plants is in various stages of implementation. It is suggested that retrofitting low-NOx burners and over fire air will help to achieve the compliances for NOx for majority of the existing plants and for other SCR/SNCR may be required. Implementation of these APCDs will also have impact on other emissions like Hg, PAH, and other trace elements. However, implementation of regulations has high capital and operating cost, and may put power sector in economic stress which lead to increased cost of production.

References

Agarwalla H, Das TB, Senapati R, Gangopadhyay M, Masto RE, Kumar M, Singh V. Mercury in coal from south eastern coalfield and mercury partitioning at sub-critical coal-fired power plant. Journal of Material Cycles and Waste Management; 2023. p. 1–10.

Agarwalla H, Senapati RN, Das TB. Mercury emissions and partitioning from Indian coal-fired power plants. J Environ Sci. 2021;100:28–33.

Bell RD, Buckingham FP. An overview of technologies for reduction of oxides of nitrogen from combustion furnaces. MPR Associates; 2010.

Bhati P. Status of implementation of new environmental norms and way forward. Tech. Rep. Centre for Science and Environment; 2016.

CEA report 2020, Growth of Electricity sector in India from 1947–2020.

CEA. CO_2 baseline database for the Indian Power Sector, 2021.

Chakraborty A, Gupta T, Tripathi SN. Chemical composition and characteristics of ambient aerosols and rainwater residues during Indian summer monsoon: insight from aerosol mass spectrometry. Atmos Environ. 2016;136:144–55.

Chen TM, Gokhale J, Shofer S, Kuschner WG. Outdoor air pollution: nitrogen dioxide, sulfur dioxide, and carbon monoxide health effects. Am J Med Sci. 2007;333:249–56.

Cohen AJ, Ross Anderson H, Ostro B, Pandey KD, Krzyzanowski M, Künzli N, Gutschmidt K, Pope A, Romieu I, Samet JM. The global burden of disease due to outdoor air pollution. J Toxicol Environ Health A. 2005;68:1301–7.

Cropper M, Cuib R, Guttikunda S, Hultman N, Jawahar P, Park Y, Yao X, Song X. The mortality impacts of current and planned coal-fired power plants in India. PNAS. 2021;118(5):e2017936118.

Das TB, Senapati RN, Agarwalla H. Mercury in Indian thermal coals. Bulletin of Environmental Contamination and Toxicology. 2020;105(3):502–12.

Das TB, Gouricharan T, Sharma KK, Choudhury A. Evaluation of reduction potential of selected heavy metals from an Indian coal by conventional coal cleaning. Int J Coal Prep Util. 2013;33:300–12.

Dunn S, Flavin C. Moving the climate change agenda forward. In: State of the world 2002. Special world summit edition. NY: WW Norton and Company; 2002. p. 25–50.

EIA (2016) International energy outlook independent statics and analysis. US Energy Information. Report No.: DOE/EIA-0484.

Energy Statistics. Ministry of Statistics and Programme Implementation. Government of India; 2021.

Garaga R, Sahu SK, Kota SH. A review of air quality modelling studies in India: local and regional scale. Curr Pollut Rep. 2018;4:59–73.

Garg A, Kapshe M, Shukla PR, Ghosh D. Large point source (LPS) emissions from India: regional and sectoral analysis. Atmos Environ. 2002;36:213–24.

Ghosh SB, Das MC, Roy RR, Banerjee NN. Mercury in Indian coal. Indian J Chem Technol. 1994;1:237.

Giang A, Stokes LC, Streets DG, Corbitt ES, Selin NE. Impacts of the Minamata convention on mercury emissions and global deposition from coal-fired power generation in Asia. Environ Sci Technol. 2015;49:5326–35.

Guo H, Kota H, Chen K, Sahu SK, Hu J, Ying Q, Wang Y, Zhang H. Source contributions and potential reductions to health effects of particulate matter in India. Atmos Chem Phys. 2018;18:15219–29.

Gupta A, Spears D. Health externalities of India's expansion of coal plants: evidence from a national panel of 40,000 households. J Environ Econ Manag. 2017;86:262–76.

Hesterberg TW, Bunn WB, McClellan RO, Hamade AK, Long CM, Valberg PA. Critical review of the human data on short-term nitrogen dioxide (NO2) exposures: evidence for NO2 no-effect levels. Crit Rev Toxicol. 2009;39:743–81.

Hoek G, Krishnan RM, Beelen R, Peters A, Ostro B, Brunekreef B, Kaufman JD. Long-term air pollution exposure and cardio-respiratory mortality: a review. Environ Health. 2013;12(1):43.

Houghton JT, Ding Y, Griggs DJ, Noguer M, Linden PJV, Dai X, et al. Climate change 2001, The scientific basis: intergovernmental panel on climate change (IPCC). Cambridge, UK: Cambridge University Press; 2001.

Huang RJ, Zhang Y, Bozzetti C, et al. High secondary aerosol contribution to particulate pollution during haze events in China. Nature. 2014;514:218–22.

International Energy Agency (IEA) Greenhouse Gas R&D Program. Carbon dioxide capture from power plant station, UK; 1995.

Itahashi S, Yumimoto K, Kurokawa J, Morino Y, Nagashima T, Miyazaki K, Maki T, Ohara T. Inverse estimation of NOx emissions over China and India 2005–2016: contrasting recent trends and future perspectives. Environ Res Lett. 2019;14:124020.

Janssens-Maenhout G, et al. HTAP_v2.2: a mosaic of regional and global emission grid maps for 2008 and 2010 to study hemispheric transport of air pollution Atmos. Chem Phys. 2015;15:11411–32.

Karagas MR, Choi AL, Oken E, Horvat M, Schoeny R, Kamai E. Evidence on the human health effects of low-level methylmercury exposure. Environ Health Perspect. 2012;120(6):799–806.

Li C, McLinden C, Fioletov V, Krotkov N, Carn S, Joiner J, Streets D, He H, Ren X, Li Z, Dickerson RR. India is overtaking China as the World's largest emitter of anthropogenic sulfur dioxide. Sci Rep. 2017;7:14304.

Lu J, Ren X. Analysis and discussion on formation and control of primary particulate matter generated from coal-fired power plants. J Air Waste Manag Assoc. 2014;64(12):1342–51.

Lu Z, Streets DG. Increase in NOx emissions from Indian thermal power plants during 1996–2010: unit-based inventories and multi satellite observations. Environ Sci Technol. 2012;46:7463–70.

Lu Z, Streets DG, Foy B, Krotkov NA. Ozone monitoring instrument observations of interannual increases in SO2 emissions from Indian coal-fired power plants during 2005–2012. Environ Sci Technol. 2013;47(24):13993–4000.

Ministry of Environment, Forest and Climate Change, Government of India, Notification dated 7th December 2015; 7th March 2016; 28 June 2018.

Moretti AL, Jones CS. Babcock & Wilcox, advanced emissions control technologies for coal-fired power plants. Barberton, Ohio: Power Generation Group, Inc.; 2012.

Nagendra SMS, Schlink U, Müller A, Khare M. Urban air quality monitoring, modelling and human exposure assessment. Springer; 2020.

Oberschelp C, Pfister S, Raptis CE, Hellweg S. Global emission hotspots of coal power generation. Nat Sustain. 2019;2:113–21.

Omidvarborna H, Kumar A, Kim DS. NOx emissions from low-temperature combustion of biodiesel made of various feedstocks and blends. Fuel Process Technol. 2015;140:113–8.

Park KS, Seo YC, Lee SJ, Lee J. Emission and speciation of mercury from various combustion sources. Powder Technol. 2008;180:151–6.

Raghuvanshi SP, Chandra A, Raghav AK. Carbon dioxide emissions from coal based power generation in India. Energy Convers Manag. 2006;47(4):427–41.

Selin NE. Global biogeochemical cycling of mercury: a review. Annu Rev Environ Resour. 2009;34(1):43–63.

Senapati RN, Agarwalla H. Mercury in Indian thermal coals. Bull Environ Contam Toxicol. 2020;105(3):502–12.

Sugathana A, Bhangale R, Kansal V, Hulke U. How can Indian power plants cost-effectively meet the new sulphur emission standards? Policy evaluation using marginal abatement cost-curves. Energy Policy. 2018;121:124–37.

Tayyeb M, Javeda N, Irfana N, Gibbsb BM. Control of combustion-generated nitrogen oxides by selective non-catalytic reduction. Journal of Environmental Management. 2007;83(3):251–289.

UNEP. Characterization of the coal fired power sector in India, assessment of the mercury contents in coal fed to power plants and calculation of mercury emissions from the sector. UNEP Chemicals Branch; 2014.

Vega F, Alonso-Farinas B, Baena-Moreno FM, Rodríguez JA, Navarrete B. Technologies for control of Sulphur and nitrogen compounds and particulates in coal combustion and gasification. New trends in coal conversion. Woodhead Publishing; 2019. p. 141–77.

WHO. Ambient (outdoor) air quality and health; 2018., http://www.who.int/news-room/fact-sheets/detail/ambient-(outdoor)-air-quality-and-health

Zhang Q, Jiang X, Tong D, et al. Transboundary health impacts of transported global air pollution and international trade. Nature. 2017;543:705–9.

Zhang X. Emission standards and control of PM2.5 from coal-fired power plant. IEA Clean Coal Centre; 2016. p. 267.

Zhao S, Duan Y, Yao T, Liu M, Lu J, Tan H, Wang X, Wu L. Study on the mercury emission and transformation in an ultra-low emission coal-fired power plant. Fuel. 2017;199:653–61.

Zhao S, Pudasainee D, Duan Y, Gupta R, Liu M, Lu J. A review on mercury in coal combustion process: content and occurrence forms in coal, transformation, sampling methods, emission and control technologies. Prog Energy Combust Sci. 2019;73:26–64.

13 Oxidative Stress: The Particulate Linkage

Suryasikha Samal and C. S. K. Mishra

13.1 Introduction

Growing trend of air pollution has become one of the familiar environmental hazards for the society. It is a mixture of toxic compounds emerging from both natural and anthropogenic sources. Depending on the source and rate of emission the components of air pollution change dramatically. The primary human-made sources of air pollution may include vehicle emissions, natural gas to heat homes, fuel oils, and by-products of manufacturing and power generation and fumes from chemical production. In addition to that, various hazardous substances such as methane and other gases are released by decomposing organic matter in soils, and other gases include ash and smoke from volcanic eruptions, which are frequently caused by humans. Various reports around the world have consistently demonstrated that air pollution is a significant adaptable risk factor for enhanced mortality. Air pollutants can harm the health of humans, animals, plants, and buildings. The pollutants are in the form of gases, solid particles, or liquid droplets. Ozone (O_3), nitric oxide (NO), carbon monoxide (CO), nitrogen dioxide (NO_2), and sulphur dioxide (SO_2) are some of the gaseous pollutants that contribute to air pollution. The pollution is now the leading environmental cause of premature death. Cardiovascular events such as acute ischemic account for the majority of the additional mortality caused by air pollution. Aside from increased mortality, air contamination is linked to significant reductions in worker productivity. Exposure to air pollution has a wide range of negative consequences on people's health. The effects of air pollution can be divided into short-term and long-term effects. The discomfort, such as irritation to the nose, eyes, throat or skin, as well as infections like bronchitis are short-term symptoms that are transient. It can also cause headaches, nausea and dizziness. The long-term effects of pollution can last for years and cause various respiratory diseases. Long-term damage to people's nerves, brains, kidneys, liver, and other organs, birth defects are the consequences of air pollution (Brook et al. 2004; Newby et al. 2015; Zivin and Neidell 2018; Hamanaka and Mutlu 2018).

Particulate matters (PM) are now the emerging contaminant in different countries including India (Fig. 13.1). The components of PM describe a combination of small airborne particles and liquid droplets found in the air. The main sources of PM in cities include industrial buildings, automobiles, non-road equipment, and power stations that burn fossil fuels. Other sources are dust, diesel emissions and secondary particle formation from gases and vapours. Reactive metals and

S. Samal · C. S. K. Mishra (✉)
Department of Zoology, Odisha University of Agriculture and Technology, College of Basic Science and Humanities, Bhubaneswar, India

P. K. Padhy et al. (eds.), *Air Quality and Human Health*,
https://doi.org/10.1007/978-981-97-1363-9_13

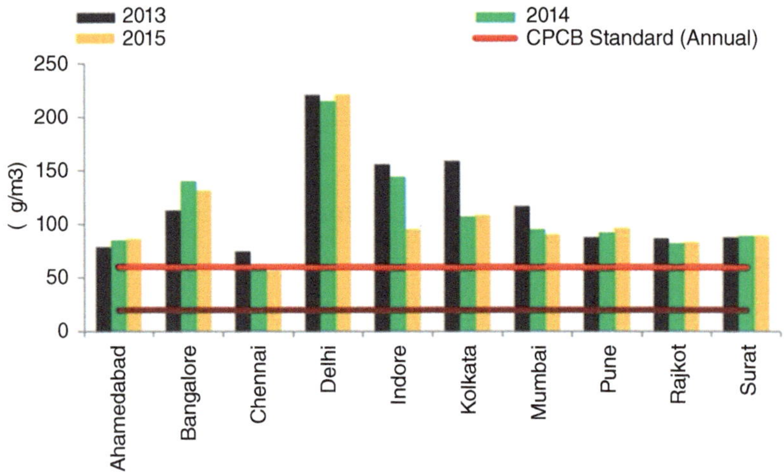

Fig. 13.1 Status of particulate matter in major cities of India (Nagpure et al. 2017)

organic compounds that have been adsorbed are present together with carbonaceous particles. Sulphates, nitrates, endotoxin, polycyclic aromatic hydrocarbons, and metals such as zinc, copper, and iron are all common constituents of PM. PM is classified according to the size of particles in various categories such as coarse particle (PM10, diameter < 10 μm), fine particle (PM2.5, diameter < 2.5 μm), and ultrafine particle (PM0.1, diameter < 0.1 μm). Coarse particles come from a variety of natural and industrial sources, and they rarely make it pass through the upper bronchus. However, the burning of fossil fuels, however, produces fine and ultrafine particles, which go deeper into the lungs than coarse particles and are more dangerous to health.

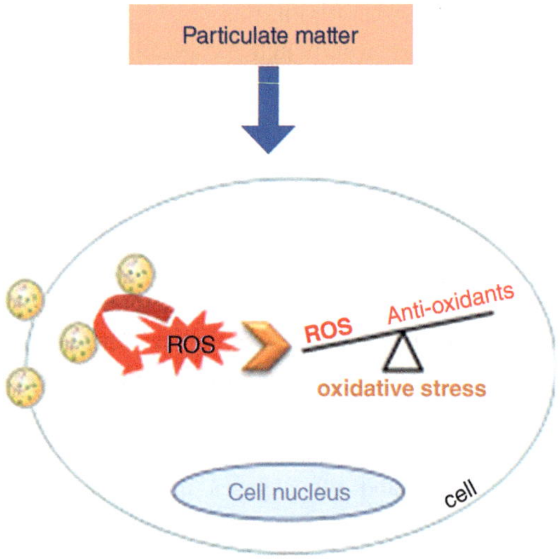

Fig. 13.2 Particulate matter induces oxidative stress. Modified from Crobeddu et al. (2017)

13.2 Generation of Reactive Oxygen Species and Oxidative Damage

Many of the biological effects of PM are thought to be exerted through the production of reactive oxygen species (ROS). Systemic markers of oxidative stress are generated when exposed to air pollution (Fig. 13.2). Various types of cells such as cells of gastrointestinal, respiratory, endothelial, lung, macrophage, cardiomyocyte, and cornea are vulnerable to air pollutants. (Miller et al. 2012; Soberanes et al. 2006; Hong et al. 2016; Hamanaka and Mutlu 2018). Chemicals and metals absorbed by PM can cause free radical production inside cells. In response to stimuli, cells can produce ROS, which act as signalling molecules. The cellular response to PM exposure has been revealed to be significantly regulated by the production of ROS.

Every air pollutant has the potential to be hazardous on the organ system in its way. Nonetheless, strong oxidants such as ozone, nitrogen oxides, and suspended particulates either directly affect lipids and proteins or indirectly trigger intracellular oxidant pathways. Reports have demonstrated that the particle's surface where nitro-PAH and PAH are adsorbed is connected to the formation of ROS. The production of ROS in biological systems is significantly influenced by the leaching of various metals like iron. Transition metals including silver, vanadium, cadmium, mercury, manganese, and copper are present in PM2.5 and can cause ROS to be generated by Fenton's reaction. These metals cause pathological disturbance in the respiratory system through oxidative stress. Sources including burning, manufacturing and release of waste water make them available as natural elements of the earth's crust. Iron also enters into the atmosphere in the form of ultrafine particles. Oxidative DNA damage can be induced in organs like kidney and lungs by particles like bound benzo(a) pyrene. In addition, volatile compounds, such as photochemical oxidants (peroxyacetyl nitrate and ozone), benzene, and secondary pollutants produced in the presence of sunlight in an atmosphere containing reactive hydrocarbons and NOx are able to enhance oxidative stress. Oxidative stress is the result of high ROS formation which leads to mitochondrial damage with elevation of NADPH oxidase isoform 4 (NOX4). Activation of inflammatory cells (neutrophils, eosinophils, and monocytes) and increased numbers of macrophages are the consequence of cellular damage. The organic carbon compounds, such as quinones and PAHs, are responsible for the oxidative potential in fine and ultrafine particles (Shukla et al. 2000; Squadrito et al. 2001; Gerlofs-Nijland et al. 2007).

Compounds like dinitro-PAH, peroxyacetyl nitrate, substituted and unsubstituted PAH, nitro-PAH, ozone, oxides of nitrogen and hydroxy radical are the major components of toxic fine particles. The metabolic activation of PAHs produces redox-active quinones, diol epoxides, DNA adducts, and radical cations catalysed by CYP450 peroxidases. ROS can be produced through the interaction of PAH, redox-cycling quinoids, and several redox-active transition metals. The organs of exposed animals experience metabolic activation of the organic fraction of DEP, a component of PAH and quinones, which results in the induction of CYP4501A1 isoform expression and an increase in ROS generation. PM causes inflammation by stimulating proinflammatory cytokines and chemokines. The generation of antioxidant enzymes through the induction of Nrf2 is the defence response of the cell against PM-induced ROS. PM inhibits antioxidant enzymes involved to protect from oxidative stress responses. Antioxidant enzyme like glutathione reductase (GR) seems to be less sensitive to particle-induced inhibition, as compared to the SOD. Moreover, glutathione peroxidase (GPx) is proved to be more sensitive to particle impacts as compared to GR, and its inhibition profile is similar to those of the SOD enzymes (Hatzis et al. 2006). This initiates extra intracellular signalling pathways that control the production of genes related to chemokines and cytokines. Fine and ultrafine structure of diesel exhaust particles (DEPs) can modulate ROS generation leading to lipid peroxidation and DNA damage due to oxidative stress (Lodovici and Bigagli 2011).

Relatively low oxidative stress activates a series of antioxidant and detoxification enzymes such as catalase and several transcription factors, such as the nuclear factor erythroid-2 (Nrf2), which can counteract ROS generation and protect the adverse biological outcomes. Failure in antioxidant response dealing with increasing ROS production leads to proinflammatory response with several cytotoxic effects. These effects are regulated by the redox-sensitive mitogen-activated protein kinase (MAPK) and NF-κB cascades. Those are responsible for the activation of certain molecules such as cytokines and chemokines, involved in inflammatory processes (Lodovici and Bigagli 2011).

13.3 Organs Affected by PM

The oxidative damage caused by PM-induced toxicity is responsible for the generation of ROS and impacts various organ systems in an adverse way. Although cardiovascular system and pulmonary system are likely to be affected by PM-caused toxicity, it may also affect the functionality of other vital organs.

13.3.1 Cardiovascular System

Deaths from air pollution are primarily caused by cardiovascular causes, with PM having a stronger adverse effect. Both short- and long-term exposure to PM2.5 levels is linked to an increased risk of ischemic heart disease, cerebrovascular disease, and heart failure. The PM in air aggravates pre-existing heart conditions and causes disease development. PM2.5 exposure enhances the risk of myocardial infarction. Exposure of PM via inhalation is correlated with enhanced pulmonary neutrophil numbers, fibrinogen, GMCSF (granulocyte-macrophage colony stimulating factor), circulating levels of C-reactive protein, IL-1b (interleukin-1b), IL-6 (interleukin-6), and TNF-a (tumour necrosis factor-a). Certain systemic markers of oxidative damage such as oxidized lipids are stimulated after exposure to PM particles (Ghio et al. 2000; Riediker et al. 2004; Tsai et al. 2012).

Study has been shown that induction of a prothrombotic state in response to PM cause arterial thrombotic (thrombotic cerebrovascular events, myocardial infarction) and venous thrombotic events. Production of haemostasis factors such as fibrinogen and other factors like Von Willebrand factor and sCD40L are due to the toxicity generated by PM. PM exposure activates antifibrinolytic pathways in addition to prothrombotic pathways. PM exposure stimulates plasminogen activator inhibitor-1 (PAI1), while tissue plasminogen activator (t-PA) activity is inhibited (Chuang et al. 2007; Lehmann et al. 2009; Signorelli et al. 2017).

13.3.2 Pulmonary System

Inhalation of PM particles causes inflammatory damage within the lung tissues. There is a clear link between redox-active chemical concentration in PM and damage in bronchial epithelial cells. Furthermore, organic chemicals adsorbed on particle surfaces stimulate inflammation in human airway epithelial cells via CYP1A1-mediated ROS production. A recent study has discovered that 1-nitropyrene in diesel exhaust causes DNA impairment in humans by forming reactive oxygen species (ROS). The fact that redox-sensitive transcription factors such as NF-κB activator protein 1 (AP-1) and CAATT/enhancer binding protein (C/EBP) regulate numerous proinflammatory genes (TNF- and IL8) are produced by exposure to PM. When human alveolar and bronchial epithelial cells are exposed to PM, several transcription factors are activated. It has been shown in primary cultures of human bronchial epithelial cells that PAH adsorbed on PM2.5 may impede repair processes of injured tissues (Ferecatu et al. 2010). Epithelial cell lines of lungs tissue exposed to PM10 cause ROS production as well as decreased glutathione and antioxidant enzyme activity, such as superoxide dismutase and glutathione reductase (Chirino et al. 2010; Lodovici and Bigagli 2011).

13.3.3 Nervous System

Once in the circulatory system, PM particles quickly exit the lungs and spread to many other organ systems, including the kidneys, lymph nodes, liver, and testes. Oxidative stress pathways cause damage to these secondary targets. The brain is prone to oxidative stress because of its high energy consumption, high cellular content of lipids and proteins, vast neural networks, and low levels of endogenous scavengers (e.g. catalase, superoxide dismutase). A study says that due to the nano size of PM, it is able to penetrate the central nervous system of animals by passing across the blood-brain barrier (Lockman et al. 2004). The permeability of mitochondrial membrane and reductions of cellular energy has been shown in PM-exposed mammalian nervous system. Significant depolarization of the microglia's mitochondrial membrane and stimulation of pro-inflammatory cytokines are the response recorded in nervous system after exposures to PM. Neurodegeneration in transgenic mice susceptible to oxidative stress can result from inhaling intense PM, according to the findings reported by Veronesi et al. (2005). Despite the fact that the fundamental mechanisms causing

PM-induced neuroinflammation and neurodegeneration are yet unknown, the study suggests that microglia play an important function in the brain's inflammatory response via innate immune pathways. Report also suggests the role of microglia and Mac-1 receptors in mediating PM-neurotoxicity (MohanKumar et al. 2008). At high quantities, diesel exhaust particles are directly hazardous to numerous cell types such as neurons, microglia, and astrocytes. DEP stimulates microglia and the release of reactive oxygen species (ROS) in vitro. Microglial superoxide mediates DEP-induced dopamine neurotoxicity. Clinical samples of nasal tissues from Mexico City people, as well as the olfactory and respiratory nasal mucosa, olfactory bulb, and cortex of feral canines living in Mexico City, were the first to show histological evidence of PM-related neurodegeneration (Calderón-Garcidueñas et al. 2002). Deteriorated cortical neurons, apoptotic glial cells, altered BBB, and neurofibrillary tangles were also found in these canine tissues. Changes in immunohistochemistry expressions of NF-kB and iNOS are suggested of oxidative stress-mediated damage (Veronesi et al. 2005; MohanKumar et al. 2008).

13.3.4 Integument System

PM has been linked to a variety of skin illnesses and problems, according to numerous studies (Williams and Kupper 1996; Krutmann et al. 2014; Araviiskaia et al. 2019). Urban PM exposure has been linked with faster extrinsic ageing in healthy skin, coarse wrinkle growth, pigment spot formation, and elastosis as clinical signs. More evidence has been reviewed from clinical, epidemiological and in vitro studies, as well as ageing mechanisms, all of which point to the generation of ROS, induction of aryl hydrocarbon receptor signalling, activation of an inflammatory cascade, and finally impairment of the skin barrier. The impact of prenatal PM exposure on epidermal barrier dysfunction was examined over the first three trimesters. PM-induced oxidative stress has been reported to cause epigenetic changes in vital DNA repair genes of the placenta, which could influence foetal immunological development (Dijkhoff et al. 2020). Because of the high particle surface reactivity, PM can promote the creation of external ROS levels by forming free radicals. PM sourced from diesel and its organic extract has been shown to have high reactivity when it comes to creating reactive oxygen species (ROS) through the catalysis of oxygen reduction by several reducing agents. Redox-active compounds (copper and quinones) have been demonstrated to cause oxidative stress by promoting the nuclear translocation of a redox-sensitive protein complex, resulting in apoptosis due to mitochondrial stress (Cervellati et al. 2020). PM could encourage endogenous ROS formation by inducing mitochondrial stress and raising the activity of ROS-producing enzymes. Reduced ATP levels in PM2.5-treated keratinocytes and fibroblasts are typical indicators of mitochondrial dysfunction. In vitro, this malfunction results in the creation of mitochondrial ROS, as well as an increase in radical production results in increasing intracellular ROS levels (Hu et al. 2017; Piao et al. 2019; Dijkhoff et al. 2020).

13.4 Future Prospective and Conclusion

The evidences suggest that exposure to PM is detrimental for living beings. Additional investigation into the signalling mediators and epigenetic regulators of PM impacts on health could result in the creation of pharmaceuticals that can lessen the detrimental effects of air pollution on health. Translational study in humans as well as animal and cell-based studies based on actual PM exposures are required for this endeavour. However, increased efforts should be made to educate the public about the harmful impacts of PM exposure. PM exposure should be considered a significant, controllable risk factor for many diseases that are potentially fatal. Future research should concentrate on mechanistic investigations in addition to efforts to limit PM production in order to better comprehend how PM generates harmful health impacts.

References

Araviiskaia E, Berardesca E, Bieber T, Gontijo G, Sanchez Viera M, Marrot L, et al. The impact of airborne pollution on skin. J Eur Acad Dermatol Venereol. 2019;33(8):1496–505.

Brook RD, Franklin B, Cascio W, Hong Y, Howard G, Lipsett M, et al. Air pollution and cardiovascular disease: a statement for healthcare professionals from the expert panel on population and prevention science of the American Heart Association. Circulation. 2004;109(21):2655–71.

Calderón-Garcidueñas L, Azzarelli B, Acuna H, Garcia R, Gambling TM, Osnaya N, et al. Air pollution and brain damage. Toxicol Pathol. 2002;30(3):373–89.

Cervellati F, Benedusi M, Manarini F, Woodby B, Russo M, Valacchi G, Pietrogrande MC. Proinflammatory properties and oxidative effects of atmospheric particle components in human keratinocytes. Chemosphere. 2020;240:124746.

Chirino YI, Sánchez-Pérez Y, Osornio-Vargas ÁR, Morales-Bárcenas R, Gutiérrez-Ruíz MC, Segura-García Y, et al. PM10 impairs the antioxidant defense system and exacerbates oxidative stress driven cell death. Toxicol Lett. 2010;193(3):209–16.

Chuang K, Chan CC, Su TC, Lee C-T, Tang C-S. The effect of urban air pollution on inflammation, oxidative stress, coagulation, and autonomic dysfunction in young adults. Am J Respir Crit Care Med. 2007;176:370–6.

Crobeddu B, Aragao-Santiago L, Bui LC, Boland S, Squiban AB. Oxidative potential of particulate matter 2.5 as predictive indicator of cellular stress. Environ Pollut. 2017;230:125–33.

Dijkhoff IM, Drasler B, Karakocak BB, Petri-Fink A, Valacchi G, Eeman M, Rothen-Rutishauser B. Impact of airborne particulate matter on skin: a systematic review from epidemiology to in vitro studies. Part Fibre Toxicol. 2020;17(1):1–28.

Ferecatu I, Borot MC, Bossard C, Leroux M, Boggetto N, Marano F, et al. Polycyclic aromatic hydrocarbon components contribute to the mitochondria-antiapoptotic effect of fine particulate matter on human bronchial epithelial cells via the aryl hydrocarbon receptor. Part Fibre Toxicol. 2010;7(1):1–14.

Gerlofs-Nijland ME, Dormans JA, Bloemen HJ, Leseman DL, Boere AJF, Kelly FJ, et al. Toxicity of coarse and fine particulate matter from sites with contrasting traffic profiles. Inhal Toxicol. 2007;19(13):1055–69.

Ghio AJ, Kim C, Devlin RB. Concentrated ambient air particles induce mild pulmonary inflammation in healthy human volunteers. Am J Respir Crit Care Med. 2000;162(3):981–8.

Hamanaka RB, Mutlu GM. Particulate matter air pollution: effects on the cardiovascular system. Front Endocrinol. 2018;9:680.

Hatzis C, Godleski JJ, González-Flecha B, Wolfson JM, Koutrakis P. Ambient particulate matter exhibits direct inhibitory effects on oxidative stress enzymes. Environ Sci Technol. 2006;40(8):2805–11.

Hong Z, Guo Z, Zhang R, Xu J, Dong W, Zhuang G, Deng C. Airborne fine particulate matter induces oxidative stress and inflammation in human nasal epithelial cells. Tohoku J Exp Med. 2016;239(2):117–25.

Hu R, Xie XY, Xu SK, Wang YN, Jiang M, Wen LR, et al. PM2. 5 exposure elicits oxidative stress responses and mitochondrial apoptosis pathway activation in HaCaT keratinocytes. Chin Med J. 2017;130(18):2205.

Krutmann J, Liu W, Li L, Pan X, Crawford M, Sore G, Seite S. Pollution and skin: from epidemiological and mechanistic studies to clinical implications. J Dermatol Sci. 2014;76(3):163–8.

Lehmann AD, Blank F, Baum O, Gehr P, Rothen-Rutishauser BM. Diesel exhaust particles modulate the tight junction protein occludin in lung cells in vitro. Part Fibre Toxicol. 2009;6(1):1–14.

Lockman PR, Koziara JM, Mumper RJ, Allen DD. Nanoparticle surface charges alter blood–brain barrier integrity and permeability. J Drug Target. 2004;12(9-10):635–41.

Lodovici M, Bigagli E. Oxidative stress and air pollution exposure. J Toxicol. 2011;2011:487074.

Miller MR, Shaw CA, Langrish JP. From particles to patients: oxidative stress and the cardiovascular effects of air pollution. Futur Cardiol. 2012;8(4):577–602.

MohanKumar SM, Campbell A, Block M, Veronesi B. Particulate matter, oxidative stress and neurotoxicity. Neurotoxicology. 2008;29(3):479–88.

Nagpure AS, Sharma KR, Sharma R, Gurjar BR, Kumar P. Association of particulate matter emissions, air quality and health risks in South Asian cities. Airborne Particles. 2017;1

Newby DE, Mannucci PM, Tell GS, Baccarelli AA, Brook RD, Donaldson K, et al. European Association for Cardiovascular Prevention and Rehabilitation; ESC Heart Failure Association. Expert position paper on air pollution and cardiovascular disease. Eur Heart J. 2015;36(2):83–93b.

Piao MJ, Kang KA, Zhen AX, Fernando PDSM, Ahn MJ, Koh YS, et al. Particulate matter 2.5 mediates cutaneous cellular injury by inducing mitochondria-associated endoplasmic reticulum stress: protective effects of ginsenoside Rb1. Antioxidants. 2019;8(9):383.

Riediker M, Cascio WE, Griggs TR, Herbst MC, Bromberg PA, Neas L, et al. Particulate matter exposure in cars is associated with cardiovascular effects in healthy young men. Am J Respir Crit Care Med. 2004;169(8):934–40.

Shukla A, Timblin C, BeruBe K, Gordon T, McKinney W, Driscoll K, et al. Inhaled particulate matter causes expression of nuclear factor (NF)-κ B–related genes and oxidant-dependent NF-κ B activation in vitro. Am J Respir Cell Mol Biol. 2000;23(2):182–7.

Signorelli SS, Ferrante M, Gaudio A, Fiore V. Deep vein thrombosis related to environment. Mol Med Rep. 2017;15(5):3445–8.

Soberanes S, Panduri V, Mutlu GM, Ghio A, Bundinger GS, Kamp DW. p53 mediates particulate matter–induced alveolar epithelial cell mitochondria-regulated apoptosis. Am J Respir Crit Care Med. 2006;174(11):1229–38.

Squadrito GL, Cueto R, Dellinger B, Pryor WA. Quinoid redox cycling as a mechanism for sustained free radical generation by inhaled airborne particulate matter. Free Radic Biol Med. 2001;31(9):1132–8.

Tsai DH, Amyai N, Marques-Vidal P, Wang JL, Riediker M, Mooser V, et al. Effects of particulate matter on inflammatory markers in the general adult population. Part Fibre Toxicol. 2012;9(1):1–9.

Veronesi B, Makwana O, Pooler M, Chen LC. Effects of subchronic exposures to concentrated ambient particles: VII. Degeneration of dopaminergic neurons in Apo E−/− mice. Inhal Toxicol. 2005;17(4-5):235–41.

Williams IR, Kupper TS. Immunity at the surface: homeostatic mechanisms of the skin immune system. Life Sci. 1996;58(18):1485–507.

Zivin JG, Neidell M. Air pollution's hidden impacts. Science. 2018;359(6371):39–40.

The Health Menace of Myriad Air Pollutants: An Indian Perspective

14

Priyadatta Satpathy, Monalin Mishra, Jyotishree Nath, R. Boopathy, and Trupti Das

14.1 Introduction

Air pollution has turned out to be one of the greatest scourges of our era. The United Nations Environment Programme (UNEP) has provided an estimation that approximately 1.1 billion people respire adulterated air globally (UNEP et al. 2003). While some interrelation between the quality of air and human diseases has been acknowledged since the ancient past, the ill effects of air pollution on health set foot into the consciousness of the world in the twentieth century. It is now the biggest environmental risk owing to its impact not only on the global climate but also on public health. Every substance or compound present in the air in concentrations high enough to threaten the health of all living beings or cause damage to the environment is designated as an air pollutant. The World Health Organization (WHO) has identified six pollutants as the major ones that pose a threat to the ecosystem. These include particulate matter, ground-level ozone (O_3), carbon monoxide (CO), sulfur oxides, nitrogen oxides, and lead (Pb). Other than these major pollutants, many substances, such as dust, fumes, hydrocarbons, volatile organic compounds, polycyclic aromatic hydrocarbons, halogen derivatives, and bioaerosols, are suspended in the air. These substances, when present at excessive concentrations, generate vulnerability to countless diseases, including malignancy (Carugno et al. 2016; Loomis et al. 2014). Human-induced changes to the atmosphere are driven largely by industrial activities, vehicle emissions, biomass burning, garbage incineration, mining activities, construction and demolition, burning of fossil fuels, agricultural production, and many others. Natural sources include volcanic eruption, microbial decay, smoke from wildfires, pollen, windblown dust, etc. In addition, meteorology has a strong influence on spatial and temporal distribution as it affects the concentration, dispersion, dilution, and other physiochemical properties of the pollutants. Out of the total deaths worldwide, air pollution accounts for deaths from lung cancer (29%), severe respiratory infections (17%), chronic obstructive pulmonary disease (43%), and cardiac arrest (24%). Epidemiological studies have indicated that the pollutants present in the ambient air have a broad spectrum of effects on human health, largely on the cardiovascular and respiratory systems. Such a situation sets up a potential risk of developing acute infections like pneumonia, asthma, and COPD, which fall

P. Satpathy · M. Mishra · J. Nath · R. Boopathy
T. Das (✉)
Aerosol & Trace gases Laboratory, Environment & Sustainability Department, CSIR-Institute of Minerals & Materials Technology (CSIR-IMMT), Bhubaneswar, Odisha, India

Academy of Scientific and Innovative Research (AcSIR), Ghaziabad, India
e-mail: truptidas@immt.res.in

P. K. Padhy et al. (eds.), *Air Quality and Human Health*,
https://doi.org/10.1007/978-981-97-1363-9_14

under non-communicable respiratory morbidities (Brashier et al. 2012). Moreover, various ailments, starting from the simple ones like irritation of the eyes, nose, and throat, headache, dizziness, and vomiting, to potentially fatal ones, namely leukemia, mesothelioma, hemolysis, allergies, heart diseases, cerebrovascular disease (stroke), pregnancy disorders, immune system defects, carcinoma, central nervous system damage, etc., are coupled with air pollutants (Haque and Singh 2017; Kim et al. 2018). It has also become one of the principal causes of infant and child mortality in emerging as well as backward countries (Brunekreef 2010; Haque and Singh 2017). Individuals with pre-existing cardiovascular and respiratory disorders are susceptible to severe health issues due to air pollution, irrespective of their level of exposure. There have been numerous reports of pregnant women and elderly people being affected by this problem (Peled 2011).

There is an asymmetric distribution of illness across the globe, with approximately two-thirds of the deaths occurring in developing countries in Asia (Anenberg et al. 2019). In this context, India has turned into a major contributor to the global onus of air pollutants due to an exponential rise in population, industrialization, and urbanization. Out of the world's ten most polluted cities, nine are in India (Who et al. 2021). Various studies have reported that India's death and disease burden due to air pollution is disproportionately high (Balakrishnan et al. 2019). As per the Global Burden of Disease 2019 report, 1.67 million deaths were accredited to air pollution, (accounting for 17.8% of the overall deaths in the country). Predominantly these mortalities (0.98 million) were caused due to an unprecedented increase in fine particulate matter and indoor air pollution (0.61 million) (Pandey et al. 2021). On account of the rising levels of air pollutants, especially $PM_{2.5}$, the Delhi government declared air pollution a public health emergency in 2019. Due to the current circumstances, air pollution is turning into a new hazard to the Indian public. Therefore, the objective of this segment is to unravel the potential health threats interrelated with deteriorating air quality in Indian cities.

14.2 Impact of Major Air Pollutants on Human Health

14.2.1 Particulate Matter

Particulate matter (PM) can be outlined as an amalgamation of solid and liquid matter suspended in the atmosphere that incorporates both organic and inorganic components such as dust, pollen, soot, droplets, etc. These particles differ in size, shape, and origin. They have been categorized principally in terms of their aerodynamic diameter, usually referred to as "particle size". These properties help in determining their transportation, removal, and deposition.

Based on the size, PM is usually divided into:

- The **coarse** mode, containing fractions ranging from 2.5 to 10 μm in size. These are produced by the disintegration of massive particles and include dust from various sources like soil, unpaved roads, mining operations, sea salt, traffic emissions, etc.
- The **fine** fraction, containing smaller particles with a size up to 2.5 μm. These smaller ones are largely shaped by gases.
- The **ultrafine** fraction, containing particles <0.1 μm. These are formed by nucleation (condensation of substances or gases within the atmosphere).

Wide-ranging approaches are available for monitoring PM concentrations. The most precise values are procured from instruments that operate based on the gravimetric method. In this method, the air is sucked across a pre-weighed filter paper and particles get collected on it, which is then removed from the sampler and weighed again. This technique has the edge that the particles gathered can be used for chemical analysis (Nussbaumer et al. 2008). Other efficient methods include instruments like sensors (based on the microbalance and electrochemical method) and optical particle counter (OPC) (based on using an illuminant and a photodetector to irradiate particles and measure the scattered light respectively) (Ministry for the Environment 2009).

Pertaining to the myriad components in the PM and to the emerging pollution scenario across the globe contributing to enormous new substances in the ambient air; studies on health implications due to exposure to PM are still quite nascent. Therefore, the characterization of PM inclusive of chemical, biological, and morphological from various parts and emission sources has to be set as one of the basic criteria of evaluation. Chemical components in the PM, both water-soluble (cations & anions) and acid-soluble (metals) along with the carbonaceous components, play a dominant role not only in regulating the chemistry of the atmosphere but also have a crucial role in altering the human metabolic processes through consistent exposure to heavy metals, nitrate, and other carbonaceous components. Similarly, the morphological characterization of the PM helps provide crucial information about the composition and source of metals and carbonaceous components in the PM. Further, the biological characterization would help in determining the presence of pathogenic microorganisms or their toxic components in ambient particulate matter. PM characterization is an important mechanism for understanding the lifetime of the pollutants in the ambient air, their emission sources, sinks, and transport potentials as well. In this chapter, we explain in brief about all the three characteristics of ambient PM.

Determining the chemical mass fraction of PM is an important step that validates the various chemical components and their sources in the ambient air. Chemical characterization of PM for determination of water-soluble ions and acid-soluble metals is carried out using ion chromatography (IC) and inductively coupled plasma-optical emission spectroscopy/mass spectrometry (ICP-OES/ICP-MS) respectively, while the elemental and organic carbon components were determined by EC-OC analyzers, using standard protocols (Mahapatra et al. 2021). The chemical components in ambient air are representative of the various emission sources that contribute to the overall PM mass. The source contribution of each component can be traced using specific tools and models (Mahapatra et al. 2021). Figure 14.1 shows the contribution of various sources to ambient PM in an urban location.

The surface morphology of ambient particulate matter (PM) plays a vital role in understanding atmospheric phenomena including oxidation reactions, hygroscopicity, gas condensation on the pre-existing particles, etc. Further, surface characteristics also provide crucial information about the metallic composition along with being a function of its source of origin, time spent in the atmosphere post-emission, and local meteorological conditions. Microstructural analysis of individual particles is carried out using a Carl Zeiss Model: Sigma Zemini supra-55, field emission scanning electron microscope (FESEM) equipped with an energy-dispersive X-ray spectrometer (EDX). Around 1 cm^2 of the sampled filter paper is used for the analysis purpose. The particles are gold coated to make the sample electrically conductive and then mounted on an elec-

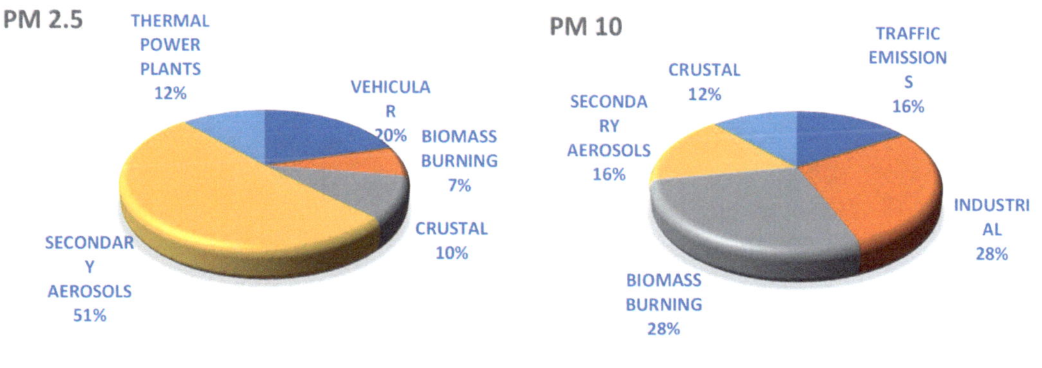

Fig. 14.1 Contribution of various sources to ambient particulate matter in an urban location (Prepared using data from Mahapatra et al. 2021)

tron microprobe stub. The instrumental detail and analytical methodology have been discussed further by Panda and Das (2017). The elemental composition of the selected particles is obtained by EDX spectra. The result showed that these particles containing organic (carboxylic acids, polycyclic aromatic hydrocarbons, aldehydes) and inorganic functional groups (sulfates, nitrates, halogens) could influence atmospheric chemistry and also alter the equilibrium affecting the soil pH. Any kind of effective air pollution strategy would need the highest possible information on particle shape, dimension, chemical composition, and surface area properties at the individual particle level. The microstructural analysis of single-particles has intensively been carried out over Bhubaneswar earlier in Panda and Das (2017). Four different major types of particles were reported based on their elemental composition and sources namely, (a) carbonaceous, (b) industrial, (c) biogenic, and (d) minerogenic particles. The present discussion includes some additional details and a general overview of the characteristics of single particles along with their possible emission sources.

High magnification images of airborne PM have been presented in Fig. 14.2 and analyzed using FESEM. Fig. 14.2a, b shows a variety of atmospheric particles collected over Bhubaneswar. The PM samples were then magnified further for a clear and in-depth view of individual aerosol particles. Most of these particles are seen to be agglomerates of granular-sized small units commonly known as soot or carbonaceous particles with carbon as their major constituent. These particles are often identified from combustion emission sources with a high surface-to-mass ratio (Popovicheva et al. 2014, 2019). The major components of these kinds of particles are high-intensity carbon with a small oxygen peak as detected in the EDX spectra.

Meteorological conditions also play a very important role in explaining and evaluating trends in PM. Previous studies have disclosed pertinent information regarding the impact of long-distance transportation and meteorology on their geographical and temporal variation (Mahapatra et al. 2018). Suspended in the atmospheric boundary layer (ABL) where wind, temperature, moisture, and intensity of solar radiation fluctuate rapidly, the PM particles may undergo physicochemical transformations triggered by the above factors coupled with the presence or absence of atmospheric reactants. PM through its surface reactions in the atmosphere predominantly affects the tropospheric ozone concentration and secondary aerosol formation (Nishanth et al. 2014). $PM_{2.5}$ is capable of acting as cloud condensation nuclei. This property facilitates the alteration of microphysical characteristics of clouds and thus in turn can regulate the weather pattern, and affect the water cycle and the overall radiation budget as well.

14.2.1.1 Health Issues on PM Exposure

As stated by the United Nations, there is a deep association between exposure to higher concentrations of PM and the multiplication of morbidity.

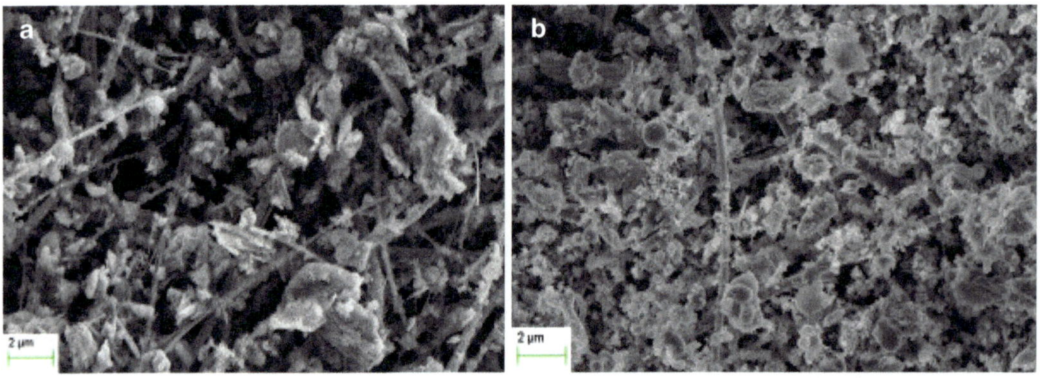

Fig. 14.2 (**a**, **b**) FESEM image of airborne PM samples collected over Bhubaneswar

PM gains access to the human anatomy through the oral or nasal tract (Li et al. 2003; Storey-Bishoff et al. 2008). Depending on the diameter of the particles, the indrawn PM settles in various segments of the respiratory tract such as the alveolar and tracheobronchial areas (Deng et al. 2020; Manojkumar and Srimuruganandam 2021). Coarse mode particles are eliminated by the nose-filtering activity which effectively stops any larger molecule from moving into our systemic circulation. Unfortunately, finer particles find their way to the lungs and end in several short and long-run effects. While many bring about native inflammation after being deposited in the lungs, soluble components enter systemic circulation leading to a rise in oxidative stress. These components may successively lead to COPD (Doiron et al. 2019; Lee et al. 2020), acute respiratory infections like pneumonia (Kim et al. 2020; Susetyo et al. 2019), hyper-stimulation of immune cells (T helper cells), DNA damage (Wunnapuk et al. 2019), and cardiovascular diseases (Martinelli et al. 2013) like stroke, reduced heart rate, etc. Chronic exposure can cause premature mortality and deteriorate the functioning of the lungs particularly for patients with pulmonary diseases like asthma (Manisalidis et al. 2020).

Frequent exposure to a high concentration of ultra-fine particles ($\geq PM_{2.5}$) can be the genesis of acute pulmonary inflammation, alveolar and interstitial edema, disruption of endothelial and epithelial cells, hemorrhage, cardiovascular problems, development of cancer, etc. (Oberdörster 2000; Peters et al. 1997; Vinzents et al. 2005). In a recent study carried out in an urban location in eastern coastal India by Mahapatra et al. (2021), it was revealed that PM consists of four different particles (dominant), i.e., alumino-silicate, metals, carbonaceous particles, and mature particles. This information is an indication that PM may originate from myriad sources. PM samples obtained from the traffic areas were loaded with soot particles whereas a partially higher concentration of metals, including chromium and other heavy metals like Cu, Ni, Pb, Al, etc. was mostly found near the construction sites. Excess carcinogenic risk is directly associated with heavy metal concentrations in ambient air. Further, measurements of endotoxin (a constituent of the exterior cell wall of the gram-negative bacteria) concentration in $PM_{2.5}$ over the same location, provide evidence of biological components in PM (Ivanova et al. 2016).

The CPCB reports of the past few years distinctly indicate a positive shift in the $PM_{2.5}$ concentrations. This detrimental inference can be directly correlated to the considerable increase in road traffic. As claimed by the State of Global Air 2020 report, in 2019, India registered the highest annual average concentration of $PM_{2.5}$ (Fig. 14.3) (Health Effects Institute, Institute for Health Metrics and Evaluation's Global Burden of Disease Project 2020). Among the cities enclosed within the National Air Quality Monitoring Program (NAMP), each daily and annual levels of more than half of them habitually exceed the threshold limit of 60 and 100 μg m^{-3} (PM_{10}) and 40 and 60 μg m^{-3} ($PM_{2.5}$) (as designated by the WHO). Considering the far-reaching health consequences of PM, 2005 WHO guidelines indicate that by reducing PM_{10} and $PM_{2.5}$ concentrations from 70 to 20 μg m^{-3} and 35 to 10 μg m^{-3} respectively, air pollution deaths can be reduced by around 15%.

14.2.1.2 COVID-19 Lockdown-Reduction of Pollutants in the Atmosphere

The countrywide lockdown announced in India to deter the rising rate of COVID-19 infection, provided a unique scenario to study the concentrations and characteristics of PM with minimal anthropogenic intervention. As per a recent study of Bhubaneswar City by Nath et al. (2021), a significant reduction of 33% in PM_{10} and 38% in $PM_{2.5}$ concentration was observed due to it. Similar reports have also showcased an exceptional reduction in the concentration of PM across all the major cities including Delhi, Mumbai, Hyderabad, Kolkata, etc. which has led to the unfolding of a clear sky nearly after a decade (Ganguly et al. 2021; Kumar et al. 2020). Thus, a decline in traffic load, shutdown of industrial activities, and reduction in mobility of the

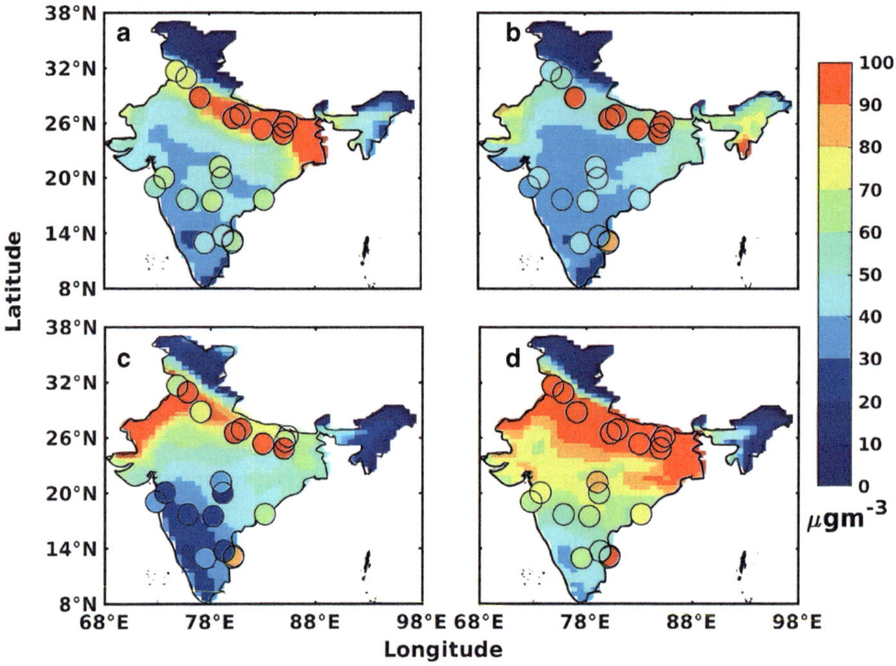

Fig. 14.3 The seasonal depiction of $PM_{2.5}$ concentration over India during (**a**) Winter, (**b**) Pre-monsoon, (**c**) Monsoon, and (**d**) Post-monsoon (circles outline the measured data from CPCB) (Navinya et al. 2020, Creative Commons Attribution BY 4.0)

population have ultimately helped in improving the overall health condition of the masses (Ivanova et al. 2016; Patra et al. 2021).

14.2.2 Trace Gases

14.2.2.1 Ozone (O_3)

The formation of ozone takes place under high voltage electric discharge of oxygen. It is a strong oxidant and arises in the stratosphere. It can also arise following the chain reactions of photochemical smog in the troposphere or during the emission of pollutants from transport vehicles or anthropogenic activities (Manisalidis et al. 2020). Moving with air masses, it can travel to distant areas from its initial source (Massachusetts Department of Public Health 2017). Thus, ozone is not inherently detrimental but tropospheric O_3 is considered a greenhouse gas because of its activity as a precursor of highly reactive hydroxyl radical (-OH) (Lal 2007). Thus, a high concentration of O_3 in the boundary layer may have some deleterious effects on biological life. The troposphere acts as a huge chamber producing ozone as a principal product in the transformation of huge quantities of trace gases from one form to another (Molina 2021) (Fig. 14.4).

According to National Ambient Air Quality Standards (NAAQS), an 8 h average is established as 100 μg m^{-3} and a 1 h average as 180 μg m^{-3} for industrial, residential, rural, and other areas (Sundaray and Bhardwaj 2019). Even though ozone plays a protective role against ultraviolet irradiation in the stratosphere, it is still harmful to the respiratory and cardiovascular systems when present in high concentrations at ground level. Therefore, it has become a major concern in recent years. Ozone uptake takes place during inhalation and affects the upper layers of skin and tear ducts of humans. Ozone exposure can trigger a plethora of health problems particularly, in children, elderly people, and asthma patients. Also, inhaled ozone can penetrate deep

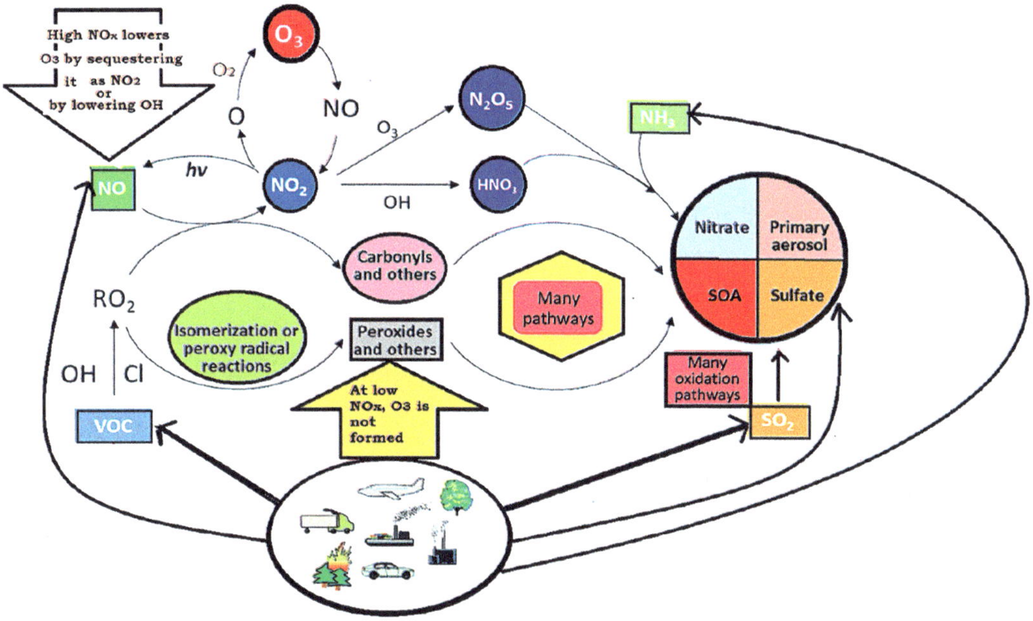

Fig. 14.4 Photolysis of ozone

into the lungs due to the low-solubility capacity of ozone (https://www.health.nsw.gov.au/environment/air/Pages/common-air-pollutants.aspx). It may lead to an increase in DNA damage in epidermal keratinocytes leading to impaired cellular function (McCarthy et al. 2013). An online analyzer is used to estimate ozone exposure in ambient air after collecting ground measurement data and fitting spatial/temporal/mathematical models (Pandey et al. 2021). Ambient ozone measurement in different parts of India has been calculated after measuring the O_3 concentration during December–January-February (DJF), March–April-May (MAM), and September–October-November (SON) as shown in Fig. 14.5.

Severe irritation in the throat, nose, eyes, or ears (irritation in the eye at around 100 ppb zone and coughing at around 2 ppm zone) and irreversible accelerated lung damage are common effects on human health. Thus, ozone has both acute and chronic health effects.

COVID-19 Lockdown Effect on Ozone

During the lockdown period in the year 2020 due to COVID-19, the concentration of O_3 has shown a sizeable reduction in some parts of the country in contrast to other parts where it has shown an increment in the daily average of the pollutant compared to its pre-lockdown days. Cities such as Kolkata and Maharashtra showed a decrement of 42% and 18% in the O_3 concentration respectively (Chowdhuri et al. 2020; Sahu et al. 2020). In contrast to this, cities such as Delhi, Utter Pradesh, and Haryana have shown an increment in concentration (Dumka et al. 2021). Over Hyderabad, two times enhancement of O_3 concentration was observed compared to normal days. Similarly, an increment of 3% was observed for the city of Bhubaneswar during the lockdown (Panda et al. 2021a, b). The reduction of O_3 concentration can be attributed to the pause in various anthropogenic activities but the enhancement can be related to insufficient NO concentration in the atmosphere resulting in the accumulation of O_3 in the troposphere engendering various health issues.

14.2.2.2 Oxides of Nitrogen (NOx)

Oxides of nitrogen ($NO_x = NO + NO_2$) are traffic-related pollutants, as they are emitted from motor engines of vehicles and are most reliable on local sources for emission (Manisalidis et al. 2020).

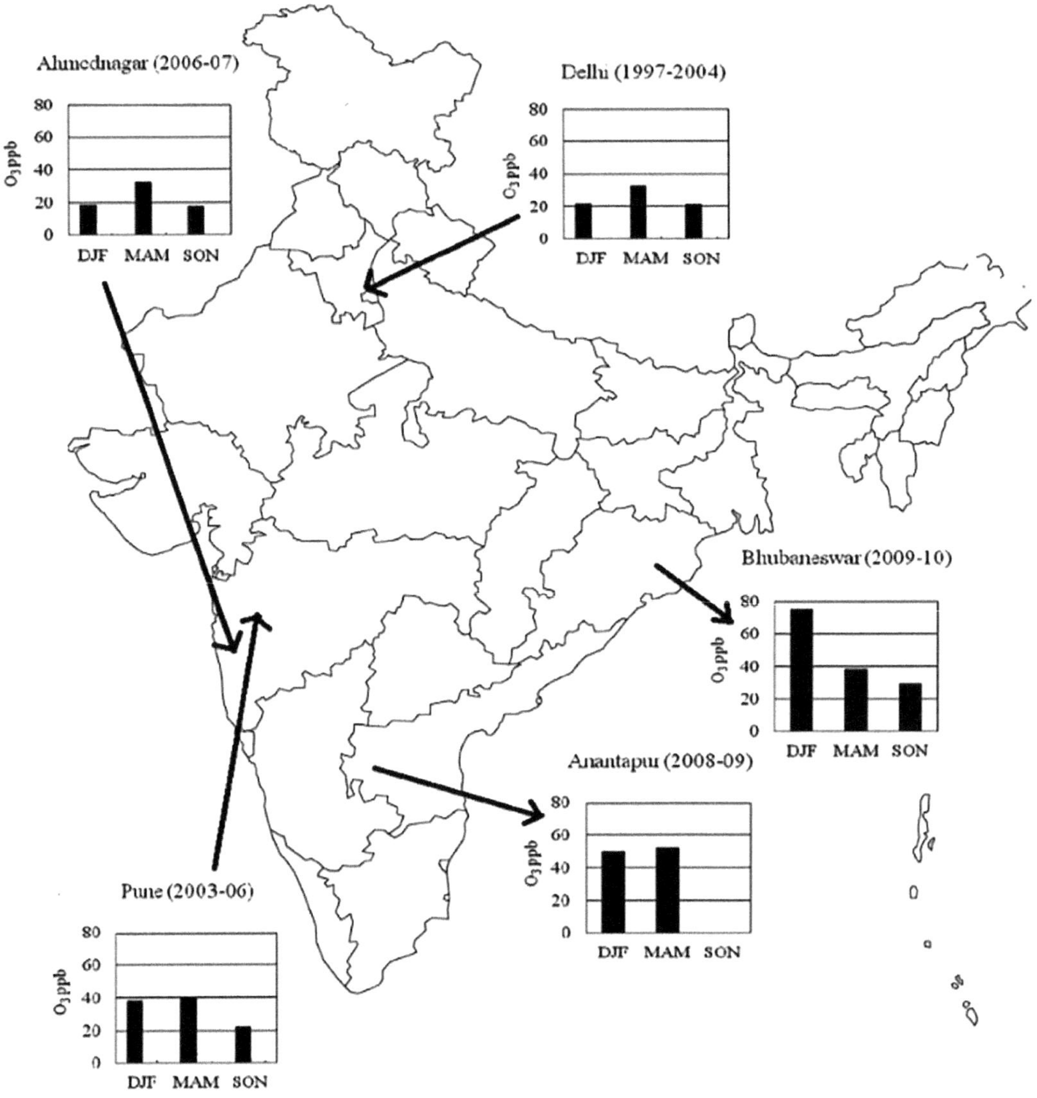

Fig. 14.5 Concentrations of ozone in different parts of India for different seasons (prepared using data from Oksanen et al. 2013)

They are irritants to the respiratory system as when high levels of NO_2 are inhaled, they very easily penetrate deep into the lungs, inducing common respiratory diseases such as coughing, wheezing, dyspnea, bronchospasm, and even pulmonary edema. It is likely to be seen that concentrations over 0.2 ppm tend to produce such adverse effects in humans, while concentrations higher than 2.0 ppm affect directly the T-lymphocytes, particularly the CD^{8+} cells and NK cells which are meant for producing our immune response. According to the studies, exposure to high levels of nitrogen dioxide for a long period can be responsible for chronic lung disease as well as has the power to impair the sense of smell. Symptoms of irritation in the eye, throat, and nose have been also registered other than the respiratory ones (Chen et al. 2007). NO_2 also has the propensity to react with other agents to form nitric acid along with some organic nitrates, those contributing to the formation of acid rain, in addition to being an obstructive and choking gas (https://www.health.nsw.gov.au/environment/air/Pages/common-air-pollutants.aspx). On the other side, NO_x reacts with other chemicals in the air to form both particulate matter and ozone, either of whose effect on respiratory health problems cannot be ignored. The NAAQS reports an annual average of around 40 $\mu g\ m^{-3}$ with toxic effects placed between 10 and 30 $\mu g\ m^{-3}$ for NO_x (Sundaray and Bhardwaj 2019). Those districts of India whose at least 50% or more areas have a concentration of NO_2 above the limiting value of ~30 $\mu g\ m^{-3}$ are identified as hotspots (Ramachandran et al. 2013) (Fig. 14.6) Development of asthma and increase in respiratory infections are a contribution from long-term exposures to elevated levels of oxides of nitrogen. Children and elderly people are generally at greater risk for the health effects of NO_x.

COVID-19 Lockdown Effect on NOx

The lockdown enforced in the country due to the global pandemic of COVID-19 has led to almost disappearance of NO_x concentration in the cities of India like Delhi, Haryana, and Punjab along with the eastern part covering Bihar, Jharkhand, West-Bengal, and Odisha which were previously marked as one of the hotspots of this toxic air pollutant (Verma and Kamyotra 2021). Precisely, a study of Bhubaneswar City in the eastern part of India during the lockdown period in the year 2020 has revealed a decrease of ~67% in the average NO_x concentration compared to the same period of its preceding year (Panda et al. 2021a, b; Sahoo et al. 2021). Thus, the result of the ban on the movement of the human race has marked a great tick toward the lessening of hazardous consequences for mankind.

14.2.2.3 Sulfur Dioxide (SO_2)

Sulfur dioxide is a harmful gas, having its emission source mainly from fossil fuel consumption or various industrial activities. Emissions from vehicles, locomotives, and ships also contribute substantially to the SO_2 concentration in the atmosphere. When these pollutants get released into the atmosphere, they undergo several reactions with oxidants like O_3 or H_2O_2 to form particulate of the sulfur which is known for being responsible for causing visibility degradation and threat to human health as well. The annual standard reported by the NAAQS for SO_2 is around 50 $\mu g\ m^{-3}$ (Sundaray and Bhardwaj 2019). Thus, short-term exposure to a higher level of SO_2 concentration may lead to major long-term effects on human life. Respiratory irritation, bronchitis, mucus production, and bronchospasm are some of the major health problems associated with sulfur dioxide emissions mainly in industrialized areas. SO_2 is a very sensory irritant and it penetrates deep into the lungs getting converted into bisulfite and in the meanwhile interacting with all the sensory receptors, causing bronchoconstriction (Chen et al. 2007; Manisalidis et al. 2020). Those with lung disease, old people, and children are the most susceptible. Moreover, redness in the skin, damage to the eyes with mucous membranes, and worsening of pre-existing cardiovascular disease have been frequently observed as effects of SO_2 emission.

COVID-19 Lockdown Effect on SO_2

Unlike other air pollutants, no significant reduction in the concentration level of SO_2 over the cities of India was observed during the lockdown days of COVID-19 due to the constant operation of the power plants and the presence of excessive coal mining as a part of essential service (Ivanova et al. 2016).

14.2.2.4 Carbon Monoxide (CO)

When combustion is incomplete in fossil fuels, it gives rise to the production of carbon monoxide having a lifetime of 2–4 months in the atmosphere. It is highly toxic and is titled as a dangerous pollutant which is infamous for its lack of identifying color and smell. Previously, this gas was extremely prevalent in houses for use in domestic heating, but then it was identified as unsafe and was replaced by natural gas. Still, it is far from extinct in the human race. Large spatial annual variability of total CO emission in India is evident from Fig. 14.7 (Nandi et al. 2020).

The annual standard for CO reported by NAAQS averaged for 1 h is 4$\mu g\ m^{-3}$ (Sundaray and Bhardwaj 2019). Various symptoms of CO

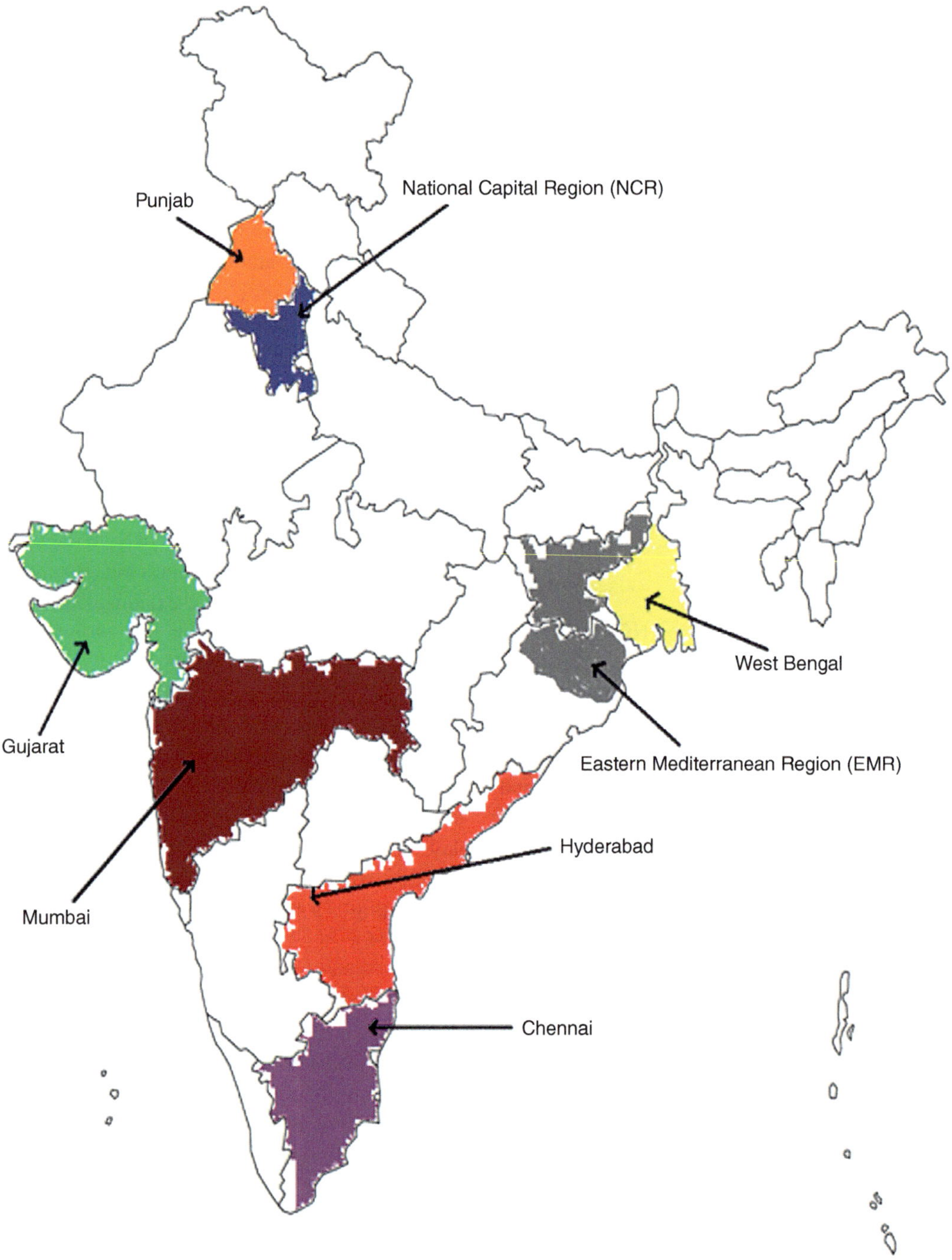

Fig. 14.6 NO_2 hotspots in India (prepared using data from Ramachandran et al. 2013; Creative Commons Attribution BY 3.0)

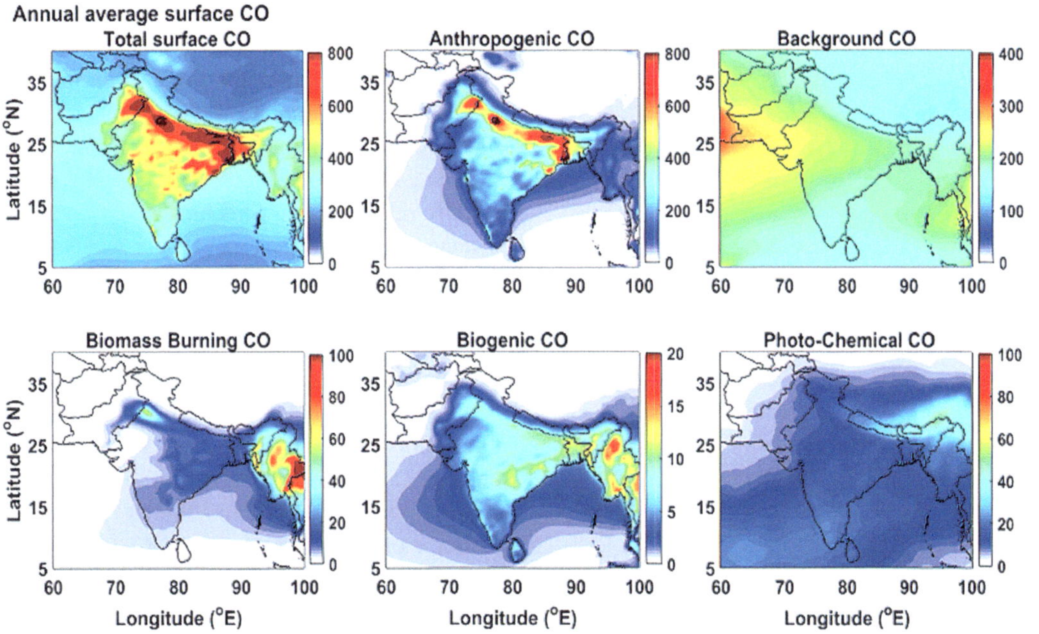

Fig. 14.7 Spatial distribution of annual average total CO mixing ratio and the absolute concentrations of different CO source types at the surface over the Indian subcontinent during 2015 (Nandi et al. 2020) [The authors permit to use the work under the terms of the CC BY 4.0 license]

poisoning include headache, vomiting, dizziness, weakness, nausea, and finally, loss of consciousness to some extent. The reason lies with the greater affinity of carbon monoxide to hemoglobin than that of oxygen. Long-term exposure of people to high levels of carbon monoxide may lead to serious poisoning in the veins. In addition, hypoxia, ischemia, and cardiovascular disease are observed due to the loss of oxygen as a result of the competitive binding of carbon monoxide which prevents the transfer of oxygen in the bloodstream to critical organs like the heart and brain. Hence, the colorless, odorless, and tasteless gas is tagged as "Silent Killer" (Chen et al. 2007; Manisalidis et al. 2020).

COVID-19 Lockdown Effect on CO

The sudden popped up pandemic situation of COVID-19, however, has led to a nationwide drop in the CO concentration during lockdown overtaking over the pre-lockdown period (Verma and Kamyotra 2021). In the year 2020, the eastern coastal city of India, Bhubaneswar, showed a significant reduction of ~14% in CO concentration during the complete lockdown compared to the same time of the previous years. Whereas the western cities of India, Pune, and Gujrat, revealed a drop of ~21% and ~ 25% respectively (Korhale et al. 2021; Panda et al. 2021a, b; Sahoo et al. 2021; Selvam et al. 2020). Hence, the effect of this air pollutant on the human species has decreased due to a less frequent chance of exposure to the silent killer gas which has shown an insignificant plummet in its concentration marked in the atmospheric layer.

Trace gases	Time-weighted average	National ambient air quality standards (μg/m^3)	Health impacts
Ozone (O_3)	8 h 1 h	100 180	Respiratory and cardiovascular dysfunctions, eye irritation, asthma, chest pain.
Nitrogen dioxide (NO_2)	Annual 24 h	40 80	Damage to liver, lungs, spleen and blood, visibility impairment
Sulfur dioxide (SO_2)	Annual 24 h	50 80	Respiratory and central nervous system involvement, eye irritation, breathing difficulty, heart diseases
Carbon monoxide (CO)	8 h 1 h	02 04	Central nervous system and cardiovascular damages, vision problems

14.2.3 VOCs

Volatile organic compounds represent a broad class of chemicals that have high vapor pressure at room temperature.

VOCs are largely divided into aromatic and halogenated hydrocarbons, oxygen-containing compounds, nitrogenous and sulfurous compounds, and several other categories (Kansal 2009). Emissions of VOCs are straight away associated with industrial operations, traffic outrushes, and countless indoor and outdoor sources. As mentioned by the US Environmental Protection Agency (USEPA), 21 among 33 precarious air pollutants are correlated to the automobile sector which is considered a substantial source of VOCs in the urban environment (Gaur et al. 2016).

Most VOCs prevail in higher concentrations within the indoor domain rather than outdoor air. Biomass burning and domestic fuel consumption are crucial sources contributing to around 17% of the overall annual VOC emissions in India (Fleming et al. 2018; Verma et al. 2019). VOCs are also released from personal care products, for instance, perfumes, cleaning agents, paints, printing machines, glues, and building materials such as plywood, particleboard, etc. (Hodgson 1991).

Though not listed under the criteria pollutants, EPA has classified them as toxic to the environment and human health. WHO has specified threshold values for some of the VOCs centered on their harmfulness, apart from odor nuisance and cancer (Srivastava et al. 2006).

Some of the most common VOCs found in the home and their threshold limit values (TLV) are listed below:

VOCs	TLV (ppm)
Ethanol	1000
Formaldehyde	0.1–0.3
Acetone	750–1000
Benzene	0.1
Dichlorobenzene	25–50

Considering the adverse effects of VOCs on human health due to inhalation, several studies have been conducted to assess their concentration levels in ambient air. At levels found in the metropolitan environment, most of the VOCs have been recorded to be noxious (Mukund et al. 1996). VOCs cover a wide range of chemicals that can give rise to headaches, fatigue, hypersensitivity, visual disorder, nausea, dizziness, irritation in the eye, nose, and throat, choking, and dermal infections. Higher concentrations may cause gene mutation, acute and chronic respiratory diseases, as well as impairment of the lungs, liver, kidney, or central nervous system (Kwon et al. 2018; Sun et al. 2019; Zhang et al. 2018). Moreover, VOC emissions show carcinogenic effects on various organs (Sarigiannis et al. 2011). Endless exposure to VOCs results in distinguished reactions with the air shaft epidermis and mucosa membrane which are strongly correlated with pulmonary diseases like asthma (Alford and Kumar 2021; Wieslander et al. 1996).

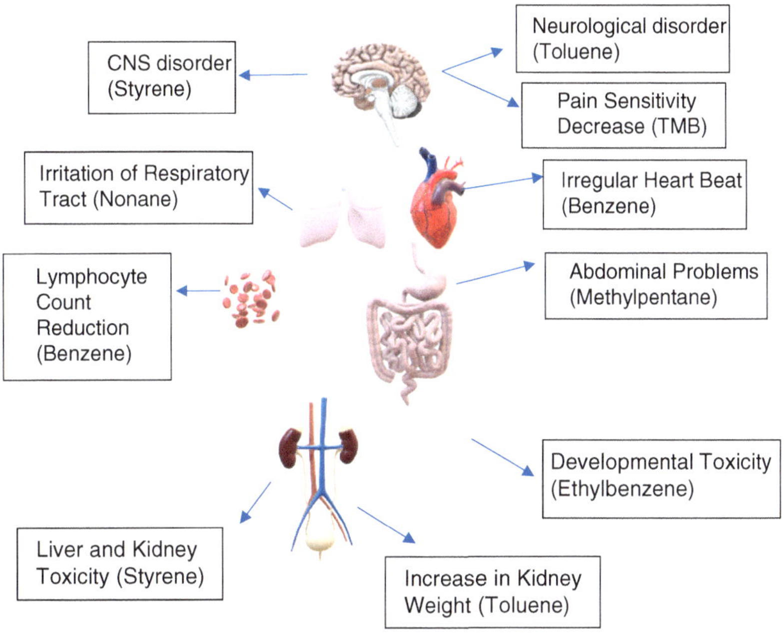

Fig. 14.8 Effect of VOCs on different organs of the human body

VOCs, such as benzene, propylene glycol, and formaldehyde sometimes lead to nasopharyngeal cancer on account of their distinguished reactivity with the epithelial lining of the airway and mucous membrane (Mølhave et al. 1997). Aromatic compounds like toluene and xylene are suspected carcinogens and may lead to leukemia (Sharma et al. 2013). Even exposure to a minor amount of VOCs in day-to-day life may trigger the risk of respiratory infections like asthma and allergies (Zhang et al. 2018) (Fig. 14.8).

14.2.4 PAHs

Polycyclic aromatic hydrocarbons are an outsized family of semi-volatile organic compounds with more than one aromatic ring. Some of the major sources of PAHs include biomass burning, combustion of fossil fuels, internal combustion engines, garbage incineration, and countless industrial activities such as coke and aluminum production, oil refining, etc.

The USEPA has included 16 out of 200 organic compounds in the checklist of priority pollutant. In India, the estimated annual emission of PAHs is approximately 90 Gg y^{-1} (Zhang and Tao 2009), whereas the safety level of PAHs according to the Occupational Safety and Health Administration (OSHA) in the workplace is 0.2 mg m^{-3}. PAHs have become a pivotal health concern due to their genotoxicity, carcinogenic attributes, and recalcitrance. The increased consciousness about the inimical effects of PAHs on the ecosystem as well as on health has led to a surge in detailed scrutiny of these detrimental compounds.

Humans are exposed to PAHs through several routes from different sources but the most important pathway of exposure from air includes inhalation. Though inhalation includes both indoor and outdoor exposure, yet indoor air would be a pertinent source as people spend 80–93% of their time indoors (Brunekreef et al. 2005). This may imply that in particular, women and children are vulnerable because of the comparatively longer time spent at home. The endorsed value for the average inhalation rate is 11.3 m^3 day^{-1} (women) and 15.2 m^3 day^{-1} (men). However, it has been noticed that this limit exceeds in maximum areas (USEPA 1997).

PAHs are promptly dispersed in various tissues with a propensity to accumulate in body fat. Being lipid-soluble, they are effortlessly ingested

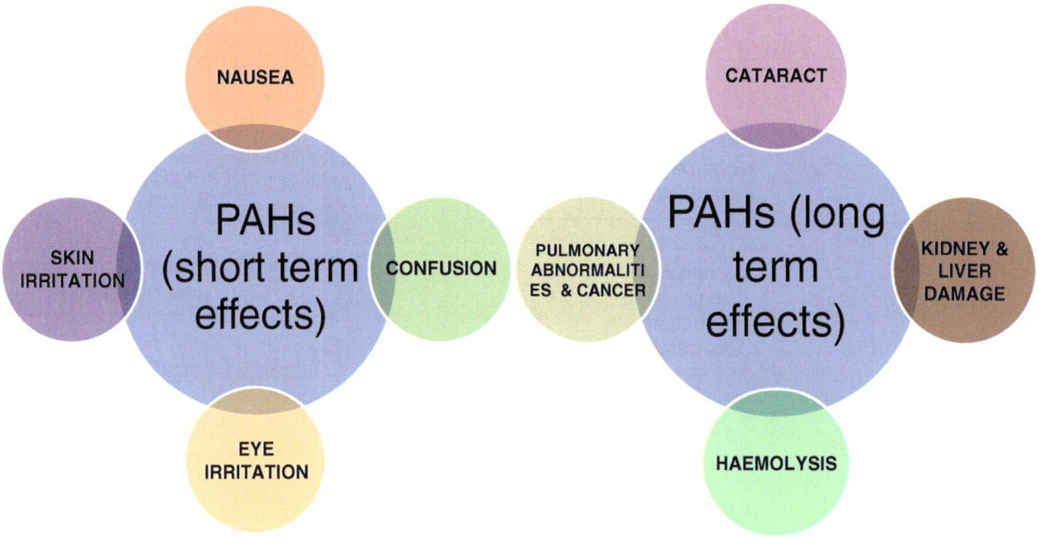

Fig. 14.9 Adverse effects of PAHs on human health

into the gastrointestinal tract of humans. Metabolism of PAHs occurs via the cytochrome P450 monooxygenase enzyme group. These enzymes help in oxidizing PAHs into epoxides, and few of these are known to be ultimate carcinogens (Abdel-Shafy and Mansour 2016). Short-term effects include nausea, dizziness, vomiting, diarrhea, illusion, etc. Prolonged exposure to PAHs can cause cataracts, kidney, and liver damage along with an escalating threat of skin, lung, bladder, and gastrointestinal cancer. PAHs can traverse through the placenta and exercise their deleterious effects on the evolving fetus (Drwal et al. 2019). Although the concentration of lighter molecular weight PAHs in the atmosphere is found to be more, yet the adverse health effect has always been associated with their heavier counterparts (Masih et al. 2020). They have the ability to bind to the DNA and alter ordinary cells into cancerous cells (Marston et al. 2001; Wohak et al. 2016). PAHs are capable of intervening with the hormones and enzymes produced by the thyroid gland. They also affect the ion transporters and the enzymes involved in signal transduction. All these effects along with their repercussions on the immune system have further raised concern (Ghosal et al. 2016; Låg et al. 2020). Numerous researches have revealed other responses of PAHs including respiratory diseases, inflammation, immune system damage, oxidative stress, etc (Låg et al. 2020; Marston et al. 2001) (Fig. 14.9).

14.2.5 Heavy Metals

Metals are substances formed naturally within the earth's crust having high electrical conductivity, malleability, and luster. They can change their valence states via redox reactions. Heavy metals have a relatively high density as compared to water (> 4 g cm^{-3}) and a high degree of toxicity even at low concentrations (ppb levels).

Heavy metals are emitted to the atmosphere from various anthropogenic sources like power generation, smelting, incineration, and traffic emissions (internal combustion engine). Other major sources include mining and industrial activities. Potentially toxic chemicals present in the air include lead, cadmium, zinc, mercury, etc.

Occupational Safety and Health Administration (OSHA) offers regulatory limits for various hazardous heavy metals. This has been mentioned in the table below along with their major health impacts.

Heavy metal	Time limit	Concentration ($\mu g/m^3$)	Effects on human health
Arsenic	8 h shifts and 40-hour work weeks	10	Bronchitis, dermatitis
Barium	–	500	Cardiac arrest, gastrointestinal dysfunction, respiratory failure
Cadmium	–	5	Renal dysfunction, lung cancer
Chromium	–	1000	Nervous system damage
Lead	–	0.5	Mental retardation in children, lung, kidney and liver cancer
Mercury	–	100	Nervous system damage, protoplasm poisoning, tremors
Selenium	–	200	Respiratory disorder
Silver	–	10	Respiratory disorder, throat and skin irritation

Exposure to metals can occur through numerous passages. The most hazardous pathway for all heavy metals is through ingestion, followed by dermal absorption and inhalation respectively (Kumari et al. 2021). Metals can be inhaled directly as dust, fume, etc., or through vaporization (e.g., mercury vapor during the manufacture of fluorescent lamps). The quantity of absorption within the alimentary canal differs widely, counting on the chemical configuration of the metal and the age and nutritional quality of an individual. Once a metal is engrossed, it disperses within the tissues and organs. Metals have gravitated to stockpile in some storage sites, namely the liver, bones, and kidneys, for years. These metals can cause several ailments which include respiratory problems such as asthma, COPD, bronchiolitis, lung carcinoma, acute myocardial infarction, central nervous system dysfunctions, and cutaneous diseases. Chronic exposure to metals such as cadmium and lead can cause hypertension and renal failure respectively. Some metals like copper, iron, cadmium, chromium, nickel, mercury, and vanadium can yield reactive radicals which end in DNA mutilation, diminution of protein sulfhydryl, and lipid peroxidation alongside other consequences (Briffa et al. 2020). Heavy metals tend to bio-accumulate through the food chain (Hu et al. 2020) which causes cancer and toxicity of several organs like the heart, brain, etc. This successively leads to an increased death rate (Du et al. 2013). An individual with a high dose of metal toxicity typically has very general symptoms, such as weakness or headache, or maybe no symptoms in the least. This makes the diagnosis of toxicity very difficult in a clinical setting unless a clinician has appropriate knowledge and training (Fig. 14.10).

14.2.6 Bioaerosols

Among the various segments of PM, the biological fraction known as bioaerosols is of special concern because of its potential to cause respiratory and infectious diseases in humans, plants, and animals. Bioaerosol primarily consists of particles of different biological origins, such as living and dead organisms, which include bacteria, fungi, archaea, protozoa, viruses, and their by-products; dispersal units, which consist of fungal spores, plant debris and pollens, and many more with aerodynamic sizes ranging from 0.01 μm to 100 μm (Després et al. 2012). These are released into the atmosphere either naturally or through various anthropogenic activities. In urban areas, anthropogenic sources such as landfills, composting, reuse of solid waste, biological wastewater treatment, industrial activities, and transport of solid/biological waste are the dominant factors for the origin of bioaerosols (Du et al. 2018). However, they do not always occur as an independent component in ambient air rather they form clusters that are attached to suspended particulate matter contributing up to 25–50% of the total PM_{10} mass concentration whereas 4–11% of the total $PM_{2.5}$ mass (Fröhlich-Nowoisky et al. 2016).

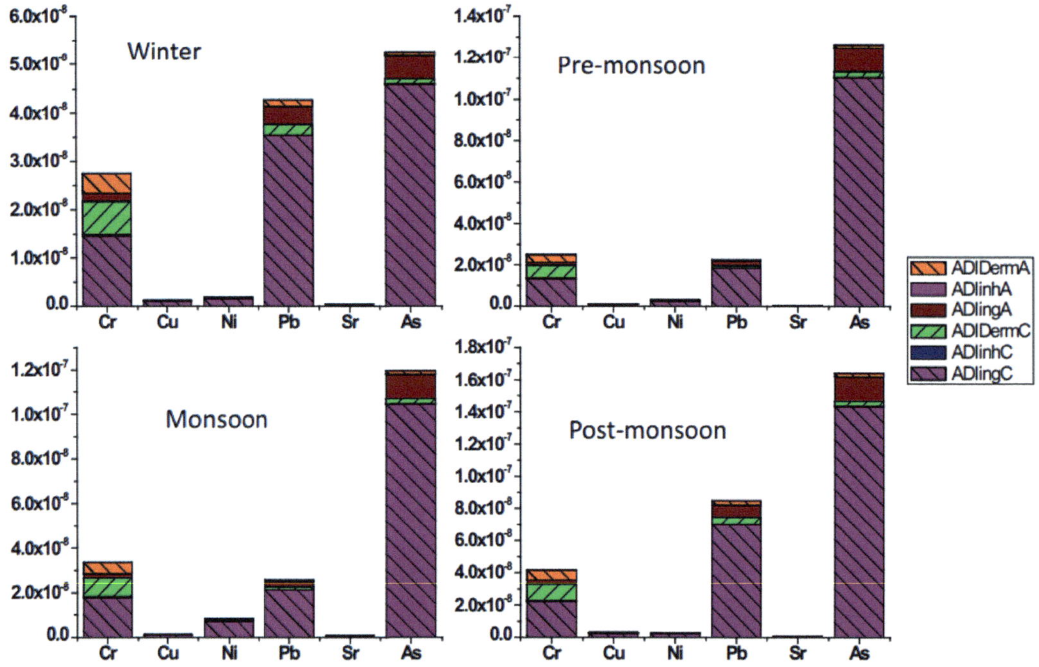

Fig. 14.10 Potential health risk assessment results of metals for different seasons in Bhubaneswar (C:-Children, A:-Adult) (Taken from Panda et al. 2021a, b)

The process of aerosolization, hydrodynamic as well as kinetic factors provides the biological entities a transportation medium through which it can be easily conveyed to different environments from the source of origin which implies that health complications associated with airborne bioaerosols are uniformly over-polluted and non-polluted regions on being overexposed or continuously exposed (Karmakar et al. 2020).

Studies on bioaerosols concluded that they contain a significant amount of pathogenic microorganisms, specifically bacteria and fungi which are responsible for several respiratory diseases such as asthma, tuberculosis, chronic obstructive pulmonary disease (COPD), allergic rhinitis, eczema, influenza, legionellosis, and anthrax, as well as some life-threatening diseases such as COPD and cancer (Humbal et al. 2018). Although there is no concrete evidence available that explains the specific mode of reaction between bioaerosols and human health, airway inflammation, as well as oxidative stress, are considered the major mechanisms of the diseases caused due to bioaerosols. Additionally, the growth of fungus yields more than 100 types of microbial volatile organic compounds, primarily the derivatives of amines, alcohols, esters, hydrocarbons, and many more, which result in irritation of the skin, eyes, and upper respiratory tract. There are reports showing the variation in the concentration of microorganisms at different places in different seasons in India. Interestingly, Madhwal and Nair reported the concentration of bacteria and fungus to be highest in monsoons because rainfall provides a humid condition for the growth of microbes in ambient air, and also because bioaerosols, when carried to the upper atmosphere, act as cloud condensation nuclei and descend back to the atmosphere with rainfall, hence increasing the concentration during monsoons whereas the chilling temperature of winters makes the atmosphere unfavorable for the growth and propagation of microbes (Madhwal et al. 2020; Nair 2021).

Nonetheless, apart from pathogenic fungi and bacteria, the microbial toxins, specifically the endotoxin, which is produced during the multiplication, growth, and death of bacteria and is present on the outermost layer of the gram-negative bacteria, are of great concern. Exposure to these endotoxins causes chills, fevers, joint pains, headaches, dry coughs, decrease in the leukocyte count in the blood, bronchial asthma, upper airway inflammation, and even death (Aghaei et al. 2020). Similarly, mycotoxin, a low molecular weight secondary metabolite of fungus, can reach the alveoli of the lungs, affecting the respiration process.

Bioaerosol sampling requires efficient sampling techniques to collect microbes from ambient air, as their concentration is not uniform. Sampling methods are divided into active and passive sampling. Active sampling utilizes a pump to draw air into the sampling unit, which can be related to particulate sampling. Filtration, impaction, liquid impingers, electrostatic samplers, and bio samplers are examples of active sampling. Whereas in passive sampling, petri plates containing the desired media are exposed to a stipulated time, which is then incubated at desired temperatures for microbial growth. A wide range of analytical techniques are available, which have their requirement, advantages as well as disadvantages (Burdsall et al. 2021). However, a valid standard sampling protocol is still missing.

Assessment of microbial concentration, as well as its organic and inorganic components, are crucial goals to determine the toxicological profile and its impact on human health. There is very little information on air micro-biome in comparison to other aspects of pollution, which requires a multi-disciplinary approach for a solution. Despite, being a very important area, no field campaign has been reported yet. The outfall of standard methods, environmental threshold limits, adequate information, and databases entangles the interpretation of the outcomes. The current COVID-19 pandemic is proof that an in-depth study is required to know more about the biological entities of the atmosphere to reduce their emission into the atmosphere as well as the health risk hazards.

14.3 Limitations and Remedial Strategies

Various studies have demonstrated a surge in mortality and morbidity due to the severity of diseases caused by both primary and secondary pollutants present in the atmosphere. Nevertheless, there are several limitations in the research available in this field. While the adverse effects of pollutants have been accessed on a whole, their actual composition is yet to be identified. Further examination is required to elucidate the time curve of pollutant-induced outcomes. Analysis defining susceptible populations along with true individual exposure will also add to the pre-existing knowledge (Anderson and Thundiyil 2012). Notably, the concentrations of air pollutants measured in different parts of the country and their repercussions on human health have not been linked to each other. Due to the absence of regular monitoring of the correlation between public health and air pollution exposure, the financial benefits associated with a small improvement in exposure are unknown. Thus, long-term epidemiological and cohort studies are the need of the hour for monitoring and control of harmful effects of air pollution (Dey et al. 2012).

An advanced wave of initiatives is imperatively needed for emerging nations like India to hinder the crippling intensity of air pollution. A collection of tentative strategies and guidelines for the formulation of a well-planned comprehensive framework with an absolute focus on a healthy and sustainable environment is given.

- Proper optimization of the transportation sector and implementation of BS-VI norms.
- Planned distribution of industries to keep pace with urban growth.
- Use of clean energy (e.g., LPG/CNG) as a replacement for fossil fuel.
- Prohibition of biomass burning particularly in autumn and winter.
- Development of techniques to improve conditions of both paved and unpaved roads.

14.4 Conclusion

The review distinctly affirms the pernicious effects of pollution on the environment as well as on human beings. The complex reciprocal interactions between climate change and air quality have important implications that need to be addressed immediately. Apart from jeopardizing health, air pollution has adverse effects on the economy of a nation. The sustainable development goals (SDGs) therefore urge for the limitation of ailments and mortalities occurring on account of air pollution. Reviews and assessments demonstrate that coordinated planning can help in achieving potential health benefits. Hence, managing air pollution is indispensable and should be the prime concern of higher authorities. The introduction of an effective environmental protection mechanism is of paramount importance. Awareness about the control measures in terms of benefits of the society needs to be promoted among the public. Overall, there is a grave necessity for an extensive air quality management strategy for the refinement of health standards and the refurbishment of the environmental status of the country.

Acknowledgments The authors acknowledge Dr. Upasana Panda (former PhD scholar CSIR-IMMT) for working on the microstructural analysis of particulate matter in CSIR-IMMT represented in Fig. 14.2 in this book chapter.

The authors are also thankful to ISRO-GBP (ATCTM and ARFI) for funding the research work carried out in CSIR-IMMT.

The authors would also like to acknowledge the Head, E&S Depat and the Director CSIR-IMMT for the encouragement.

References

Abdel-Shafy HI, Mansour MSM. A review on polycyclic aromatic hydrocarbons: source, environmental impact, effect on human health and remediation. Egypt J Pet. 2016;25:107–23. https://doi.org/10.1016/j.ejpe.2015.03.011.

Aghaei M, Yaghmaeian K, Hassanvand MS, Hedayati MH, Yousefian F, Janjani H, Nabizadeh R, Yunesian M. Exposure to endotoxins and respiratory health in composting facilities. Ecotoxicol Environ Saf. 2020;202:110907. https://doi.org/10.1016/j.ecoenv.2020.110907.

Alford KL, Kumar N. Pulmonary health effects of indoor volatile organic compounds—a meta-analysis. Int J Environ Res Public Health. 2021;18:1–12. https://doi.org/10.3390/ijerph18041578.

Anderson JO, Thundiyil JG. Clearing the air : a review of the effects of particulate matter air pollution on human health; 2012, 166–175. https://doi.org/10.1007/s13181-011-0203-1.

Anenberg S, Miller J, Henze D, Minjares R. A global snapshot of the air pollution-related health impacts of transportation sector emissions in 2010 and 2015. Int Counc Clean Transpl. 2019:55.

Balakrishnan K, Dey S, Gupta T, Dhaliwal RS, Brauer M, Cohen AJ, Stanaway JD, Beig G, Joshi TK, Aggarwal AN, Sabde Y, Sadhu H, Frostad J, Causey K, Godwin W, Shukla DK, Kumar GA, Varghese CM, Muraleedharan P, Agrawal A, Anjana RM, Bhansali A, Bhardwaj D, Burkart K, Cercy K, Chakma JK, Chowdhury S, Christopher DJ, Dutta E, Furtado M, Ghosh S, Ghoshal AG, Glenn SD, Guleria R, Gupta R, Jeemon P, Kant R, Kant S, Kaur T, Koul PA, Krish V, Krishna B, Larson SL, Madhipatla K, Mahesh PA, Mohan V, Mukhopadhyay S, Mutreja P, Naik N, Nair S, Nguyen G, Odell CM, Pandian JD, Prabhakaran D, Prabhakaran P, Roy A, Salvi S, Sambandam S, Saraf D, Sharma M, Shrivastava A, Singh V, Tandon N, Thomas NJ, Torre A, Xavier D, Yadav G, Singh S, Shekhar C, Vos T, Dandona R, Reddy KS, Lim SS, Murray CJL, Venkatesh S, Dandona L. The impact of air pollution on deaths, disease burden, and life expectancy across the states of India: the Global Burden of Disease Study 2017. Lancet Planet Health. 2019;3:e26–39. https://doi.org/10.1016/S2542-5196(18)30261-4.

Brashier B, Londhe J, Madas S, Vincent V, Salvi S. Prevalence of self-reported respiratory symptoms, asthma and chronic bronchitis in slum area of a rapidly developing Indian City. Open J Respir Dis. 2012;02:73–81. https://doi.org/10.4236/ojrd.2012.23011.

Briffa J, Sinagra E, Blundell R. Heavy metal pollution in the environment and their toxicological effects on humans. Heliyon. 2020;6:e04691. https://doi.org/10.1016/j.heliyon.2020.e04691.

Brunekreef B. Air pollution and human health: from local to global issues. Procedia Soc Behav Sci. 2010;2:6661–9. https://doi.org/10.1016/j.sbspro.2010.05.010.

Brunekreef B, Janssen NAH, de Hartog JJ, Oldenwening M, Meliefste K, Hoek G, Lanki T, Timonen KL, Vallius M, Pekkanen J, Van Grieken R. Personal, indoor, and outdoor exposures to PM2.5 and its components for groups of cardiovascular patients in Amsterdam and Helsinki. Res Rep Health Eff Inst. 2005;

Burdsall AC, Xing Y, Cooper CW, Harper WF. Bioaerosol emissions from activated sludge basins: characterization, release, and attenuation. Sci Total Environ. 2021;753:141852. https://doi.org/10.1016/j.scitotenv.2020.141852.

Carugno M, Consonni D, Randi G, Catelan D, Grisotto L, Bertazzi PA, Biggeri A, Baccini M. Air pollution exposure, cause-specific deaths and hospitalizations in a highly polluted italian region. Environ Res. 2016;147:415–24. https://doi.org/10.1016/j.envres.2016.03.003.

Chen TM, Gokhale J, Shofer S, Kuschner WG. Outdoor air pollution: nitrogen dioxide, sulfur dioxide, and carbon monoxide health effects. Am J Med Sci. 2007;333:249–56. https://doi.org/10.1097/MAJ.0b013e31803b900f.

Chowdhuri I, Pal SC, Saha A, Chakrabortty R, Ghosh M, Roy P. Significant decrease of lightning activities during COVID-19 lockdown period over Kolkata megacity in India. Sci Total Environ. 2020;747:141321. https://doi.org/10.1016/j.scitotenv.2020.141321.

Deng Y, Liu W, Liu K, Fang YY, Shang J, Zhou L, Wang K, Leng F, Wei S, Chen L, Liu HG. Clinical characteristics of fatal and recovered cases of coronavirus disease 2019 in Wuhan, China: a retrospective study. Chin Med J. 2020;133:1261–7. https://doi.org/10.1097/CM9.0000000000000824.

Després VR, Alex Huffman J, Burrows SM, Hoose C, Safatov AS, Buryak G, Fröhlich-Nowoisky J, Elbert W, Andreae MO, Pöschl U, Jaenicke R. Primary biological aerosol particles in the atmosphere: a review. Tellus Ser B Chem Phys Meteorol. 2012;64:15598. https://doi.org/10.3402/tellusb.v64i0.15598.

Dey S, Di L, Van Donkelaar A, Tripathi SN, Gupta T, Mohan M. Remote sensing of environment variability of outdoor fi ne particulate (PM 2 . 5) concentration in the Indian Subcontinent : A remote sensing approach. Remote Sens Environ. 2012;127:153–61. https://doi.org/10.1016/j.rse.2012.08.021.

Doiron D, de Hoogh K, Probst-Hensch N, Fortier I, Cai Y, de Matteis S, Hansell AL. Air pollution, lung function and COPD: results from the population-based UK Biobank study. Eur Respir J. 2019;54:1802140. https://doi.org/10.1183/13993003.02140-2018.

Drwal E, Rak A, Gregoraszczuk EL. Review: polycyclic aromatic hydrocarbons (PAHs)—action on placental function and health risks in future life of newborns. Toxicology. 2019;411:133–42. https://doi.org/10.1016/j.tox.2018.10.003.

Du P, Du R, Ren W, Lu Z, Zhang Y, Fu P. Variations of bacteria and fungi in PM2.5 in Beijing, China. Atmos Environ. 2018;172:55–64. https://doi.org/10.1016/j.atmosenv.2017.10.048.

Du Y, Gao B, Zhou H, Ju X, Hao H, Yin S. Health risk assessment of heavy metals in road dusts in urban parks of Beijing, China. Procedia Environ Sci. 2013;18:299–309. https://doi.org/10.1016/j.proenv.2013.04.039.

Dumka UC, Kaskaoutis DG, Verma S, Ningombam SS, Kumar S, Ghosh S. Silver linings in the dark clouds of COVID-19: improvement of air quality over India and Delhi metropolitan area from measurements and WRF-CHIMERE model simulations. Atmos Pollut Res. 2021;12:225–42. https://doi.org/10.1016/j.apr.2020.11.005.

Fleming LT, Weltman R, Yadav A, Edwards RD, Arora NK, Pillarisetti A, Meinardi S, Smith KR, Blake DR, Nizkorodov SA. Emissions from village cookstoves in Haryana, India, and their potential impacts on air quality. Atmos Chem Phys. 2018;18:15169–82. https://doi.org/10.5194/acp-18-15169-2018.

Fröhlich-Nowoisky J, Kampf CJ, Weber B, Huffman JA, Pöhlker C, Andreae MO, Lang-Yona N, Burrows SM, Gunthe SS, Elbert W, Su H, Hoor P, Thines E, Hoffmann T, Després VR, Pöschl U. Bioaerosols in the earth system: climate, health, and ecosystem interactions. Atmos Res. 2016;182:346–76. https://doi.org/10.1016/j.atmosres.2016.07.018.

Ganguly R, Sharma D, Kumar P. Short-term impacts of air pollutants in three megacities of India during COVID-19 lockdown. Environ Dev Sustain. 2021;23:18204. https://doi.org/10.1007/s10668-021-01434-9.

Gaur M, Singh R, Shukla A. Volatile organic compounds in India: concentration and sources. J Civ Environ Eng. 2016;06 https://doi.org/10.4172/2165-784x.1000251.

Ghosal D, Ghosh S, Dutta TK, Ahn Y. Current state of knowledge in microbial degradation of polycyclic aromatic hydrocarbons (PAHs): a review. Front Microbiol. 2016;7 https://doi.org/10.3389/fmicb.2016.01369.

Haque MS, Singh RB. Air pollution and human health in Kolkata, India: a case study. Climate. 2017;5:1–16. https://doi.org/10.3390/cli5040077.

Health Effects Institute, Institute for Health Metrics and Evaluation's Global Burden of Desease project, 2020. State of Global Air 2020.

Hodgson MJ. Indoor air quality: volatile organic compounds (VOC). Allergy Proc. 1991;12:371–83.

Hu B, Xue J, Zhou Y, Shao S, Fu Z, Li Y, Chen S, Qi L, Shi Z. Modelling bioaccumulation of heavy metals in soil-crop ecosystems and identifying its controlling factors using machine learning. Environ Pollut. 2020;262:114308. https://doi.org/10.1016/j.envpol.2020.114308.

Humbal C, Gautam S, Trivedi U. A review on recent progress in observations, and health effects of bioaerosols. Environ Int. 2018;118:189–93. https://doi.org/10.1016/j.envint.2018.05.053.

Ivanova N, Gugleva V, Dobreva M, Pehlivanov I, Stefanov S, Andonova V. We are IntechOpen , the world ' s leading publisher of Open Access books Built by scientists, for scientists TOP 1%. Intech i; 2016. 13

Cities under National Air Quality Monitoring Programme (NAMP), Different monitoring modes & facility under NAMP

Kansal A. Sources and reactivity of NMHCs and VOCs in the atmosphere: a review. J Hazard Mater. 2009;166:17–26. https://doi.org/10.1016/j.jhazmat.2008.11.048.

Karmakar B, SenGupta K, Kaur A, Roy A, Gupta Bhattacharya S. Fungal bio-aerosol in multiple micro-environments from eastern India: source, distribution, and health hazards. SN Appl Sci. 2020;2 https://doi.org/10.1007/s42452-020-2323-1.

Kim D, Chen Z, Zhou L-F, Huang S-X. Air pollutants and early origins of respiratory diseases. Chronic Dis Transl Med. 2018;4:75–94. https://doi.org/10.1016/j.cdtm.2018.03.003.

Kim KN, Kim S, Lim YH, Song IG, Hong YC. Effects of short-term fine particulate matter exposure on acute respiratory infection in children. Int J Hyg Environ Health. 2020;229:113571. https://doi.org/10.1016/j.ijheh.2020.113571.

Korhale N, Anand V, Beig G. Disparity in ozone trends under COVID-19 lockdown in a closely located coastal and hillocky metropolis of India. Air Qual Atmos Health. 2021;14:533–42. https://doi.org/10.1007/s11869-020-00958-9.

Kumar P, Hama S, Omidvarborna H, Sharma A, Sahani J, Abhijith KV, Debele SE, Zavala-Reyes JC, Barwise Y, Tiwari A. Temporary reduction in fine particulate matter due to 'anthropogenic emissions switch-off' during COVID-19 lockdown in Indian cities. Sustain Cities Soc. 2020;62:102382. https://doi.org/10.1016/j.scs.2020.102382.

Kumari S, Jain MK, Elumalai SP. Assessment of pollution and health risks of heavy metals in particulate matter and road dust along the road network of Dhanbad, India. J Health Pollut. 2021;11 https://doi.org/10.5696/2156-9614-11.29.210305.

Kwon JW, Park HW, Kim WJ, Kim MG, Lee SJ. Exposure to volatile organic compounds and airway inflammation. Environ Heal A Glob Access Sci Source. 2018;17:1–8. https://doi.org/10.1186/s12940-018-0410-1.

Låg M, Øvrevik J, Refsnes M, Holme JA. Potential role of polycyclic aromatic hydrocarbons in air pollution-induced non-malignant respiratory diseases. Respir Res. 2020;21:299. https://doi.org/10.1186/s12931-020-01563-1.

Lal S. Trace gases over the Indian region. Indian J Radio Space Phys. 2007;36:556–70.

Lee YM, Lee JH, Kim HC, Ha E. Effects of PM10 on mortality in pure COPD and asthma-COPD overlap: difference in exposure duration, gender, and smoking status. Sci Rep. 2020;10:1–10. https://doi.org/10.1038/s41598-020-59246-2.

Li N, Hao M, Phalen RF, Hinds WC, Nel AE. Particulate air pollutants and asthma: a paradigm for the role of oxidative stress in PM-induced adverse health effects. Clin Immunol. 2003;109:250–65. https://doi.org/10.1016/j.clim.2003.08.006.

Loomis D, Huang W, Chen G. The International Agency for Research on Cancer (IARC) evaluation of the carcinogenicity of outdoor air pollution: focus on China. Chin J Cancer. 2014;33:189–96. https://doi.org/10.5732/cjc.014.10028.

Madhwal S, Prabhu V, Sundriyal S, Shridhar V. Ambient bioaerosol distribution and associated health risks at a high traffic density junction at Dehradun city, India. Environ Monit Assess. 2020;192:196. https://doi.org/10.1007/s10661-020-8158-9.

Mahapatra PS, Panda U, Mallik C, Boopathy R, Jain S, Sharma SK, Mandal TK, Senapati S, Satpathy P, Panda S, Das T. Chemical, microstructural, and biological characterization of wintertime PM2.5 during a land campaign study in a coastal city of eastern India. Atmos Pollut Res. 2021;12:101164. https://doi.org/10.1016/j.apr.2021.101164.

Mahapatra PS, Sinha PR, Boopathy R, Das T, Mohanty S, Sahu SC, Gurjar BR. Seasonal progression of atmospheric particulate matter over an urban coastal region in peninsular India: role of local meteorology and long-range transport. Atmos Res. 2018;199:145–58. https://doi.org/10.1016/j.atmosres.2017.09.001.

Manisalidis I, Stavropoulou E, Stavropoulos A, Bezirtzoglou E. Environmental and health impacts of air pollution: a review. Front Public Health. 2020;8:1–13. https://doi.org/10.3389/fpubh.2020.00014.

Manojkumar N, Srimuruganandam B. Health effects of particulate matter in major Indian cities. Int J Environ Health Res. 2021;31:258–70. https://doi.org/10.1080/09603123.2019.1651257.

Marston CP, Pereira C, Ferguson J, Fischer K, Hedstrom O, Dashwood WM, Baird WM. Effect of a complex environmental mixture from coal tar containing polycyclic aromatic hydrocarbons (PAH) on the tumor initiation, PAH-DNA binding and metabolic activation of carcinogenic PAH in mouse epidermis. Carcinogenesis. 2001;22:1077–86. https://doi.org/10.1093/carcin/22.7.1077.

Martinelli N, Olivieri O, Girelli D. Air particulate matter and cardiovascular disease: a narrative review. Eur J Intern Med. 2013;24:295–302. https://doi.org/10.1016/j.ejim.2013.04.001.

Masih J, Dyavarchetty S, Nair A, Taneja A, Singhvi R, Team R. EPA Public Access; 2020, https://doi.org/10.1016/j.eti.2018.10.012.Concentration.

Massachusetts Department of Public Health. 2017 Massachusetts State Health Assessment; 2017, 1–312.

McCarthy JT, Pelle E, Dong K, Brahmbhatt K, Yarosh D, Pernodet N. Effects of ozone in normal human epidermal keratinocytes. Exp Dermatol. 2013;22:360–1. https://doi.org/10.1111/exd.12125.

Ministry for the Environment. Good practice guide for air quality monitoring and data management; 2009, 2009.

Mølhave L, Clausen G, Berglund B, De Ceaurriz J, Kettrup A, Lindvall T, Maroni M, Pickering AC, Risse U, Rothweiler H, Seifert B, Younes M. Total volatile organic compounds (TVOC) in indoor air quality investigations. Indoor Air. 1997;7:225–40. https://doi.org/10.1111/j.1600-0668.1997.00002.x.

Molina LT. Introductory lecture: air quality in megacities. Faraday Discuss. 2021;226:9–52. https://doi.org/10.1039/d0fd00123f.

Mukund R, Kelly TJ, Spicer CW. Source attribution of ambient air toxic and other VOCs in Columbus, Ohio. Atmos Environ. 1996;30:3457–70. https://doi.org/10.1016/1352-2310(95)00487-4.

Nair AT. Bioaerosols in the landfill environment: an overview of microbial diversity and potential health hazards. Aerobiologia (Bologna). 2021;37:185–203. https://doi.org/10.1007/s10453-021-09693-9.

Nandi I, Srivastava S, Yarragunta Y, Kumar R, Mitra D. Distribution of surface carbon monoxide over the Indian subcontinent: investigation of source contributions using WRF-Chem. Atmos Environ. 2020;243:117838. https://doi.org/10.1016/j.atmosenv.2020.117838.

Nath J, Panda S, Patra SS, Ramasamy B, Das T. Variation of black carbon and particulate matter in Bhubaneswar during the pre-monsoon: possible impact of meteorology and COVID-19 lockdown. Curr Sci. 2021;120:313–21. https://doi.org/10.18520/cs/v120/i2/313-321.

Navinya C, Patidar G, Phuleria HC. Examining effects of the COVID-19 national lockdown on ambient air quality across urban India. Aerosol Air Qual Res. 2020;20:1759–71. https://doi.org/10.4209/aaqr.2020.05.0256.

Nishanth T, Praseed KM, Kumar MKS, Valsaraj KT. Influence of ozone precursors and PM10 on the variation of surface O3 over Kannur, India. Atmos Res. 2014;138:112–24. https://doi.org/10.1016/j.atmosres.2013.10.022.

Nussbaumer T, Czasch C, Klippel N, Johansson L, Tullin C. Particulate emissions from biomass combustion in IEA countries, 16th European biomass conference and exhibition; 2008.

Oberdörster G. Pulmonary effects of inhaled ultrafine particles. Int Arch Occup Environ Health. 2000;74:1–8. https://doi.org/10.1007/s004200000185.

Oksanen E, Pandey V, Pandey AK, Keski-Saari S, Kontunen-Soppela S, Sharma C. Impacts of increasing ozone on Indian plants. Environ Pollut. 2013;177:189–200. https://doi.org/10.1016/j.envpol.2013.02.010.

Panda S, Mallik C, Nath J, Das T, Ramasamy B. A study on variation of atmospheric pollutants over Bhubaneswar during imposition of nationwide lockdown in India for the COVID-19 pandemic. Air Qual Atmos Health. 2021a;14:97–108. https://doi.org/10.1007/s11869-020-00916-5.

Panda U, Boopathy R, Gadhavi HS, Renuka K, Gunthe SS, Das T. Metals in coarse ambient aerosol as markers for source apportionment and their health risk assessment over an eastern coastal urban atmosphere in India. Environ Monit Assess. 2021b;193:311. https://doi.org/10.1007/s10661-021-09057-3.

Panda U, Das T. Micro-structural analysis of individual aerosol coarse particles during different seasons at an eastern coastal site in India. Atmos Pollut Res. 2017;8:196–207. https://doi.org/10.1016/j.apr.2016.08.012.

Pandey A, Brauer M, Cropper ML, Balakrishnan K, Mathur P, Dey S, Turkgulu B, Kumar GA, Khare M, Beig G, Gupta T, Krishnankutty RP, Causey K, Cohen AJ, Bhargava S, Aggarwal AN, Agrawal A, Awasthi S, Bennitt F, Bhagwat S, Bhanumati P, Burkart K, Chakma JK, Chiles TC, Chowdhury S, Christopher DJ, Dey S, Fisher S, Fraumeni B, Fuller R, Ghoshal AG, Golechha MJ, Gupta PC, Gupta R, Gupta R, Gupta S, Guttikunda S, Hanrahan D, Harikrishnan S, Jeemon P, Joshi TK, Kant R, Kant S, Kaur T, Koul PA, Kumar P, Kumar R, Larson SL, Lodha R, Madhipatla KK, Mahesh PA, Malhotra R, Managi S, Martin K, Mathai M, Mathew JL, Mehrotra R, Mohan BVM, Mohan V, Mukhopadhyay S, Mutreja P, Naik N, Nair S, Pandian JD, Pant P, Perianayagam A, Prabhakaran D, Prabhakaran P, Rath GK, Ravi S, Roy A, Sabde YD, Salvi S, Sambandam S, Sharma B, Sharma M, Sharma S, Sharma RS, Shrivastava A, Singh S, Singh V, Smith R, Stanaway JD, Taghian G, Tandon N, Thakur JS, Thomas NJ, Toteja GS, Varghese CM, Venkataraman C, Venugopal KN, Walker KD, Watson AY, Wozniak S, Xavier D, Yadama GN, Yadav G, Shukla DK, Bekedam HJ, Reddy KS, Guleria R, Vos T, Lim SS, Dandona R, Kumar S, Kumar P, Landrigan PJ, Dandona L. Health and economic impact of air pollution in the states of India: the global burden of disease study 2019. Lancet Planet Health. 2021;5:e25–38. https://doi.org/10.1016/S2542-5196(20)30298-9.

Patra A, Chaudhary P, Ravi KS. Adverse impact of Covid-19 on anatomical sciences teachers of India and proposed ways to handle this predicament. Anat Sci Educ. 2021;14:163–5. https://doi.org/10.1002/ase.2052.

Peled R. Air pollution exposure: who is at high risk? Atmos Environ. 2011;45:1781–5. https://doi.org/10.1016/j.atmosenv.2011.01.001.

Peters A, Wichmann HE, Tuch T, Heinrich J, Heyder J. Respiratory effects are associated with the number of ultrafine particles. Am J Respir Crit Care Med. 1997;155:1376–83. https://doi.org/10.1164/ajrccm.155.4.9105082.

Popovicheva O, Kistler M, Kireeva E, Persiantseva N, Timofeev M, Kopeikin V, Kasper-Giebl A. Physicochemical characterization of smoke aerosol during large-scale wildfires: extreme event of August 2010 in Moscow. Atmos Environ. 2014;96:405–14. https://doi.org/10.1016/j.atmosenv.2014.03.026.

Popovicheva O, Timofeev M, Persiantseva N, Jefferson MA, Johnson M, Rogak SN, Baldelli A. Microstructure and chemical composition of particles from small-scale gas flaring. Aerosol Air Qual Res. 2019;19:2205–21. https://doi.org/10.4209/aaqr.2019.04.0177.

Ramachandran A, Jain NK, Sharma SA, Pallipad J. Recent trends in tropospheric NO2 over India observed by SCIAMACHY: identification of hot spots. Atmos Pollut Res. 2013;4:354–61. https://doi.org/10.5094/APR.2013.040.

Sahoo PK, Mangla S, Pathak AK, Salãmao GN, Sarkar D. Pre-to-post lockdown impact on air quality and the role of environmental factors in spreading the COVID-19 cases—a study from a worst-hit state of India. Int J Biometeorol. 2021;65:205–22. https://doi.org/10.1007/s00484-020-02019-3.

Sahu SK, Tyagi B, Beig G, Mangaraj P, Pradhan C, Khuntia S, Singh V. Significant change in air quality parameters during the year 2020 over 1st smart city of India: Bhubaneswar. SN Appl Sci. 2020;2:1–8. https://doi.org/10.1007/s42452-020-03831-7.

Sarigiannis DA, Karakitsios SP, Gotti A, Liakos IL, Katsoyiannis A. Exposure to major volatile organic compounds and carbonyls in European indoor environments and associated health risk. Environ Int. 2011;37:743–65. https://doi.org/10.1016/j.envint.2011.01.005.

Selvam S, Muthukumar P, Venkatramanan S, Roy PD, Manikanda Bharath K, Jesuraja K. SARS-CoV-2 pandemic lockdown: effects on air quality in the industrialized Gujarat state of India. Sci Total Environ. 2020;737:140391. https://doi.org/10.1016/j.scitotenv.2020.140391.

Sharma SB, Jain S, Khirwadkar P, Kulkarni S. The effects of air pollution on the environment and human health. Indian J Res Pharm Biotechnol. 2013;1:2320–3471.

Srivastava A, Joseph AE, Devotta S. Volatile organic compounds in ambient air of Mumbai - India. Atmos Environ. 2006;40:892–903. https://doi.org/10.1016/j.atmosenv.2005.10.045.

Storey-Bishoff J, Noga M, Finlay WH. Deposition of micrometer-sized aerosol particles in infant nasal airway replicas. J Aerosol Sci. 2008;39:1055–65. https://doi.org/10.1016/j.jaerosci.2008.07.011.

Sun J, Shen Z, Zhang L, Zhang Y, Zhang T, Lei Y, Niu X, Zhang Q, Dang W, Han W, Cao J, Xu H, Liu P, Li X. Volatile organic compounds emissions from traditional and clean domestic heating appliances in Guanzhong plain, China: emission factors, source profiles, and effects on regional air quality. Environ Int. 2019;133:105252. https://doi.org/10.1016/j.envint.2019.105252.

Sundaray SNK, Bhardwaj SR. NCAP National Clean Air Programme. India: Minist. Environ. For. Clim. Chang. Gov; 2019.

Susetyo SH, Assomadi AF, Hermana J. Res Environ Sci Eng. 2019;1:25–9.

UNEP, UNICEF, WHO. Children in the new millennium: Envi- ronmental impact on health. Geneva, Switzerland: World Health Organization; 2003. 2003, World Health Organization

USEPA. Exposure factors handbook; EPA, Environmental Protection Agency. Washington, DC: U.S. Environ. Prot. Agency; 1997. EPA/600/R-090/052F 1–1466

Verma M, Pervez S, Majumdar D, Chakrabarty R, Pervez YF. Emission estimation of aromatic and halogenated VOCs from household solid fuel burning practices. Int J Environ Sci Technol. 2019;16:2683–92. https://doi.org/10.1007/s13762-018-1920-7.

Verma RL, Kamyotra JS. Impacts of COVID-19 on air quality in India. Aerosol Air Qual Res. 2021;21:200482. https://doi.org/10.4209/aaqr.200482.

Vinzents PS, Møller P, Sørensen M, Knudsen LE, Hertel O, Jensen FP, Schibye B, Loft S. Personal exposure to ultrafine particles and oxidative DNA damage. Environ Health Perspect. 2005;113:1485–90. https://doi.org/10.1289/ehp.7562.

Who T, Quality A, Update G, Asia WHOS, Pacific W. Ambient (outdoor) air pollution 1–9 2021.

Wieslander G, Norbäck D, Björnsson E, Janson C, Boman G. Asthma and the indoor environment: the significance of emission of formaldehyde and volatile organic compounds from newly painted indoor surfaces. Int Arch Occup Environ Health. 1996;69:115–24. https://doi.org/10.1007/s004200050125.

Wohak LE, Krais AM, Kucab JE, Stertmann J, Øvrebø S, Seidel A, Phillips DH, Arlt VM. Carcinogenic polycyclic aromatic hydrocarbons induce CYP1A1 in human cells via a p53-dependent mechanism. Arch Toxicol. 2016;90:291–304. https://doi.org/10.1007/s00204-014-1409-1.

Wunnapuk K, Pothirat C, Manokeaw S, Phetsuk N, Chaiwong W, Phuackchantuck R, Prapamontol T. PM10-related DNA damage, cytokinetic defects, and cell death in COPD patients from Chiang Dao district, Chiang Mai, Thailand. Environ Sci Pollut Res. 2019;26:25326–40. https://doi.org/10.1007/s11356-019-05641-w.

Zhang N, Sun Q, Li W, Li H, Han L, Huang P. Detection of volatile organic compounds in exhaled breath for early diagnosis of lung cancer. Tumor. 2018;38:874–82. https://doi.org/10.3781/j.issn.1000-7431.2018.33.287.

Zhang Y, Tao S. Global atmospheric emission inventory of polycyclic aromatic hydrocarbons (PAHs) for 2004. Atmos Environ. 2009;43:812–9. https://doi.org/10.1016/j.atmosenv.2008.10.050.

Strategic Urban Air Quality Improvement: Perspectives on Public Health

15

Biswajit Patra, Surya Narayan Pradhan, and R. Paulraj

15.1 Introduction

Fine particulate matter ($PM_{2.5}$) is the principal constituent behind the formulation of regional air quality index (Quarmby et al. 2019). The yearly normal convergence of $PM_{2.5}$ in urban areas is genuinely increasing around the world. Epidemiological investigations have shown a causal connection between prolonged $PM_{2.5}$ exposure and increased rate of respiratory sicknesses, cellular breakdown in the lungs, and stroke (Liu et al. 2021a, b, c; Li et al. 2021; Farkas et al. 2021). Rate of outpatient visit and hospitalization of cardiovascular and cerebrovascular illnesses is increased after brief repeated exposure to PM $_{2.5}$. As of now, an expanding number of analysts are worried about the imminent dangers of $PM_{2.5}$ pollution and are mostly zeroing in on the reason for $PM_{2.5}$ pollution (Deary and Griffiths 2021; Guarino et al. 2021; Mendoza et al. 2021). Enhanced level of $PM_{2.5}$ aggravates cardiovascular and cerebrovascular illnesses (ischaemic coronary illness [IHD] and stroke) leading to untimely deaths (Korhonen et al. 2021). Liberalization and globalization bring about nonstop industrialization, and urbanization bothers air contamination. Without successful air contamination control measures, public health is put at risk (de Vries et al. 2021; Schaefer et al. 2021). The combined impact of unplanned urbanization and improper air pollution control accounts for 44–48% change in $PM_{2.5}$ (Kim et al. 2013). Various studies have been conducted to investigate the distinctions between $PM_{2.5}$ reduction strategy in metropolitan cities and air quality improvement (Alari et al. 2021; Margolis et al. 2021). Disorganized urbanization and indiscriminate use of motorized transport lead to severe rise in various lifestyle-related metabolic disorders like hyperglycaemia, heftiness, dyslipidaemia, and hypertension (Chen et al. 2021a, b; Gkrillas et al. 2021; Lemarchand et al. 2021). Subsequently, urban air quality is a prime concern since air contamination has been found to affect a number of health risks like respiratory sicknesses and skin disorders (Liang et al. 2021). Since air contamination has been perceived as one of the prominent threats to human health and climate resilience, numerous key ecological policies are focused on it (Liu et al. 2021c). In developed economies, the air is contaminated with toxic gases, mixtures of nitrogen and sulphur oxides and particulate matter (Stenson et al. 2021; Timonen et al. 2021). There are distinctive theoretical ways to deal with real-time air quality like ground stations of observing organizations, satellite telemonitoring and the

B. Patra · S. N. Pradhan (✉)
School of Life Sciences, Sambalpur University, Sambalpur, Odisha, India

R. Paulraj
School of Environmental Sciences, Jawaharlal Nehru University, New Delhi, India

P. K. Padhy et al. (eds.), *Air Quality and Human Health*,
https://doi.org/10.1007/978-981-97-1363-9_15

utilization of biomonitoring (Xu et al. 2021). The capacity of plants to gather pollutants in the various organs or tissues (roots, natural products, leaves or barks) has been misused worldwide since the sixties of last century. A few anthropogenic exercises can cause significant increments in suspended molecule loads in marine conditions, for instance digging for marine navigation purpose, subsea development including the development of seaward windfarms, run-off from farming, release of metropolitan effluents, land-based mining and subsea mining exercises. Direct and indirect human interventions adversely affect the ecosystem by mixing with pollutants in the water, air and soil (Almeida 2021; Gardon et al. 2021). Air quality disturbance is correlated with various decline in socioeconomic developmental factors (Di Virgilio et al. 2021; Kocak et al. 2021). Public reports have referenced the near danger of hospitalizations and mortality associated with fine particulate air pollution. Fine and ultrafine particulate matter arrive at the circulatory system via the tracheal system. Urban air contamination largely affects the air quality and health of resident citizens (Newell et al. 2017; Evans et al. 2021; Maphutha et al. 2021). Trans-border spread of particulate pollution in winter season poses immediate health risk to the thickly populated cities as it remains in the respirable zone for longer time (Nageen et al. 2021; Sun et al. 2021). Overpopulation and uncontrolled urbanization as well as industrialization issues become major problems for developing countries too. The utilization of cooking wood and cow dung for domestic cooking adds more indoor air pollutants (Carlsten et al. 2020). Around three billion individuals all throughout the planet are utilizing the plant biomass, wood and other excreta of animals for their daily cooking needs. The ladies of the family in agricultural nations inhale soots and other particulate matter emitted during household cooking (Lee et al. 2020; Deary and Griffiths 2021). India, one of the Asia's most populated countries, is also facing rise in particulate air pollutants (Tong 2019). Rate of rise in cardiovascular sicknesses is significantly higher in developed nations as compared to developing ones. New Delhi, the national capital territory of India, is one of the most polluted metropolitan areas in India. Many flight trips all through New Delhi International Airport are frequently dropped or diverted to other nearby airports because of the decreased visibility required for flight operation due to excessive particulate matters in air in winter seasons (Carlsen et al. 2019). India is also witnessing rapid rise in quick industrialization, urbanization and use of vehicles in transport. Biomass burning for cooking is rampant in India and in Nepal leading to increased urban air pollution (Rivas-Santiago et al. 2015). Besides, very significant degrees of $PM_{2.5}$ contamination are noticed in Mexico City and Rio de Janeiro, trailed by Milan, Ankara, Melbourne, Tokyo, and Moscow. Various think tanks, non-governmental organizations and agencies have made protests and raised voice of resistance regarding the rising levels of urban air pollution and implementation of stricter health advisories time to time (Bascom et al. 1996; Boogaard and van Erp 2019).

15.2 Urban Air Pollution and Climate Change

Urban air pollution and its subsidiary events influence climatic conditions globally. Environmental fluctuations and climatic factors influence the span, timing and force of respiratory illness episodes throughout the globe (Castro-Díez et al. 2021). Parasitic or viral sicknesses borne by some insects are prone to deviation in environmental parameters. Even rise in atmospheric temperature, the breeding behaviour of mosquitoes is influenced (Chen et al. 2021a). Also, water-warming event prompts the incidence of high rate of waterborne diseases because of environmental changes. Incidences of cholera, poliomyelitis, tick-borne encephalitis, and malaria have been multiplied because of the alteration of climatic conditions in Europe as reported by researchers.

15.3 Effect of Particulate Matter on Human Health

Particulate matter ($PM_{2.5}$) is usually created in the atmosphere as an end product of various physico-chemical reactions occurring between diverse pollutants. The alveolar penetration of particles is closely related to their micrometre size and weight. Biological contaminants associated with $PM_{2.5}$ include microorganisms (bacteria, viruses, fungi, moulds and bacterial spores), cat allergens, house dust and allergens, and pollen whereas dust samples include suspended atmospheric dust, settling dust, and heavy dust (Czubaj-Kowal et al. 2021). PM $_{2.5}$, due to its minute size, does cause more serious health effects. These aforementioned fine particles are the main constituent of the "haze" formation in different metropolitan areas (Evans et al. 2021). Ozone uptake usually occurs by inhalation. Ozone affects the upper layers of the skin and the tear ducts. A study of short-term exposure of mice to high levels of ozone showed malondialdehyde formation in the upper skin (epidermis) but also depletion in vitamins C and E (Gulia et al. 2015). It is likely that ozone levels are not interfering with the skin barrier function and integrity to predispose towards skin disease.

15.4 Effect of CO and NO_2 on Human Health

The biochemical affinity of carbon monoxide to haemoglobin is unexpectedly high as much as 250 times greater than that for oxygen. Carboxyhaemoglobin is a highly stable end product created when haemoglobin reacts with carbon monoxide (Hwang et al. 2020). Hypoxia, ischaemia, and cardiovascular sickness are observed because of nonformation of oxyhaemoglobin. Apart from lower respiratory tract infections, eye, throat, and nose aggravation have been noticed (Cai et al. 2016). Nitrogen oxide is generated due to engine emission during motorized transport. The fumes of CO and NO_2 penetrate inside the lung, inciting respiratory irritations, wheezing, dyspnoea, bronchospasm, and surprisingly aspiratory oedema when breathed in at high levels (Guan et al. 2020).

15.5 Effect of SO_2 and Lead on Human Health

Sulphur dioxide gas is produced basically from petroleum-based engine functioning or mechanical activities (Nhung et al. 2017). The yearly emission norm for SO_2 is 0.03 ppm. Health effects of SO_2 exposure include reddening of skin, inflammation of eyes and mucous layers, and aggravation of cardiovascular sickness. Lead is a major metal used in various modern industries and also transmitted from some petroleum engine motors, secondary storage cells, radiators, mining incinerators, and waste water treatment plants (Yang et al. 2018). Lead induces serious health hazards upon bodily penetration. Through channels of breath, ingestion, and dermal retention, it makes entry into the human body. Lead after entry via the breathing route rapids transmits the entire body because of the blood circulatory system and associated organ systems (Liu et al. 2020). Neonates and babies are more prone to the neurotoxic effects of lead as it hampers learning behaviour, impairs memory functions and retardation of mental growth.

15.6 Polyaromatic Hydrocarbons and Volatile Organic Compound Effects on Health System

PAH is readily found in coal and tar products. It is originated from the incomplete combustion of organic matter. Benzopyrene, acenaphthylene, anthracene, and fluoranthene are the prominent PAH having potent toxicogenicity and carcinogenicity (Santi et al. 2016). Lung cancer is induced because of PAH exposure in the long run. Rise in cancer incidences in human is attributed to prom-

Table 15.1 Different pollutants and their source on different diseases

Pollutant	Source	Diseases	References
NO_2	Kerosene, fossil fuel, vehicle, gas	Asthma, lungs infection	Quarmby et al. (2019)
Volatile organic compound	Petroleum, automobile emissions	Allergic, breathlessness	Newell et al. (2017)
Particulate matter	Dusts, vehicle emissions	Bronchitis, skin irritation, nose and blocking	Li et al. (2020)
SO_2	Petroleum, coal	Respiratory tract infection	Liu et al. (2020)
Ozone	Industrial gas	Lungs infection	Landrigan (2017)

inent volatile organic compounds (VOCs), such as toluene, benzene, ethylbenzene, and xylene. Temporary PAH exposure is associated with irritation of eyes, inflammation of nose, throat, and mucosal layers, whereas long-term exposure creates chronic health hazards, even hospitalizations too (Liu et al. 2021a) (Table 15.1).

15.7 Effect of Other Pollutants on Human Health

Human respiratory issues are firmly connected with the nasal inhalation of fine particulate matters. Upon entry through the nostrils, they reach the deeper the alveolar cells and respiratory epithelium. The extent of lung infection lies with the quantum of alveolar deposition. Variation in geographic position, season, and time also causes alteration in $PM_{2.5}$ deposition (Liu et al. 2014). Long-term deposition of PM in the atmosphere creates a steep hike in the ambient air quality. Incidences of respiratory illnesses, asthma and other sicknesses are highly reported in areas of PM pollution (Qian et al. 2016). Persistent obstructive aspiratory sickness might be initiated following air contamination, expanding dreariness and mortality. Neurological impacts have been seen in grown-ups and youngsters after prolonged exposure to PM pollution. Cerebrum irritation was seen in canines living in an exceptionally contaminated region in Mexico for a longer time exposure. There are more chances of atherosclerosis and oxidative stress upon prolonged air pollution exposure in urban areas. Even in the later stage of respiratory infection, lung and associated organs are paralysed (Janson et al. 2020; Li et al. 2020). Neuroinflammation is also associated with PM pollution. Symptomatic dermal irritation and inflammation such as skin maturation, psoriasis, urticaria, dermatitis, and atopic dermatitis. Newborn babies are at the receiving end as several studies reported inadequate fetal growth, improper birth weight, and low lung function rate among the babies. Asymptomatic inflammation of eye is also noticed. Moreover, eye is also partly affected by chronic air pollution (Asghary et al. 2015; Kuźma et al. 2020; Les Bujanda et al. 2021) (Fig. 15.1).

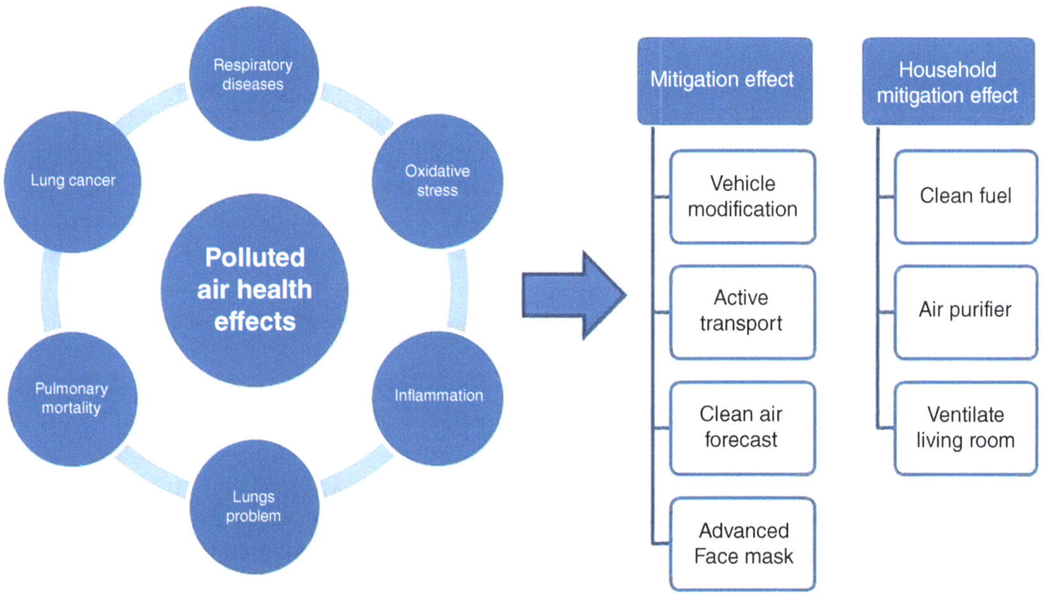

Fig. 15.1 Graphical representation showing the health effect associated with pollution and their mitigation effect

15.8 Improving Air Quality by Technology

Restricting vehicular emission can definitely stabilize regional air quality to a larger extent. Prioritizing sensitive areas such as schools, residential areas, and hospitals as minimal zones of vehicular emission can be a checkpoint for excessive personal vehicle use. In European Union countries, low emission zone has been implemented across 12 European nations including England, Italy, Sweden, and Holland (Mishra et al. 2020; Evans et al. 2021; Maphutha et al. 2021). London's low emissivity zone is limited to a large portion of the space inside the M25 ring street which checks vehicular congestion making it a role model for air quality stabilization (Newell et al. 2017; Ngoc et al. 2017; Moshammer et al. 2021). Implementation of ultra-low emission zone was made effective from April 2019 in the Congestion Charging region. Stricter norms following emission standards will be applicable to vehicles, bikes, and vans in the ultra-low emission zone (Palacios-Mateo et al. 2021). Banning excessive use of personal vehicle for routine office work and promoting other alternative modes of surface transport like cycling, e-vehicles and public transport has made tremendous improvement in metropolitan air quality. Cycling is one of the most efficient means of transport which cuts down air pollution, hence improving air quality. As per the recommendation of European nations' air quality standards, many speed limit areas are notified time to time with respect to rise in pollutant load (Cepeda et al. 2017). Measures have been taken to create more public awareness about ill-effects of $PM_{2.5}$ pollution. Regular media advertisements are necessitated to reach out the masses (Landrigan 2017). Bottom-up approach is justified for effective air pollution mitigation. Few nations, including the Netherland, Japan and England. Delhi has adopted the model of water guns to wipe out air pollutants but no such visible progress was observed over the long run. Various technological advancements have been exercised to make diesel and petroleum vehicles more environment friendly. There is some proof that blending of ethanol with petrol or diesel emits less NOx as compared to conventional fuel used in engines (Yin et al. 2020; Khomenko et al. 2021; Pandey et al. 2021). Government issues advisory regarding banning of old vehicles having high emission with some penalty if used while allowance of newer vehicles equipped with lesser emissivity engines (WHO 2013; Von Schneidemesser et al. 2020; Acharya and Vinoj 2021). Adoption of electric vehicles has definitely stabilized the polluted urban air quality. Avenues for more electric charging stations have been commissioned for the effective implementation of e-vehicles (Anderson 2017; Das and Horton 2018; Neira et al. 2018). The afforestation drive can multiply the greenish vegetation sphere in metropolitan environments causing drastic change in air quality (Pelgrims et al. 2021; Schlutow et al. 2021; Sicard et al. 2021). Proper town planning with adequate space for green space can enrich the oxygen saturation hence purifying the polluting air (Carlsen et al. 2019; Sly et al. 2021; Wei et al. 2021). Most nations lack the rational scientific advisory body to regulate the episodic rise of air pollutants while implementing of national ambient air quality standard. Hence proper policy formulation to tackle indiscriminate rise in air pollution is missing (Bukowy-Bieryłło 2021; Fakunle et al. 2021; Turrisi et al. 2021). Several innovations in air pollution reduction technology in urban areas have significantly improved urban air quality (Vandyck et al. 2018; Knapp et al. 2021; Siddharthan et al. 2021; Vlachogiannis et al. 2021). Installation of more numbers of electric charging stations has been envisaged in future urban planning. Incentivizing electric vehicle use can be a green initiative of maintaining regional air quality (Aguilera et al. 2021).

15.9 Discussion

The uncontrolled urbanization and use of motorized transport around the globe caused the alarming rise of traffic-related particulate matter emission and air contamination. Release of metals and metalloids in exhaust fumes of petrol and diesel-operated motorized engines used in transport not only pollutes the atmospheric air but also affects human respiratory health causing severe chronic health disorders (Anderson 2017; Shams et al. 2021; Singh et al. 2021). Eco-toxicity associated with air pollution impacts the living organism present in the biosphere including humans (Shi et al. 2020). Promulgation of sound public policy on air pollution management may be undertaken with the participation of administrators, clinicians, lawyers, environmentalists, and common public as stakeholders (Giani et al. 2020). Campaign about air pollution and its mitigation strategies should be adopted by Governments to transmit actual real-time air quality data through various social media platforms including print and mass media.

15.10 Conclusion

Here we attempted to analyse the health effects of air pollution and the processes and technologies to beat off air pollution globally. Ride-sharing and carpooling are adopted in many developed countries to minimize traffic congestion as well as pollution. Encouragement of the use of alternative forms of public transport like bus, train, and metro is highly recommended for drastic change in PM pollution. Even promotion of cycling, and electric vehicles will restrict the air pollution level in the long run (Balakrishnan et al. 2019). There is no concrete evidence of any single technology for the complete reduction of air pollution (Zhang et al. 2020). Hence, the collaboration of joint initiatives is instrumental to achieve drastic minimization in pollutant load both on short- and long-term basis. Moreover, emphasis should be taken at the individual level to use public transport maximally to attain pollution free environment. Local actions taken at the individual level can efficiently address the global issue of particulate matter accumulation and climate change broadly.Competing InterestThe authors declare that they have no competing financial interests.

References

Acharya A, Vinoj V. Investigation of June 2020 Giant Saharan dust storm using remote sensing observations and model reanalysis. 2021; https://doi.org/10.20944/preprints202107.0100.v1.

Aguilera R, Corringham T, Gershunov A, Benmarhnia T. Wildfire smoke impacts respiratory health more

than fine particles from other sources: observational evidence from Southern California. Nat Commun. 2021;12:1493. https://doi.org/10.1038/s41467-021-21708-0.
Alari A, Schwarz L, Zabrocki L, Le Nir G, Chaix B, Benmarhnia T. The effects of an air quality alert program on premature mortality: a difference-in-differences evaluation in the region of Paris. Environ Int. 2021;156:106583. https://doi.org/10.1016/j.envint.2021.106583.
Almeida SM. Air quality mapping and visualisation : An affordable solution based on a vehicle-mounted sensor network. 2021:315. https://doi.org/10.1016/j.jclepro.2021.128194.
Anderson HR. Implications for the science of air pollution and health. Lancet Respir Med. 2017;5:916–8. https://doi.org/10.1016/S2213-2600(17)30396-X.
Asghary M, Raoof JB, Hamidi-Asl E, Ojani R. A novel electrochemical genosensor based on banana and nano-gold modified electrode using tyrosinase enzyme as indicator. J Nanosci Nanotechnol. 2015;15:3394–404. https://doi.org/10.1166/jnn.2015.10211.
Balakrishnan K, Dey S, Gupta T, Dhaliwal RS, Brauer M, Cohen AJ, et al. The impact of air pollution on deaths, disease burden, and life expectancy across the states of India: the Global Burden of Disease Study 2017. Lancet Planet Health. 2019;3:e26–39. https://doi.org/10.1016/S2542-5196(18)30261-4.
Bascom R, Bromberg PA, Costa DA, Devlin R, Dockery DW, Frampton MW, et al. Health effects of outdoor air pollution. Am J Respir Crit Care Med. 1996;153:3–50. https://doi.org/10.1164/ajrccm.153.1.8542133.
Boogaard H, van Erp AM. Assessing health effects of air quality actions: what's next? Lancet Public Health. 2019;4:e4–5. https://doi.org/10.1016/S2468-2667(18)30235-4.
Bukowy-Bieryłło Z. Long-term differentiating primary human airway epithelial cell cultures: how far are we? *Cell Commun.* Signals. 2021;19:1–18. https://doi.org/10.1186/s12964-021-00740-z.
Cai Y, Zhang B, Ke W, Feng B, Lin H, Xiao J, et al. Associations of short-term and long-term exposure to ambient air pollutants with hypertension: a systematic review and meta-analysis. Hypertension. 2016;68:62–70. https://doi.org/10.1161/HYPERTENSIONAHA.116.07218.
Carlsen HK, Valdimarsdóttir U, Briem H, Dominici F, Finnbjornsdottir RG, Jóhannsson T, et al. Severe volcanic SO 2 exposure and respiratory morbidities in the Icelandic population—a register study. 2019:1–12. https://doi.org/10.1101/19013474.
Carlsten C, Salvi S, Wong GWK, Chung KF. Personal strategies to minimise effects of air pollution on respiratory health: advice for providers, patients and the public. Eur Respir J. 2020;55:1902056. https://doi.org/10.1183/13993003.02056-2019.
Castro-Díez P, Alonso Á, Saldaña-López A, Granda E. Effects of widespread non-native trees on regulating ecosystem services. Sci Total Environ. 2021;778:146141. https://doi.org/10.1016/j.scitotenv.2021.146141.
Cepeda M, Schoufour J, Freak-Poli R, Koolhaas CM, Dhana K, Bramer WM, et al. Levels of ambient air pollution according to mode of transport: a systematic review. Lancet Public Health. 2017;2:e23–34. https://doi.org/10.1016/S2468-2667(16)30021-4.
Chen B, Jia P, Han J. Role of indoor aerosols for COVID-19 viral transmission: a review. Environ Chem Lett. 2021a;19:1953–70. https://doi.org/10.1007/s10311-020-01174-8.
Chen HG, Lu Q, Tu ZZ, Chen YJ, Sun B, Hou J, et al. Identifying windows of susceptibility to essential elements for semen quality among 1428 healthy men screened as potential sperm donors. Environ Int. 2021b;155:106586. https://doi.org/10.1016/j.envint.2021.106586.
Czubaj-Kowal M, Kurzawa R, Mazurek H, Sokołowski M, Friediger T, Polak M, et al. Relationship between air pollution and the concentration of nitric oxide in the exhaled air (Feno) in 8–9-year-old school children in krakow. Int J Environ Res Public Health. 2021;18 https://doi.org/10.3390/ijerph18136690.
Das P, Horton R. Pollution, health, and the planet: time for decisive action. Lancet. 2018;391:407–8. https://doi.org/10.1016/S0140-6736(17)32588-6.
Deary ME, Griffiths SD. A novel approach to the development of 1-hour threshold concentrations for exposure to particulate matter during episodic air pollution events. J Hazard Mater. 2021;418:126334. https://doi.org/10.1016/j.jhazmat.2021.126334.
Di Virgilio G, Hart MA, Maharaj AM, Jiang N. Air quality impacts of the 2019-2020 black summer wildfires on Australian schools. Atmos Environ. 2021;261:118450. https://doi.org/10.1016/j.atmosenv.2021.118450.
Evans JS, Rojas-Bracho L, Hammitt JK, Dockery DW. Mortality benefits and control costs of improving air quality in Mexico City: the case of heavy duty diesel vehicles. Risk Anal. 2021;41:661–77. https://doi.org/10.1111/risa.13655.
Fakunle AG, Jafta N, Naidoo RN, Smit LAM. Association of indoor microbial aerosols with respiratory symptoms among under - five children : a systematic review and meta-analysis. Environ Health. 2021:1–16. https://doi.org/10.1186/s12940-021-00759-2.
Farkas J, Nordtug T, Svendheim LH, Amico ED, Davies EJ, Ciesielski T, et al. Effects of mine tailing exposure on early life stages of cod (Gadus morhua) and haddock (Melanogrammus aeglefinus). Environ Res. 2021;200 https://doi.org/10.1016/j.envres.2021.111447.
Gardon T, El Rakwe M, Paul-Pont I, Le Luyer J, Thomas L, Prado E, et al. Microplastics contamination in pearl-farming lagoons of French Polynesia. J Hazard Mater. 2021;419:126396. https://doi.org/10.1016/j.jhazmat.2021.126396.
Giani P, Castruccio S, Anav A, Howard D, Hu W, Crippa P. Short-term and long-term health impacts of air pollution reductions from COVID-19 lockdowns in China and Europe: a modelling study. Lancet Planet

Health. 2020;4:e474–82. https://doi.org/10.1016/S2542-5196(20)30224-2.

Gkrillas A, Dirven H, Papadopoulou E, Andreassen M, Husøy T. Exposure estimates of phthalates and DINCH from foods and personal care products in comparison with biomonitoring data in 24-hour urine from the Norwegian EuroMix biomonitoring study. Environ Int. 2021;155:106598. https://doi.org/10.1016/j.envint.2021.106598.

Guan Q, Chen S, Wang B, Dou X, Lu Y, Liang J, et al. Effects of particulate matter exposure on semen quality: a retrospective cohort study. Ecotoxicol Environ Saf. 2020;193:110319. https://doi.org/10.1016/j.ecoenv.2020.110319.

Guarino F, Improta G, Triassi M, Castiglione S, Cicatelli A. Air quality biomonitoring through Olea europaea L.: The study case of "Land of pyres.". Chemosphere. 2021;282:131052. https://doi.org/10.1016/j.chemosphere.2021.131052.

Gulia S, Shiva Nagendra SM, Khare M, Khanna I. Urban air quality management—a review. Atmos Pollut Res. 2015;6:286–304. https://doi.org/10.5094/APR.2015.033.

Hwang J, Kwon J, Yi H, Bae HJ, Jang M, Kim N. Association between long-term exposure to air pollutants and cardiopulmonary mortality rates in South Korea. BMC Public Health. 2020;20:1–8. https://doi.org/10.1186/s12889-020-09521-8.

Janson E, Johannessen A, Holm M, Franklin K, Holst GJ, Gislason T, et al. Insomnia associated with traffic noise and proximity to traffic-a cross-sectional study of the respiratory health in northern Europe III population. J Clin Sleep Med. 2020;16:545–52. https://doi.org/10.5664/jcsm.8274.

Khomenko S, Cirach M, Pereira-Barboza E, Mueller N, Barrera-Gómez J, Rojas-Rueda D, et al. Premature mortality due to air pollution in European cities: a health impact assessment. Lancet Planet Health. 2021;5:e121–34. https://doi.org/10.1016/S2542-5196(20)30272-2.

Kim KH, Jahan SA, Kabir E. A review on human health perspective of air pollution with respect to allergies and asthma. Environ Int. 2013;59:41–52. https://doi.org/10.1016/j.envint.2013.05.007.

Knapp J, Eberle B, Bernhard M, Theiler L, Pietsch U, Albrecht R. Analysis of tracheal intubation in out-of-hospital helicopter emergency medicine recorded by video laryngoscopy. Scand J Trauma Resusc Emerg Med. 2021;29:1–10. https://doi.org/10.1186/s13049-021-00863-9.

Kocak TK, Gurram S, Bertini RL, Stuart AL. Impacts of a metropolitan-scale freeway expansion program on air pollution and equity. J Transp Health. 2021;22:101114. https://doi.org/10.1016/j.jth.2021.101114.

Korhonen A, Relvas H, Miranda AI, Ferreira J, Lopes D, Rafael S, et al. Analysis of spatial factors, time-activity and infiltration on outdoor generated PM2.5 exposures of school children in five European cities. Sci Total Environ. 2021;785:147111. https://doi.org/10.1016/j.scitotenv.2021.147111.

Kuźma Ł, Struniawski K, Pogorzelski S, Bachórzewska-Gajewska H, Dobrzycki S. Gender differences in association between air pollution and daily mortality in the Capital of the Green Lungs of Poland–population-based study with 2,953,000 person-years of follow-up. J Clin Med. 2020;9:2351. https://doi.org/10.3390/jcm9082351.

Landrigan PJ. Air pollution and health. Lancet Public Health. 2017;2:e4–5. https://doi.org/10.1016/S2468-2667(16)30023-8.

Lee KK, Bing R, Kiang J, Bashir S, Spath N, Stelzle D, et al. Adverse health effects associated with household air pollution: a systematic review, meta-analysis, and burden estimation study. Lancet Glob Health. 2020;8:e1427–34. https://doi.org/10.1016/S2214-109X(20)30343-0.

Lemarchand C, Gabet S, Cénée S, Tvardik N, Slama R, Guénel P. Breast cancer risk in relation to ambient concentrations of nitrogen dioxide and particulate matter: results of a population-based case-control study corrected for potential selection bias (the CECILE study). Environ Int. 2021;155:106604. https://doi.org/10.1016/j.envint.2021.106604.

Les Bujanda I, Loureiro-Amigo J, Bastons FC, Guerra IE, Sánchez JA, Murgadella-Sancho A, et al. Treatment of COVID-19 pneumonia with glucocorticoids (CORTIVID): a structured summary of a study protocol for a randomised controlled trial. Trials. 2021;22:20–2. https://doi.org/10.1186/s13063-020-04999-4.

Li T, Tang X, Liu Y, Li Y, He B. Dietary patterns and metabolic syndrome among urbanized Tibetans: a cross-sectional study. Environ Res. 2021;200:111354. https://doi.org/10.1016/j.envres.2021.111354.

Li W, Bertisch SM, Mostofsky E, Vgontzas A, Mittleman MA. Associations of daily weather and ambient air pollution with objectively assessed sleep duration and fragmentation: a prospective cohort study. Sleep Med. 2020;75:181–7. https://doi.org/10.1016/j.sleep.2020.06.029.

Liang J, Liu J, Zhan Y, Zhou S, Xue CX, Sun C, et al. Succession of marine bacteria in response to Ulva prolifera-derived dissolved organic matter. Environ Int. 2021;155:106687. https://doi.org/10.1016/j.envint.2021.106687.

Liu J, Ghastine L, Um P, Rovit E, Wu T. Environmental exposures and sleep outcomes: a review of evidence, potential mechanisms, and implications. Environ Res. 2021a;196:110406. https://doi.org/10.1016/j.envres.2020.110406.

Liu J, Wu T, Liu Q, Wu S, Chen JC. Air pollution exposure and adverse sleep health across the life course: a systematic review. Environ Pollut. 2020;262:114263. https://doi.org/10.1016/j.envpol.2020.114263.

Liu L, Kauri LM, Mahmud M, Weichenthal S, Cakmak S, Shutt R, et al. Exposure to air pollution near a steel plant and effects on cardiovascular physiology: a randomized crossover study. Int J Hyg Environ

Health. 2014;217:279–86. https://doi.org/10.1016/j.ijheh.2013.06.007.

Liu Q, Zhang Z, Shao C, Zhao R, Guan Y, Chen C. Spatio-temporal variation and driving factors analysis of PM 2 . 5 health risks in Chinese cities. Ecol Indic. 2021b;129:107937. https://doi.org/10.1016/j.ecolind.2021.107937.

Liu Y, Gong S, Ye L, Li J, Liu C, Chen D, et al. Organophosphate (OP) diesters and a review of sources, chemical properties, environmental occurrence, adverse effects, and future directions. Environ Int. 2021c;155:106691. https://doi.org/10.1016/j.envint.2021.106691.

Maphutha J, Twilley D, Lall N. Inhibition of phosphatidylinositol 3-kinase (PI3K) enzyme and human skin carcinoma cell growth by Combretum apiculatum Sond. South African J Bot. 2021;140:95–102. https://doi.org/10.1016/j.sajb.2021.03.031.

Margolis AE, Pagliaccio D, Ramphal B, Banker S, Thomas L, Robinson M, et al. Prenatal environmental tobacco smoke exposure alters children's cognitive control circuitry: a preliminary study. Environ Int. 2021;155:106516. https://doi.org/10.1016/j.envint.2021.106516.

Mendoza DL, Benney TM, Bares R, Crosman ET. Intra-city variability of fine particulate matter during COVID-19 lockdown: a case study from Park City, Utah. Environ Res. 2021;201:111471. https://doi.org/10.1016/j.envres.2021.111471.

Mishra R, Krishnamoorthy P, Gangamma S, Raut AA, Kumar H. Particulate matter (PM10) enhances RNA virus infection through modulation of innate immune responses. Environ Pollut. 2020;266:115148. https://doi.org/10.1016/j.envpol.2020.115148.

Moshammer H, Poteser M, Hutter HP. COVID-19 and air pollution in Vienna—a time series approach. Wien Klin Wochenschr. 2021;133:951. https://doi.org/10.1007/s00508-021-01881-4.

Nageen Y, Asemoloye MD, Põlme S, Wang X, Xu S, Ramteke PW, et al. Analysis of culturable airborne fungi in outdoor environments in Tianjin, China. BMC Microbiol. 2021;21:1–10. https://doi.org/10.1186/s12866-021-02205-2.

Neira M, Pfeiffer M, Campbell-Lendrum D, Prüss-Ustün A. Towards a healthier and safer environment. Lancet. 2018;391:408–10. https://doi.org/10.1016/S0140-6736(17)32545-X.

Newell K, Kartsonaki C, Lam KBH, Kurmi OP. Cardiorespiratory health effects of particulate ambient air pollution exposure in low-income and middle-income countries: a systematic review and meta-analysis. Lancet Planet Health. 2017;1:e360–7. https://doi.org/10.1016/S2542-5196(17)30166-3.

Ngoc LTN, Park D, Lee Y, Lee YC. Systematic review and meta-analysis of human skin diseases due to particulate matter. Int J Environ Res Public Health. 2017;14 https://doi.org/10.3390/ijerph14121458.

Nhung NTT, Amini H, Schindler C, Kutlar Joss M, Dien TM, Probst-Hensch N, et al. Short-term association between ambient air pollution and pneumonia in children: a systematic review and meta-analysis of time-series and case-crossover studies. Environ Pollut. 2017;230:1000–8. https://doi.org/10.1016/j.envpol.2017.07.063.

Palacios-Mateo C, van der Meer Y, Seide G. Analysis of the polyester clothing value chain to identify key intervention points for sustainability. Environ Sci Eur. 2021;33:2. https://doi.org/10.1186/s12302-020-00447-x.

Pandey A, Brauer M, Cropper ML, Balakrishnan K, Mathur P, Dey S, et al. Health and economic impact of air pollution in the states of India: the global burden of disease study 2019. Lancet Planet Health. 2021;5:e25–38. https://doi.org/10.1016/S2542-5196(20)30298-9.

Pelgrims I, Devleesschauwer B, Guyot M, Keune H, Nawrot TS, Remmen R, et al. Association between urban environment and mental health in Brussels, Belgium. BMC Public Health. 2021;21:1–18. https://doi.org/10.1186/s12889-021-10557-7.

Qian Z, Liang S, Yang S, Trevathan E, Huang Z, Yang R, et al. Ambient air pollution and preterm birth: a prospective birth cohort study in Wuhan, China. Int J Hyg Environ Health. 2016;219:195–203. https://doi.org/10.1016/j.ijheh.2015.11.003.

Quarmby S, Santos G, Mathias M. Air quality strategies and technologies: a rapid review of the international evidence. Sustain For. 2019;11:1–18. https://doi.org/10.3390/su11102757.

Rivas-Santiago CE, Sarkar S, Cantarella P, Osornio-Vargas Á, Quintana-Belmares R, Meng Q, et al. Air pollution particulate matter alters antimycobacterial respiratory epithelium innate immunity. Infect Immun. 2015;83:2507–17. https://doi.org/10.1128/IAI.03018-14.

Santi D, Vezzani S, Granata AR, Roli L, De Santis MC, Ongaro C, et al. Sperm quality and environment: a retrospective, cohort study in a northern province of Italy. Environ Res. 2016;150:144–53. https://doi.org/10.1016/j.envres.2016.05.053.

Schaefer M, Salari HE, Köckler H, Thinh NX. Assessing local heat stress and air quality with the use of remote sensing and pedestrian perception in urban microclimate simulations. Sci Total Environ. 2021;794:148709. https://doi.org/10.1016/j.scitotenv.2021.148709.

Schlutow A, Schröder W, Scheuschner T. Assessing the relevance of atmospheric heavy metal deposition with regard to ecosystem integrity and human health in Germany. Environ Sci Eur. 2021;33 https://doi.org/10.1186/s12302-020-00391-w.

Shams SR, Jahani A, Kalantary S, Moeinaddini M, Khorasani N. Artificial intelligence accuracy assessment in NO2 concentration forecasting of metropolises air. Sci Rep. 2021;11:1–9. https://doi.org/10.1038/s41598-021-81455-6.

Shi J, Chi C, Gong X, Chen C, Yu W, Huang J, et al. Examining health disparities and characteristics in general practice utilization: Based on outpatient data from 2014–2018 in Shanghai. BMC Fam Pract. 2020;21:1–12. https://doi.org/10.1186/s12875-020-01146-5.

Sicard P, Agathokleous E, De Marco A, Paoletti E, Calatayud V. Urban population exposure to air pollution in Europe over the last decades. Environ Sci Eur. 2021;33:28. https://doi.org/10.1186/s12302-020-00450-2.

Siddharthan T, Pollard SL, Jackson P, Robertson NM, Wosu AC, Rahman N, et al. Effectiveness of low-dose theophylline for the management of biomass-associated COPD (LODOT-BCOPD): study protocol for a randomized controlled trial. Trials. 2021;22:1–9. https://doi.org/10.1186/s13063-021-05163-2.

Singh RK, Drews M, De la Sen M, Srivastava PK, Trisasongko BH, Kumar M, et al. Highlighting the compound risk of COVID-19 and environmental pollutants using geospatial technology. Sci Rep. 2021;11:1–12. https://doi.org/10.1038/s41598-021-87877-6.

Sly PD, Trottier BA, Bulka CM, Cormier SA, Fobil J, Fry RC, et al. The interplay between environmental exposures and COVID-19 risks in the health of children. Environ Health. 2021;20:1–10. https://doi.org/10.1186/s12940-021-00716-z.

Stenson C, Wheeler AJ, Carver A, Donaire-Gonzalez D, Alvarado-Molina M, Nieuwenhuijsen M, et al. The impact of traffic-related air pollution on child and adolescent academic performance: a systematic review. Environ Int. 2021;155:106696. https://doi.org/10.1016/j.envint.2021.106696.

Sun D, Liu Y, Zhang J, Liu J, Wu Z, Liu M, et al. Long-term effects of fine particulate matter exposure on the progression of arterial stiffness. Environ Health. 2021;20:1–7. https://doi.org/10.1186/s12940-020-00688-6.

Timonen H, Mylläri F, Simonen P, Aurela M, Maasikmets M, Bloss M, et al. Household solid waste combustion with wood increases particulate trace metal and lung deposited surface area emissions. J Environ Manag. 2021;293:112793. https://doi.org/10.1016/j.jenvman.2021.112793.

Tong S. Air pollution and disease burden. Lancet Planet Health. 2019;3:e49–50. https://doi.org/10.1016/S2542-5196(18)30288-2.

Turrisi TB, Bittel KM, West AB, Hojjatinia S, Hojjatinia S, Mama SK, et al. Seasons, weather, and device-measured movement behaviors: a scoping review from 2006 to 2020. Int J Behav Nutr Phys Act. 2021;18:1–26. https://doi.org/10.1186/s12966-021-01091-1.

Vandyck T, Keramidas K, Kitous A, Spadaro JV, Van Dingenen R, Holland M, et al. Air quality co-benefits for human health and agriculture counter-balance costs to meet Paris agreement pledges. Nat Commun. 2018;9:1–11. https://doi.org/10.1038/s41467-018-06885-9.

Vlachogiannis DM, Xu Y, Jin L, González MC. Correlation networks of air particulate matter (PM 2.5): a comparative study. Appl Netw Sci. 2021;6:32. https://doi.org/10.1007/s41109-021-00373-8.

Von Schneidemesser E, Driscoll C, Rieder HE, Schiferl LD. How will air quality effects on human health, crops and ecosystems change in the future? Philos Trans R Soc A Math Phys Eng Sci. 2020;378:20190330. https://doi.org/10.1098/rsta.2019.0330.

de Vries W, Schulte-Uebbing L, Kros H, Voogd JC, Louwagie G. Spatially explicit boundaries for agricultural nitrogen inputs in the European Union to meet air and water quality targets. Sci Total Environ. 2021;786:147283. https://doi.org/10.1016/j.scitotenv.2021.147283.

Wei Y, Yazdi MD, Di Q, Requia WJ, Dominici F, Zanobetti A, et al. Emulating causal dose-response relations between air pollutants and mortality in the Medicare population. Environ Health. 2021;20:1–10. https://doi.org/10.1186/s12940-021-00742-x.

WHO. Review of evidence on health aspects of air pollution—REVIHAAP Project: Technical Report, vol. 49. Copenhagen, Denmark: World Health Organisation; 2013. p. 35–42. Available at: http://www.euro.who.int/__data/assets/pdf_file/0004/193108/REVIHAAP-Final-technical-report-final-version.pdf

Xu JL, Lin X, Hugelier S, Herrero-Langreo A, Gowen AA. Spectral imaging for characterization and detection of plastic substances in branded teabags. J Hazard Mater. 2021;418:126328. https://doi.org/10.1016/j.jhazmat.2021.126328.

Yang BY, Qian Z, Howard SW, Vaughn MG, Fan SJ, Liu KK, et al. Global association between ambient air pollution and blood pressure: a systematic review and meta-analysis. Environ Pollut. 2018;235:576–88. https://doi.org/10.1016/j.envpol.2018.01.001.

Yin P, Brauer M, Cohen AJ, Wang H, Li J, Burnett RT, et al. The effect of air pollution on deaths, disease burden, and life expectancy across China and its provinces, 1990–2017: an analysis for the global burden of disease study 2017. Lancet Planet Health. 2020;4:e386–98. https://doi.org/10.1016/S2542-5196(20)30161-3.

Zhang X, Song Z, Li Y, Wang H, Zhang S, Reid AM, et al. Cytotoxic and antiangiogenetic xanthones inhibiting tumor proliferation and metastasis from Garcinia xipshuanbannaensis. J Nat Prod. 2020;84:1515. https://doi.org/10.1021/acs.jnatprod.0c01354.

GPSR Compliance

The European Union's (EU) General Product Safety Regulation (GPSR) is a set of rules that requires consumer products to be safe and our obligations to ensure this.

If you have any concerns about our products, you can contact us on ProductSafety@springernature.com

In case Publisher is established outside the EU, the EU authorized representative is:

Springer Nature Customer Service Center GmbH
Europaplatz 3
69115 Heidelberg, Germany

Batch number: 10732809

Printed by Printforce, the Netherlands